THE NATURE AND FUNCTION
OF SCIENTIFIC THEORIES

Volume 4   *University of Pittsburgh Series*
              in the Philosophy of Science

The Nature & Function

Editor

ROBERT G. COLODNY

GROVER MAXWELL

MARY HESSE

ABNER SHIMONY

WESLEY C. SALMON

NORWOOD RUSSELL HANSON

PAUL K. FEYERABEND

of Scientific Theories

Essays in Contemporary
Science and Philosophy

University of Pittsburgh Press

The artwork in "A Picture Theory of Theory Meaning" was drawn by Howard N. Ziegler.

The material about Duhem on pages 52–54 has been adapted from "Duhem, Quine, and a New Empiricism" by Mary Hesse in *Knowledge and Necessity*, Royal Institute of Philosophy Lectures, Vol. 3, 1968–69 (London: Macmillan, 1970). Used by permission of Macmillan and Co., Ltd.

Grateful acknowledgment is made to the following for permission to quote material in "Problems of Empiricism, Part II":

Abelard-Schuman Limited, for quotations from *The Naming of the Telescope* by Edward Rosen. Copyright 1947 by Henry Schuman, Inc. By permission of Abelard-Schuman Limited.

Basic Books, Inc., for quotations from "Complexities, advances, and misconceptions in the development of the science of vision: what is being discovered?" by Ronchi in *Scientific Change* edited by Alistair Crombie, Basic Books, Inc., Publishers, New York, 1963.

Harvard University Press, for quotations from *A Source Book in Greek Science* by M. R. Cohen and I. E. Drabkin. Copyright 1948.

Springer-Verlag and C. Truesdell, for the quotation from "A Program Towards Rediscovering the Rational Mechanics of the Age of Reason" by C. Truesdell in *Archive for the History of Exact Sciences,* Vol. 1. Copyright 1960. This article has been reprinted, with corrections, in *Essays in the History of Mechanics* by C. Truesdell (Berlin, Heidelberg, New York: Springer-Verlag, 1968).

University of California Press, for quotations from *A Dialogue Concerning the Two Chief World Systems* by Galileo Galilei, trans. by Stillman Drake. Copyright 1953. Reprinted by permission of The Regents of the University of California.

Library of Congress Catalog Card Number 70–123094

ISBN 0–8229–3211–3

Henry M. Snyder & Co., Inc., London

Manufactured in the United States of America

In memory of Norwood Russell Hanson

Contents

Preface

IN MAY OF 1965 a workshop type of conference was held at the Center for Philosophy of Science at the University of Pittsburgh. The discussion sessions focused on seven related papers presented in chronological order as follows:

> Paul K. Feyerabend, "Outline of a Model for the Acquisition and Improvement of Knowledge"
> Wilfrid Sellars, "Realism Versus Instrumentalism"
> Abner Shimony, "Proposals for a Naturalistic Epistemology"
> Grover Maxwell, "Remarks on Perception and Theoretical Entities"
> Norwood Russell Hanson, "A Picture Theory of Theory Meaning"
> Adolf Grünbaum, "The Scientific Assertibility of Statements About the Past and About the Future"
> Wesley C. Salmon, "Deductive and Inductive Explanation"

The present volume includes updated versions of five of these seven papers. A paper by Mary Hesse, based on a 1966 lecture which she gave in the Center's annual series of public lectures, is likewise included because of its topical relevance to the overall theme of this book.

The Center is greatly indebted to Dr. Charles H. Peake, provost of the University of Pittsburgh, for support which helped to make the 1965 workshop-conference possible. And the participants in the work of the Center warmly appreciate the frequent assistance rendered by the late Dorothy Colodny to her husband in his editing of the volumes in the University of Pittsburgh Series in the Philosophy of Science.

> ADOLF GRÜNBAUM
> Andrew Mellon Professor of Philosophy and
> Director of the Center for Philosophy of Science,
> University of Pittsburgh

ROBERT G. COLODNY
University of Pittsburgh

Introduction

> Who would have said, a few years ago, that we could ever know of what substances stars are made whose light may have been longer in reaching us than the human race has existed? Who can be sure of what we shall now know in a few hundred years? Who can guess what would be the result of continuing the pursuit of science for ten thousand years, with the activity of the last hundred? And if it were to go on for a million, or a billion, or any number of years you please, how is it possible to say that there is any question which might not ultimately be solved?
> —Charles S. Peirce
> *Values in a Universe of Chance* [1]

IN THE OLD GREEK, theory, theoria, implied vision—a spectacle. A theoretical scientist was in this sense a supreme visionary, a fabricator of pictures of the world, an artisan working with ideas. Democritus and Empedocles were magnificent in such labors, as was the poet Plato. Parmenides and Zeno upset these artistic efforts to reduce the complexity of the world to picture language, and Aristotle, particularly the craftsman of *Posterior Analytics,* set vigorous rules for theorizing that have influenced scientists and philosophers ever since.

In our own post-Faustian times, when scientific theories are freighted with power over nature and when communities of scientists constitute an incredibly important "national resource," the nature of these scientists' most important activity—theorizing about the world—becomes of concern to all thinking members of the global community. This indubitable fact arises not only from the expectation of more control over the brute powers of nature nor even from the hope of intellectual clarity concerning the fine grain of the world, but rather from the realization that the world outside the laboratories and observatories will be irreversibly altered in its attitudes about how human reason intersects the fabric of the cosmos and conveys its conclusions in human languages—both the natural language of the common man and of common sense and the beautifully flexible, artificial symbolic systems of mathematics and logic.

There is a subtle feedback from the total activity of scientists that radically affects the cognitive as well as the emotive behavior of all other inhabitants of the Republic of Science and Letters. These other workers, to the extent that they are more than passive spectators of the drama provided by the scientists, correctly perceive that there are procedures followed by the scientific community which, it is presumed, are in part responsible for the immense fecundity of scientific efforts. There exists, therefore, a more or less conscious desire to universalize these procedures so as to maximize their domain of efficacy. In the modern world, i.e., since the scientific revolution of the seventeenth century, the principal carriers of this class of insights have been, in the main, the philosophers of science. Here one wishes to distinguish between popularizers of scientific theories and careful analyzers of scientific methodologies. Newton, Locke, Descartes, Kant, Whewell, Mill, Comte, Duhem, Mach, and Poincaré would obviously belong to the latter category. The last phase of this three hundred year history was dominated by various forms of logical positivism, a mode of philosophizing that was spread from imperial Vienna in the second and third decades of the twentieth century and that, like Vienna-based psychoanalysis, won more cultural subjects than the arms or diplomacy of the Hapsburgs.

We are now in the post-positivist phase of the philosophy of science, wherein the exaggerations and warped focus of this amorphous school are clearly perceived [2] and the more traditional quest of philosophers of science is being renewed. This refocusing of attention is manifested by both the content and the tone of the essays that make up this volume.

There is evidence that the career of logical positivism and related philosophical activity, despite great technical virtuosity, has become counterproductive because its leading practitioners ignored the history of philosophy and convinced themselves that the conceptual history of science was irrelevant to the philosophy of science. This error has been documented in detail by the research of Laurens Laudan, [3] who shows dramatically that in philosophy as well as in social life to ignore history is to be doomed to repeat it. Thus, the extreme claims made for the principles of verifiability and falsifiability of scientific theories by the Vienna Circle and its epigones led to a repetition of the theoretical disputes of the nineteenth century. Furthermore, the rich intellectual biographies of scientists as diverse as, say, Kepler, Darwin, Planck, and Einstein could have, if exploited, added new dimensions to the understanding of the creative processes of scientific theory-building.

Readers of this volume will discover that the six philosophers who herein engage in a coordinated effort to clarify the contemporary status of scientific theories are concerned not only with the linguistic, logical, epistemic, and ontological elements of theory-building and knowledge claims, but that each finds it intellectually rewarding to examine historical antecedents and to reflect upon the broad cultural matrix within which actual scientific work takes place.

Thus Maxwell, in his plea for a realist rather than an instrumentalist interpretation of science, draws upon later and ignored works of Bertrand Russell and suggests that this implied reorientation of philosophy will remove serious inconsistencies from our understanding of the claims both of common sense and highly developed scientific hypotheses.

Mary Hesse, in a far-ranging critique of one of the core ideas of traditional positivism concerning the relation between observational and theoretical languages, remarks that "the account which follows is by no means original; in fact, versions of it have been lying around in the literature for so long that it may even sound trite." Her references are not only to Kuhn and Chomsky, Quine and Popper, but also to Duhem, to the facts of Aristotelian biology and the language problems of quantum theory. Thus the results of her inquiry not only shed light on the technical problems of the relations of various predicates to the physical world but enable us to understand some of the problems of scientific discourse in any age whatsoever.

With the essays of Abner Shimony and Wesley Salmon, the thorny problems of inductive inference and scientific explanation are raised to a new level. Shimony's remarkable analysis of the relationship between the reliability of common sense and the authenticity of the world view of science leads him to radically new conclusions concerning the nature of human perception of the external world. Besides containing certain affinities with the conclusions of Hesse, Maxwell, and Salmon, Shimony's solutions are innovative not only in the use he makes of probability theory but in the broad evolutionary frame in which he embeds homo faber and homo sapiens. Elements of the thought of Charles S. Peirce are acknowledged here, but, like all potentially revolutionary leaps, the primary appeal is to the future, to the architects of a new naturalistic epistemology to come who will be working when one will no longer have to repeat

Shimony's observation that "the mind-body problem is a vast desert of ignorance in the sciences of psychology and physiology."

Salmon's essay, the latest in a series of magisterial works on the problem of inductive logic, accomplishes the extremely difficult task of removing long held ambiguities from the theory of inductive logic by directly confronting traditional orthodoxies on the relations between deductive and inductive argument. His discoveries shed new light on the foundations of statistical explanation.

With Hanson's paper, unfortunately the last that is to be published by this gifted polemecist, an oft overlooked aspect of scientific theories is probed in detail. The representational aspect of such constructs with their isomorphic qualities is herein examined in detail. Ironically, the examples are drawn from the relatively new science of aerodynamics in which Hanson was both a theoretician and fatally a practitioner.

Paul K. Feyerabend's concluding essay moves directly to the issues central in this introduction—the relationship between the actuality of conceptual revolutions in the history of science and certain traditional philosophical pictures or theories about scientific theory-building. Acknowledging debts to Hanson, to Thomas Kuhn, and to Ronchi, Feyerabend with characteristic audacity argues for the necessity of extracting methodological insight from concrete case studies (herein Galileo) and drives home the conclusion that the continued growth of science, its freedom from ossification promoted by unyielding orthodoxy, requires the most daring use of hypotheses that may be formulated despite the presumed weight of countervailing physical evidence.

The foregoing remarks should indicate that philosophers of science have moved far beyond polemics rooted in "classical positions," that they are again pioneer explorers on the frontiers of their discipline. As C. H. Waddington said, a "scientific theory cannot remain a mere structure within the world of logic, but must have implications for action and that in two different ways. In the first place, it must involve the consequence that if you do so and so, such and such results will follow. That is to say it must give, or at least offer the possibility of controlling the process; and secondly—and this is a point not so often mentioned by those who discuss the nature of scientific theories—its value is quite dependent on its power of suggesting the next step in scientific advance." [4]

Finally, the goal that governed the production of this volume is indicated by Höffding's remark to Emile Myerson: "Science will discover (what many philosophers have seen) that here again, truth is in the action, in the movement of thought not in a particular given position." [5]

NOTES

1. The thematic quotes by Peirce in this volume are taken from *Values in a Universe of Chance, The Selected Writings of Charles S. Peirce,* ed. Philip P. Wiener (New York: Doubleday, 1958).
2. See Peter Achinstein and Stephen F. Barker, eds., *The Legacy of Logical Positivism; Studies in the Philosophy of Science* (Baltimore: Johns Hopkins Press, 1969). Of particular interest are the essays by Dudley Shapere, "Notes Toward a Post-Positivistic Interpretation of Science," and Carl G. Hempel, "Logical Positivism and the Social Sciences."
3. Reference is to his paper "From Testability to Meaning and Back Again," which will be published in a subsequent volume in the series sponsored by the Center for Philosophy of Science at the University of Pittsburgh. An extensive bibliography "Theories of Scientific Method from Plato to Mach" appeared in *History of Science,* VII (1968).
4. C. H. Waddington, *The Nature of Life* (New York: Atheneum, 1962), p. 12.
5. Quoted by L. Rosenfeld in "Velocity of Light and the Evolution of Electrodynamics," 4, supp. no. 5, ser. X, *Nuovo Cimento,* p. 11. This essay is a model of philosophical and historical clarity and exemplifies many of the points made programatically by the authors of this volume.

THE NATURE AND FUNCTION
OF SCIENTIFIC THEORIES

GROVER MAXWELL

University of Minnesota

Theories, Perception, and Structural Realism

> It used to be supposed by empiricists that the justification of [nondeductive] inference rests upon induction. Unfortunately, it can be proved that induction by simple enumeration, if conducted without regard to common sense, leads very much more often to error than to truth. And if a principle needs common sense before it can be safely used, it is not the sort of principle that can satisfy a logician. We must, therefore, look for a principle other than induction if we are to accept the broad outlines of science, and of common sense in so far as it is not refutable. This is a very large problem.
>
> —Bertrand Russell
> *My Philosophical Development*

> *Owing to the world being such as it is,* certain occurrences are sometimes, *in fact,* evidence for certain others; and *owing to animals being adapted to their environment,* occurrences which are, *in fact,* evidence of others tend to arouse expectation of those others.
>
> —Bertrand Russell
> *Human Knowledge:*
> *Its Scope and Limits* (italics added)[1]

DISCUSSIONS OF PROBLEMS about the meaning of theoretical terms, the status of theoretical (unobservable) entities, and the like seldom include any detailed consideration of traditional philosophical problems about perception and knowledge of the external world or the problem about induction raised by Hume and now extended to the general problem of

---

Support of research by the National Science Foundation, the Carnegie Foundation, and the Minnesota Center for Philosophy of Science at the University of Minnesota is gratefully acknowledged.

4 : *Grover Maxwell*

the confirmation of knowledge claims. Although these may be mentioned briefly, and certain rough analogies of traditional problems with problems about scientific theories may be affirmed or denied, they seem to be regarded as having only incidental interest for the philosophy and methodology of science. The most notable exception occurs in the later writings of Bertrand Russell (circa 1925 and thereafter). Incredible as it is, this monumentally important work by "the philosopher of the century" has, virtually without exception, been either totally neglected or completely misunderstood. As Russell himself has remarked, "It is true that nobody has accepted what seems to me the solution [to certain problems about physics, perception, and the relation of mind and matter], but I believe and hope that this is only because my theory has not been understood." [2] I hope that his statement now admits of at least one exception, for although the views that are defended here are not identical with Russell's and were, to a considerable extent, developed before I understood his theory, the debt to him is great and is gratefully acknowledged.[3]

Although explicit consideration of these traditional philosophical issues is rarely encountered in philosophical discussions of scientific theories, such discussions almost always presuppose or tacitly make use of one or another solution to each such problem; and this determines, to a great extent, what kinds of views about scientific theories will be held and defended. It is the purpose of this essay to consider several of such important cases in point and also to explore implications that run in the opposite direction: that is, to discover what light can be thrown upon these traditional philosophical areas by considerations from the philosophy and history of science as well as from the actual content of current scientific theory.

According to some influential contemporary views, the very task that I have set for myself is philosophically illegitimate. Since these contemporary views are themselves cases in point for my main thesis, let us begin the investigation with them. Specifically, their objections are against drawing philosophical conclusions from scientific premises. Their arguments may be reconstructed as follows: (1) knowledge claims are of two and only two kinds. (2) On the one hand, there are those which are contingent and thus have factual content, and, on the other, there are those which are necessary and are thus, in some important sense, factually empty. (3) The former are at least sometimes testable or confirmable or disconfirmable by means of experience and observation, while the latter depend for their truth or falsity wholly upon matters that are purely logical, conceptual, or linguistic in nature. The premises so far stated can be, and of course have been, challenged, but I have no inclination to do so myself and have stated them only for the sake of completeness. The argument continues, claiming that (4) contingent statements are always, "in principle," testable, or confirmable or disconfirmable by experience or observation and that (5) they are, therefore, exclusively matters for either *scientific* inquiry or everyday observations and inferences. (6) Matters that fall within the province of science or everyday observation and inference must be left to science and common sense and, therefore, (7) they are not to be treated as philosophical. It follows that (8) philosophy must restrict itself to matters that are purely logical or purely conceptual or purely linguistic (excluding contingent, scientific linguistics, of course) in nature (end of the argument). (There is a variation on this argument that admits a third category of statements, "metaphysical" ones, but this need not concern us here.)

The parts of this argument that seem to require the closest scrutiny are premise (4) and the inference therefrom to (5) and eventually to (8). I shall argue that the acceptability of this premise and the legitimacy of the subsequent inference depend crucially upon what is taken to count as *testability* or *confirmability* and *disconfirmability*. Therefore, whether or not one subscribes to this argument will depend to a large extent upon one's views about induction, confirmation, testability, and related matters. Naturally, most philosophers have not done extensive research in confirmation theory, but it plays such a fundamental role in epistemic matters that it seems reasonable to assume that they hold, explicitly or tacitly, some sort of view about the relation of evidence to that which is evidenced. I shall argue that one's positions on each of a number of important philosophical and methodological issues in addition to the one now being considered depend crucially upon the theory of confirmation to which one subscribes.

The classification of confirmation theories that follows is not intended to be exhaustive and is somewhat oversimplified, but it is not, I believe, seriously distorted. One kind of confirmation theory will be called "strict inductivism."[4] According to it, confirmation can only be accomplished by inductive arguments of a very simple kind such as induction by simple enumeration (including the "straight rule," that is, infer that the relative frequency in the population will be the same as that in the sample) and Mill's methods. In other words, all reasoning must be either deductive or inductive in this simple sense. It is also generally held that although induction, unlike deduction, cannot guarantee truth transmission, deductive logic and inductive logic nevertheless share a number of important features. For one thing, it is contended, both deductive and inductive logic provide the standards for the assessment of *rationality*. Although induction may fail at times and may fail almost always in some possible worlds, to make inductive inferences, it is held, is always the rational thing to do in any possible world—just as much so as it is to make deductive inferences. Moreover, for both deductive and inductive arguments, it is form rather than content that makes inferences legitimate. Thus, inductive logic is conceived as really being *logic*, in a perfectly legitimate and important sense.

Strict inductivisim is a special case of a view that I shall call *strict confirmationism*. It holds that the confirmatory relationship between evidence and hypothesis is of a *purely logical nature*, where "logical" means purely formal as well as rational in all possible worlds. From what was said in the latter part of the previous paragraph, it is obvious that all strict inductivists are strict confirmationists. However, the converse does not hold; some strict confirmationists advocate modes of nondeductive reasoning other than the simple inductive ones mentioned above. These may sometimes also be called "inductive," but, if so, the term is being used in a broader sense to mean something like "nondeductive." One of the more important kinds of strict confirmationism differs from strict inductivism in that it permits, indeed espouses, what is usually called "hypothetico-deductive [5] reasoning." According to it, it is a mistake to believe that general knowledge is, for the most part, acquired and confirmed by inferring it from the evidence, in any usual sense of "infer," inductive or deductive. Rather, in the more interesting and important cases in both science and everyday life, theories or hypotheses are proposed as a result of as yet poorly understood creative acts of the mind. When a satisfactory hypothesis is conjoined with certain other knowledge claims (for example, auxiliary hypotheses and initial conditions), the evidence may be inferred from this conjunction, usually deductively although nondeductive inferences from the conjunction to the evidence are not excluded. Unfortunately the inference seems, at first blush, to be in the wrong direction; we are inferring something that we knew already, the evidence, from something whose epistemic status is in question and is, in fact, the very thing we wanted to confirm, the theory or hypothesis. Nevertheless, it is claimed that as more and more of such inferences yield predictions that turn out to be true, the theory or hypothesis becomes better and better confirmed. Evidential support is somehow supposed to seep up backward through the inferential linkage that goes from hypothesis to evidence and to deposit itself upon the hypothesis.

It seems impossible to deny that, both in scientific inquiry and every-day reasoning, this hypothetico-inferential kind of confirmation is frequently employed. Quite often in common-sense reasoning our main reason for holding that a certain hypothesis is true or quite likely to be true is that it *explains* the facts so well, where "it explains the facts" means something like "when conjoined with certain other statements, it yields as consequences statements expressing 'the facts.'" For example, reread any Sherlock Holmes detective story. Certainly in the more advanced sciences, especially when the theories are about unobservables, *something* like hypothetico-inferential confirmation seems to be employed almost exclusively. It is interesting and important to note that confirmation by simple inductive arguments can always be reconstructed as a special case of hypothetico-inferential confirmation. For example, the hypothesis "All crows are black," together with the initial condition "This is a crow," yields as a consequence "This is black," which, if true, confirms the hypothesis.

I have explained these elementary matters in this much detail because I am going to claim that strict confirmationism in general and strict inductivism in particular are both absolutely untenable, and I want to be as clear as possible about what I am rejecting. Clearly, however, to reject strict inductivism is not necessarily to prohibit use of simple inductive inferences; it is only to deny the claim that such inferences are legitimate because of purely logical principles and/or the claim that simple inductive inferences comprise the whole of permissible nondeductive reasoning. And certainly to reject strict confirmationism is not necessarily to reject hypothetico-inferential reasoning but, rather, to deny that the backward seepage of evidential support mentioned above is all there is to the confirmation relation. More generally, to reject strict confirmationism is to maintain that any viable theory of confirmation must be a contingent one.

I have argued the case against strict inductivism and strict confirmationism in detail elsewhere,[6] and a convincing case against them has also been made by Russell in his *Human Knowledge* and *My Philosophical Development* (see above). Therefore, I shall only summarize some of the most important arguments here. First of all, for every simple inductive argument with true premises that gives correct results (a true or probable conclusion), there can be constructed an indefinitely large number of arguments with the same logical form as the one with the acceptable conclusion and with all true premises but which give false, totally unacceptable results—results which, moreover, are incompatible with the acceptable ones. This can be demonstrated in numerous ways. I shall select one of the briefer, simpler ones although it may not be the most rigorous or convincing. First, consider the argument that cites n instances of black crows and concludes that all crows are black. We now define the predicate "whack" as follows:

> x is whack $=_{df} x$ inhabits a planet with a diurnal rotation period of 24.0000012 sidereal hours or less and is black or x inhabits a planet with a diurnal rotation period of more than 24.0000012 sidereal hours and is white.

As with Nelson Goodman's "grue," if all crows so far observed are black, then they are also whack. So we can cite n instances of whack crows and conclude that all crows are whack, so that when the period of rotation becomes more than 24,0000012 sidereal hours, say, tomorrow, why should we not feel that we have good inductive support for the prediction that all crows existing thereafter will be white? (We know that the period of diurnal rotation gradually increases due to tidal friction.)

Acknowledging the debt to Goodman, I contend that use of a predicate like "whack" instead of one like "grue" makes a stronger case, for whack is not epistemically "tainted" (it contains no reference to observation); nor does it refer to specific temporal location; it is a completely general predicate and could be applicable on *any* planet having an appropriate diurnal rotation period. Actually this and Goodman's examples are special instances of the more general and widely accepted principle that any observed or otherwise postulated regularity can be extrapolated in an infinite number of mutually incompatible ways. These examples show that simple inductive inferences are *one* means of providing these bewildering extrapolations.

Obviously there is an indefinitely large number of properties similar to *whackness* which would enable us (or, at any rate, enable God or Omniscient Jones) to construct an indefinitely large number of inductive arguments which have the same logical form and which have all premises true but give us an unacceptable conclusion about crow color. It will do no good to object that everyone knows that small changes in the diurnal rotation period of ones home planet has nothing to do with feather pigmentation so that whackness is a silly, irrelevant property to begin with, for I will certainly agree. However, I must then go on to inquire how it is that everyone knows this. It is certainly a general knowledge claim and thus, according to strict inductivism, must be supported by simple inductive arguments. But again, for each such argument with an acceptable conclusion there will be, again, the indefinitely large number of arguments with logical form identical to it and with all true premises but with unacceptable conclusions. The attempt to reply in terms of "cross inductions" fails; strict inductivism provides no means for supposing that *any* acceptable network of "cross inductions" could ever get started, for we have seen that logic alone does not provide a means for distinguishing acceptable inductions from the indefinitely more numerous unacceptable ones. So although we do know that whackness and grueness are silly, irrelevant properties, we know it for reasons other than strict inductivist ones.

We have already seen that hypothetico-inferential confirmation is stronger than simple inductive confirmation, since the former contains the latter as a special case and the converse is not true. This is sufficient to demonstrate that since simple inductive confirmation cannot be a matter of logic alone, neither can hypothetico-inferential confirmation. This can be shown by other means. For example, no matter how much evidence we may accumulate (by observation, experiment, etc.), there will always be an indefinitely large number of mutually incompatible hypotheses each of which, when conjoined with propositions expressing appropriate "background knowledge," will yield, as consequences, propositions expressing all of the evidence. Thus, at any point in the search for knowledge, there exist numerous mutually incompatible hypotheses all of which, it would seem, are equally well confirmed hypothetico-inferentially by the evidence at hand. True, we are not usually intelligent enough or highly motivated enough to think of more than one or two such hypotheses, but the indefinitely large number are all known to Omniscient Jones and are all logically on a par. As far as pure logic is concerned and thus as far as strict confirmationism is concerned, there is no reason to suppose that the ones upon which we happen to stumble are any more likely to be true or closer to the truth than any of the other numerous ones known to O.J.

As far as constructive alternatives to strict confirmationism are concerned, the purposes of this paper will be served by pointing out that we do, of course, sometimes make what seem to be fairly successful selections among competing hypotheses all of which are on a par regarding evidence and logic [7] but that, on the other hand, there is often disagreement, even in the advanced sciences, that remains even as more and more experimental evidence is accumulated and that may span decades or even centuries. Simplicity is often mentioned as a criterion and, certainly, is often helpful, as well as being an intrinsically desirable property. But I do not think that it is by any means a sufficient or a necessary criterion. Examples from the history of science show that simple theories sometimes give way to more complicated ones even though the former could have been "saved" with only a little bit of "simple ad hoc-ing." Certainly there are no purely logical reasons for supposing that a simpler hypothesis is more likely to be true or closer to the truth.

Now let us consider again premise (4) in the argument cited earlier against drawing philosophical conclusions from scientific premises. This premise claims that contingent statements are always in principle testable, or confirmable or disconfirmable by observation or experience. It is easy to see that if the premise is to avoid a *reductio,* "testable," "confirmable," and "disconfirmable" must be intended in the strict confirmationist sense. For if nonlogical assumptions were involved in confirmation, they would either have to be confirmed in the strict confirmationist sense or we would be off to a vicious infinite regress so that any actual instance of confirmation would be impossible except, perhaps, those expressing observations of the moment. It is, perhaps, a moot point as to whether these are confirmable or whether they require no confirmation. And if confirmation is impossible, no propositions would be confirmable. As Russell has pointed out, it is still possible, indeed "logically impeccable" to restrict oneself to experience and logic and to refuse to assert anything that goes beyond observation of the moment. But, as he says of such a skeptic, "I shall not call him irrational but I shall have profound doubts about his sincerity." The untenability of strict confirmationism, then, refutes the kind of argument proffered above for restricting philosophy to purely logical, conceptual, or linguistic matters, since it refutes the crucial premise (4).

We are now prepared to consider the claim that the difference between instrumentalism and realism as regards scientific theories must be only a linguistic one—as Nagel has put it, "a conflict over preferred modes of speech." This position does not *commit* one to strict confirmationism but the converse seems to hold. If one subscribed to simple confirmationism either explicitly or tacitly, this would explain his philosophical motivation for holding this view about the difference between instrumentalism and realism. For the instrumentalist position is almost always stated so as to guarantee that no difference between it and realism can be established by means of observation and purely logical considerations. It seems to me that there is abundant evidence for my explanation of this kind of philosophical motivation. For example, Carnap makes no bones about it. He says, "The statement asserting the reality of the external world (realism) as well as its negation in various forms, e.g., solipsism and several forms of idealism [the principle remains the same if we insert "instrumentalism" here], in the traditional controversies are *pseudo-statements,* i.e., devoid of cognitive content." [8] And, he continues, "The view that [such] . . . sentences . . . are non-cognitive was based on Wittgenstein's principle of verifiability . . . [*which*] *was later replaced by the more liberal principle of confirmability*" (my italics).[9] In other words, Carnap's contention clearly seems to be that since there are no differences that can be established by strict confirmationist methods between realism and its negation, there is no contingent difference, indeed nothing other than a verbal difference, between them. In view of the total impotence of strict confirmationism in selecting among the myriad of mutually incompatible possible knowledge claims in the sciences and everyday life as well as in philosophy, there seems no reason not to yield to our philosophical common sense and hold that the difference between realism and instrumentalism is substantial, indeed, contingent. Realism is true if and only if it is contingently true that the unobservable entities like those referred to by scientific theories exist and false if they do not. If it is false, then perhaps instrumentalism, or operationism, or some other alternative is true.

It might be objected that we have still neglected an important difference between the realism-instrumentalism controversy and the case of competing theories in the sciences. For with competing scientific theories at any given time, it is always possible in principle that *future* evidence will count more for or against one than the other, whereas, the objection continues, this is not true for the realism-instrumentalism issue. There are two replies. First, if we use the more common meaning of "observable," then as our instruments become more highly developed, in some cases (for example, as progress is made in microscopy), things that were once unobservable become observable. Surely this *evidence* disconfirms instrumentalism and confirms realism. Second, regarding the more general and important theories in the advanced sciences, it is always possible to "save" any theory in the face of any evidence by changing the "auxiliary" hypotheses and the assumptions about the unobservable initial conditions. So there are cases where it is, in principle, impossible for strict confirmationism to make a selection between competing scientific theories no matter how much evidence accumulates. If it is objected that some of such theories can be "saved" only by making the auxiliary assumptions hopelessly complicated and ad hoc and at the expense of having them explain less than their competitors, I shall be delighted. For surely as our theoretical knowledge increases in scope and power, the competitors of realism become more and more convoluted and ad hoc and explain less than realism. For one thing, they do not explain why the theories which they maintain are mere, cognitively meaningless instruments are so successful, how it is that they can make such powerful, successful predictions. Realism explains this very simply by pointing out that the predictions are consequences of the true (or close to true) propositions that comprise the theories.

Since I have quoted Nagel, I should, in all fairness, say that I do not know whether or not he is a strict confirmationist. At any rate the reasons he gives for his position on this issue are not strict confirmationist ones. I have commented on his argument in the passage that follows from "The Ontological Status of Theoretical Entities":

There follows here a brief and what I hope is a not too inaccurate summary of his [Nagel's] argument. Various criteria of "real" or "exist" (runs the argument) are employed by scientists, philosophers, etc., in their considerations of the "reality problem." (Among these criteria—some of them competing, some compatible with each other—are public perceivability, being mentioned in a generally accepted law, being mentioned in more than one law, being mentioned in a "causal" law, and being invariant "under some stipulated set of transformations, projections, or perspectives.") Since, then (it continues) any two disputants will, in all probability, be using "real" or "exist" in two different senses, such disputes are merely verbal. Now someone might anticipate the forthcoming objections to this argument by pointing out that the word "criteria" is a troublesome one and that perhaps, for Nagel, the connection between criteria and reality or existence is a contingent one rather than one based on meaning. But a moment's reflection makes it obvious that for Nagel's argument to have force, "criteria" must be taken in the latter sense; and, indeed, Nagel explicitly speaks for the connection between criteria and the "senses [sic] of 'real' or 'exist'." Before proceeding to a criticism of these arguments, let me point out that Professor Gustav Bergmann, completely independently, treats ontological questions in a similar manner. Rather than criteria, he speaks of "patterns," although he does say that he "could instead have spoken of criteria," and he makes explicit reference to various "uses" of 'exist'.

There are two main points that I wish to make regarding this kind of approach to ontological issues. First, it seems to me that it commits the old mistake of confusing meaning with evidence. To be sure, the fact that a kind of entity is mentioned in well-confirmed laws or that such entities are publicly perceptible, etc.—such facts are evidence (very good evidence!) for the existence or "reality" of the entities in question. But I cannot see how a prima-facie—or any other kind of—case can be made for taking such conditions as *defining characteristics* of *existence*.

The second point is even more serious. One would hope that (Professor

Norman Malcolm notwithstanding) over nine hunderd years of debate and analysis have made it clear that existence is *not* a property. Now surely the characteristics of being mentioned in well-confirmed laws, being publicly perceptible, etc., *are* properties of sorts; and if these comprised part of the meaning of 'exists,' then 'existence' would be a predicate (and existence a property).

Thus, it is seen that the issue between instrumentalism and realism can be made into a merely verbal one *only* by twisting the meanings of 'existence' and 'reality', not only beyond their "ordinary" meaning but, also, far beyond any *reasonable* meanings which these terms might be given. [And, I would add today, beyond any meaning ever intended by either party of the realism-instrumentalism controversy.] [10]

It is not my purpose here to give my constructive arguments for realism, which can be found elsewhere,[11] but to examine the implications of certain traditional philosophical issues for this problem and, in so doing, remove some of the grounds for opposing realism. My aim to begin by making a case for the legitimacy of the *problem* by showing that there is no viable reason for supposing it to be a "pseudo-issue" has made it necessary to treat some of the more difficult and complex issues ahead of some of the simpler ones. Let us now turn to some of these. The implications of strict inductivism are straightforward and extreme. Recall that the premises in inductive arguments countenanced by strict inductivists consist of singular statements which either are totally about the directly observables or have been confirmed on the basis of direct observation by simple inductive arguments, and also that the form of the permitted simple inductive arguments insures that the conclusions do not refer to properties or kinds of entities unless they are also referred to in the premises. This makes confirmation of any statement referring to anything unobservable completely impossible. The strict inductivist's contingent knowledge is limited to whatever he takes his observables to be, as is his epistemically viable ontology. If he takes direct observables to be private experiences, he will tend to be a subjective idealist or some variety of phenomenalist and, if consistent moreover, a solipsist. Obviously, he must hold an instrumentalist or some similar view of scientific theories. If he holds that observables must be publicly observable, he will tend to be a direct realist and a naïve realist[12] as well as a behaviorist. If he is consistent, he will disavow any knowledge of the private experiences of others *as well as those of his own.* I cannot withhold a certain wry admiration for philosophers, psychologists, and social scientists who openly espouse these strange views. Since they find my views and some of Russell's that I defend even stranger, I hope that they will tender me the same forbearance. But one cannot dismiss so cavalierly the consequences of strict inductivism (and to a lesser degree other kinds of strict confirmationism) for scientific methodology, particularly in psychology and the social sciences. I have no doubt that the behaviorist makes the invalid inference from "Behavior is what we *observe*" to "Behavior is what we can *know about*" because he is, tacitly at least, a strict inductivist, and it does not occur to him that, by means of hypothetico-inferential reasoning, we can use behavior as evidence for confirming hypotheses about private experiences. Once the total impotence of strict inductivism and the necessity for something like hypothetico-inferential reasoning (freed from the millstone of strict confirmationism, of course)—if complete skepticism is to be avoided—are recognized, the behaviorist tragicomedy that has reduced large segments of psychology and the social sciences to trivial poll-taking, head-counting, and lever-pressing can be seen for what it is—no more grounded in reality than the search for the philosopher's stone or the Salem witch trials!

The abandonment of strict inductivism, then, removes much of the grounds for opposing realism, not only realism regarding the theoretical entities of physics, chemistry, and the like, but realism concerning the external world in general and realism regarding the private experience of others and ourselves. Except in the more indirect way discussed earlier, strict confirmationism that employs hypothetico-inferential reasoning does not pose difficulties for realism regarding unobservables, although we have seen that it is just as untenable as strict inductivism.

Some readers may have noticed that what I have called "strict confirmationism" is the same as what has sometimes been called "judgment empiricism," [13] a position which, unlike some varieties of empiricism, is silent or agnostic concerning the *source* of contingent, general knowledge claims, but holds that they must all be justified or confirmed by experience or observation in order to be acceptable. I mention this in order to make it clear that I recognize the gravity of my charge that strict confirmationism is invalid. It means that judgment empiricism must be abandoned; although it has already been emphasized that this does not diminish the importance of observational, experimental testing. It only means that in assessing our hypotheses on the basis of such tests, it is necessary to use assumptions that have not been in the least confirmed and that will neither be confirmed or disconfirmed by the particular test being conducted.

Having rejected the objections to realism due to judgment empiricism by rejecting the doctrine itself, we must now face a kind of objection that arises from a view sometimes called "concept empiricism." As the name implies, it holds that concepts must originate in experience. Or, in contemporary terms, the meaning of all words that are not given explicit verbal definitions must come from ostension. Two succinct statements of the position are Hume's maxim "No idea without a previous [sense] impression" and Russell's *principle of acquaintance, "Every proposition that we can understand must be composed of ingredients with which we are acquainted."* An exception is often made of logical concepts, that is, the logical connectives, variables, and quantifiers, although Russell makes a valiant attempt to avoid this exception or, at least, to reduce it to a minimum (see, for example, his *Human Knowledge,* part II). A concept empiricist may or may not be a judgment empiricist, although historically, I suppose, most of them have been. Berkeley may not have been since, as far as I know, he made no claim of experiential evidence for asserting God's role in feeding into us our ideas and insuring by His own observations that objects of sense do not cease to exist when we cease to observe them. Or, as is often charged, he may have been merely inconsistent.

Concept empiricism is unfashionable in many circles today. I do not know whether or not it is true, although, at least in the case of descriptive predicates, I am inclined to accept it and find the arguments offered against it unconvincing. One can admit, for example, that there is a syntactical component in the meaning of descriptive predicates and still maintain that the ostensive, experiential (and referential) component is necessary and crucial. And one can be a concept empiricist without being a logical atomist and without introducing sense data or other such "objects" (as opposed to *events*). Fortunately, it is not necessary to consider these matters further here or to affirm or deny concept empiricism. For the obstacles that it seems to pose for realism can be removed by procedures that for other, more cogent reasons are necessary for an adequate philosophy of science and epistemology.

At first blush these obstacles to realism seem insurmountable for the concept empiricist unless he is also a direct realist, and even then any kind of realism regarding unobservables seems impossible. The difficulties are parallel to, but should be clearly distinguished from those of the strict inductivist. The strict inductivist lacks forms of inference that permit him to confirm knowledge claims about unobservables. This difficulty disappears if strict inductivism is rejected and hypothetico-inferential reasoning (freed from strict confirmationism) is permitted as a means of confirmation. But the concept empiricist, it would seem, lacks the concepts necessary even to *express* knowledge claims about unobservables, since for him observation is necessary for meaning. It does no good to give him a means of confirmation if he can neither state nor understand anything to confirm because he lacks meaningful terms with which to do so. It is of passing interest to note that these two kinds of objections to realism, the one from strict inductivism and the one from concept empiricism, are often confused and conflated into ostensibly one objection. This is especially true when the view that is being attacked is *representative* realism. The confusion results from confusion of meaning with evidence or confusion of what a proposition asserts with how it comes to be known, the *fallacy of epistemologism*. This confusion has been very unfortunate because the two kinds of objections are very different from each other and often the proponents of realism, not being aware of the confusion, have not been able to formulate the two different kinds of replies necessary to counter the objections satisfactorily.

Russell, among others, has shown how to reconcile realism and concept empiricism. His distinction between knowledge by acquaintance and knowledge by description and his theory of descriptions (including indefinite descriptions) show how it is possible both to maintain his *principle of acquaintance* and nevertheless to formulate propositions about entities with which we are not acquainted.[14] If I say that somebody stole my car last night and if what I say is true, I have referred by means of indefinite description to the thief (or thieves), although I am not acquainted with him (them). As Russell puts it, the thief is not an ingredient of the proposition that somebody stole my car. This means, as far as I am concerned at any rate, that, although the proposition asserts that there *is* a thief, it does not say *who* he is—it does not *name* him. Reference to items with which we are not acquainted is *indirect* reference and is always accomplished as in the example given, that is, by purely logical terms (variables, quantifiers, etc.) plus terms whose *direct* referents are items of acquaintance. If we recognize, as I believe we must, that, in addition to individual things, we can refer to properties and classes by descriptions, using quantified predicate variables, then the difficulties that the concept empiricist has with realism regarding unobservables vanish. He can formulate propositions that refer to unobservable properties or to classes of unobservable things by means of existentially quantified predicate variables and other purely logical terms plus terms whose direct referents are observables. Fortunately any theory whatever can be transformed without loss of significant content into such a proposition. It is only necessary to replace the conjunction of the assertions of the theory by its Ramsey sentence; that is, each theoretical predicate (each term referring to unobservables) is replaced by an existentially quantified predicate variable, the scope of each such quantifier extending over the entire conjunction. It is well known that every observational consequence of the original theory is also a consequence of its Ramsey sentence; but it should also be emphasized that although theoretical *predicate constants* are absent and the problem of the meaning of theoretical terms that plagues the concept empiricist is absent, nevertheless every unobservable *property* and *class* referred to by the original theory is still referred to by the Ramsey sentence (if it is true). We do not know *what* these properties and classes are—we cannot "name" them with predicate constants, but if we know the Ramsey sentence, we know that the properties exist *and we know something else about them.* We know what the Ramsey sentence asserts about them by means of variables, quantifiers, logical connectives, and terms referring directly to observables. The assertion that we do not know *what* these properties are may be taken as an explication of Russell's contention that we are ignorant of the intrinsic nature of the unobservable. (We do not know *what* first order unobservable properties are, as I put it.) And the assertion that we know that the unobservable properties exist and that *they,* in turn, have the properties (second or higher order properties) attributed to them by the Ramsey sentence is approximately equivalent to Russell's contention that we *do* have knowledge of the *structural* properties of the unobservable. Temporal succession, simultaneity, and causal connection must be counted among these structural properties, for it is by virtue of them that the unobservables interact with one another and with observables and, thus, that Ramsey sentences have observable consequences.

I believe that most objections to realism either as an ontological thesis or as a methodological policy are due either to mistaken views about confirmation—specifically strict confirmationism and especially strict inductivism—or to uneasiness about the meaning of terms that are supposed to refer to unobservables. We have seen that all varieties of strict confirmationism must be rejected on independent grounds and that the second kind of objection is removed by using Russell's theory of descriptions implemented by Ramsey sentences. There may be important objections to realism either as an ontology or as a methodological policy other than these epistemological ones, but I do not know what they are. It is not the purpose here to give constructive arguments *for* realism, but I shall mention one, since it depends on the constructive theory of confirmation that I have summarized in note 7. I contend that realism is much more highly *confirmed* than any other theory that "explains the facts," because its *prior probability* is so much greater than any other such theory, and prior probability is always one of the crucial factors in selecting among competing hypotheses all of which explain current evidence. My reasons for accepting realism are of the same kind as those for accepting any scientific theory over others which also explain current evidence. That a philosophical position such as realism turns out to be a contingent theory and one, moreover, for which scientific theory and observational evidence are relevant should not surprise us once we have freed ourselves of the mistaken views about the nature of philosophy that arise from strict confirmationism.[15] Contemporary empiricists, who have tended to emphasize the factual emptiness of propositions that are certifiable on the basis of purely logical, conceptual, or linguistic considerations and that severe limits are thereby set on the potency of such considerations, should, it would seem, have been profoundly suspicious of claims that such considerations were sufficient to decide issues that have, among many other things, such pervasive and extreme implications for work-a-day scientific practice such as realism versus phenomenalism, realism versus operationism, realism regarding private experiences versus behaviorism, etc. The failure of such suspicion to develop can be understood, perhaps, by remembering that in order to reject the claim, it would be necessary to reject strict confirmationism and, thus, to abandon empiricism of the judgmental variety. What holds for these issues regarding realism holds also, I believe, for most—perhaps not all—of the important, interesting problems of epistemology and metaphysics. These problems differ only in degree and not in kind from more run-of-the-mill scientific problems in that they ask somewhat more general and fundamental questions and in that the proposed answers are more indirectly connected with evidence and they are thus more difficult to confirm or disconfirm.[16]

Let us now turn to some implications that flow in the opposite direction, that is, to implications of currently accepted scientific theories for traditional philosophical problems, specifically the puzzles about sense perception and the nature of the external world. Russell and others[17] have argued, conclusively it seems to me, that current scientific knowledge entails that the things that we commonsensically identify with tables, chairs, human bodies, etc., and take to be directly perceived, in fact, the entire physical environment or "external world," are all, strictly speaking, unobservable and that, moreover, we have no knowledge as to what their first order properties are but only knowledge about their structural properties. I shall not repeat these arguments here but only give a summary of one that I believe to be conclusive. But first, since this position and accompanying arguments, like *most* of Russell's later philosophy, have been ignored, misunderstood, or rejected as having no relevance for "philosophical" problems of perception, I must pause and request the reader to hear me out and, if interested, to see some of the references listed in note 17. I shall deal with the metaphilosophical objections presently. The only first order properties of which we have knowledge by acquaintance—the only observable first order properties —are those that are exemplified in our private experience. We have knowledge by *description* of the *higher* order properties *of* the *first* order properties of the external world, knowledge that derives from the Ramsey sentences of our theories (if they are true or reasonably close to the truth). The Ramsey sentence refers to these first order properties, as explained earlier, by means of the terms of pure logic and terms referring directly to items of our direct experience.

The main argument for this position may be summarized briefly. If our current theories in physics, neurophysiology, and psychophysiology are at all close to the truth or even if they are at all headed in the right direction, then a complete description, including a complete causal account, of everything that is involved in perception except the private experience itself would mention only such entities and events as submicroscopic particles, electromagnetic quanta, etc., and their relations and interactions with one another and with, for example, neural termini in the retina, afferent neural impulses, and patterns of neuronal activity in the brain. At no point in the entire, complete description and causal explanation is there mention of any first order property such as colors *until* we come to the private experience that results from the pattern of neuron firings in the brain.[18] It seems to me that we must conclude that colors are exemplified *only* in our private experiences and that there is no reason to believe that they are ever properties of the material objects of the external environment. *What holds for colors must also be true for all of the first order properties that we perceive directly.* We do not know what any of the first order properties of material objects are, although our (Ramseyfied) theories tell us that they exist and what some of *their* (second and higher order) properties are. This ends the summary of the scientific argument for Russell's contention that our knowledge of the external (or *physical,* or nonmental) realm is limited to its structural aspects. It also seems to be an equally strong argument for concept empiricism (excepting logical concepts), but I shall not press this point here.

It could immediately be objected that the proposition that first order properties known through the senses such as colors are not properties of physical entities is not a conclusion that follows necessarily. This is true. What does follow is that even if physical objects were colored,[19] we could not perceive, directly or otherwise, *that* they were, nor, most certainly, could we perceive *what* their colors were. We could not perceive them directly, since whatever information about the external world we get from perception results from long, complicated, and devious causal chains. And we could not "perceive them indirectly" (whatever this might mean); nor could we know by any means that such colors exist or what they are, for they are completely causally redundant. According to physics, they would play no role in determining the frequency of the electromagnetic radiation which is the *only* factor, as far as the physical object is concerned, in determining what colors we see. These considerations make the hypothesis that colors are among the first order properties of the external world as gratuitous and unwarranted as the hypothesis that invisible witches dance nightly on the University Mall and are responsible for its fiscal, administrative, and pedagogical woes. They also reveal as whistling in the dark—as attempts to evade or dismiss the problems rather than to solve them—the contention that, although perception involves complex causal chains, it is *by means of these very chains* that we perceive external objects *directly* (what could "directly" mean here?), including their first order properties, such as color, pretty much as they are. As Russell has said in a letter to A. J. Ayer, "You say that from the fact that the perceived qualities of physical objects are causally dependent upon the state of the percipient, it does not follow that the object does not really have them. This, of course, is true. What does follow is that there is no reason to suppose that it has them. From the fact that when I wear blue spectacles, things look blue, it does not follow that they are not blue, but it does follow that I have no reason to suppose they are blue." [20]

I must now face an objection that I have not been able to understand very well. According to it, a scientific description and causal account of perception is not relevant to the traditional philosophical problems about perception and reality and, thus, must be ignored when considering these problems. I do understand, as noted earlier, that some philosophers for strict confirmationist or other reasons hold that philosophy must limit itself to matters that are purely logical, conceptual, or linguistic. But what follows from this view of philosophy seems to be not that a causal account is irrelevant for the traditional problems but, rather, that what have counted as the traditional philosophical problems of perception are not truly philosophical problems—that whether or not, for example, the sensory qualities are exemplified only in private experience or as properties of external, physical objects is a scientific rather than a philosophical matter. I would be pretty much in agreement with such a conclusion. My objections would be mainly verbal; I do not favor restricting the extension of "philosophy" so narrowly; nor do I think a sharp distinction between the scientific and the philosophical is desirable. But I am afraid that the objection is usually not intended in this manner. It is true that the objector often continues by holding that the task of the philosopher is to determine what we mean by our perceptual judgments. I agree that this is interesting and important, but the task seems quite easy and the answer obvious. Their meaning is such, it seems to me, that although they are intended to refer to instances of sensory qualities, we directly experience these instances, and they, nevertheless, are properties that belong to physical objects. Whether I am right about this does not matter here. What I am concerned to maintain is that, no matter *what* our conceptual judgments mean, if they conflict with the implications of the "causal theory of perception" that we drew above, it is they, rather than the scientific account, that must be appropriately modified before they can be accepted as being, strictly speaking, true.

According to one objection to this that must be taken seriously, we come to give meaning to words referring to these qualities by learning to make true perceptual judgments; therefore, any theory that holds that our perceptual judgments are generally false entails that all terms referring to observable qualities are meaningless. Well, in the first place, it is not necessary for me to maintain that all of our perceptual judgments are false but only that those which attribute first order sensory properties to external objects are false. For example, the judgment expressed by "This is blue" may be true even though there is a mistaken belief about some of the other characteristics of whatever "this" refers to. Moreover, I must deny that we cannot learn or give meanings by making false judgments. Let me first use a common-sense example. Suppose that a child does not know what "beige" means and that one of several men in the parlor is wearing a beige scarf. Suppose that the child and his mother both mistakenly judge that the man's overcoat is brown. The next day the child asks, "What does 'beige' mean?" His mother replies that the man in the parlor yesterday who wore the brown overcoat was wearing a beige scarf. "Oh yes, I remember that scarf. I see; so the man with the brown overcoat wore a beige scarf. Now I know what 'beige' means." Even though the perceptual judgment was, strictly speaking, false, surely it sufficed for giving "beige" meaning for the child due to the correctness of the judgment that the man incorrectly judged to be wearing a brown overcoat *was* truly wearing a beige scarf.

Similarly, we may falsely judge that we are directly perceiving an external object that is blue, yet the judgment might help give someone meaning to "blue" because of the correctness of the judgment that an instance of blue is present in whatever is being judged external whether it *is* external or not. "But," the objection continues, "in all such cases, perception of an instance that truly did exemplify the quality referred to by the word in question was necessary. According to your view, we never observe anything external. What meaning can you give to 'external'? I should say that just as we perceive instances that are truly blue, we also perceive as external instances that are truly external. Otherwise we could not know what 'external' means, which is absurd." Well, a rough and ready definition of "external" in the sense that I have been using it would be "mind independent, that is, not among the entities, events, etc., present in my private experience." "External" in this sense refers to a structural property and I can, thus, understand its meaning by understanding the meaning of logical terms and terms referring directly to items of experience. There are other senses of "external," of course. The related set of perceived properties that I commonsensically label "this desk" is directly perceived to be *external* to the set that I label "my body" in my own (private) visual space. From this it may or may not be correct to *infer* that the physical entity that plays an important causal role in my perception of the first set of properties (let us call it the *physical desk*) is external to my physical body in physical space. But it would be a groundless and gross confusion to infer from this that the physical desk has anything more than a complicated causal relation to the first-mentioned set of perceived properties the "perceptual desk," which is an item of private experience, even though common sense may mistakenly identify them. It would be perhaps an even worse confusion to infer that the perceptual desk is outside my physical body in physical space. Needless to say, physical space and its properties are structural so that externality in physical space is a structural relation.

Of course, it remains logically possible to ignore or deny the "causal theory of perception." And the strange philosophical views that result from doing so, like most theories, including scientific ones, cannot be conclusively refuted. But once the untenability of strict confirmationism, including, of course, strict inductivism, is recognized, there seems to be nothing that can be said in their favor. Nor does there appear to be any good reason for refusing to recognize what is virtually forced upon us by common sense as well as by science: our perceptions are caused by a large number of things or events, some of them external to our (physical) bodies and some of them internal. We cannot hope to understand perception or to understand adequately or know what is perceived without understanding these complicated causal chains.

The argument given earlier to show that the sensory properties are exemplified only in private experience and that our knowledge of the external world is structural is sometimes held to be inconsistent and, therefore, invalid or, at best, unsound.[21] For, it is claimed, such arguments have premises that assume the existence of the physical thing, say the surface of a table, and that, moreover, describe the physical thing in a (perhaps unavoidably) naïve realist [22] fasion (that is, at least some sensory properties are predicated of them). Thus among the premises we find the assumption that naïve realism is true, which contradicts the conclusion that naïve realism is false. Even if this were true, it would not invalidate the argument, as Russell has pointed out; whence his notorious explanation that naïve realism leads to physics and physics implies that naïve realism is false; thus, if naïve realism is true, it is false; therefore it is false. This is essentially correct. The structure of the argument may be schematized as follows:

$$\frac{\begin{array}{c} S \\ N \end{array}}{\text{Therefore} \sim N}$$

where "S" stands for the conjunction of the assumptions from science used in the argument and "N" for those of naïve realism. We know from elementary logic that if this argument is valid the following one is also valid:

$$\frac{S}{\text{Therefore } N \text{ implies} \sim N} \cdot$$

But "N implies $\sim N$" is equivalent to "$\sim N$." So the argument is equivalent to

$$\frac{S}{\sim N} \cdot$$

It might be said that this shows that if the argument *is* valid, it should be possible to present it in the form directly above, so that no naïve realist assertions appear in the premises. This is true, and the task is easily accomplished. We need only to replace the conjunction of the premises with its Ramsey sentence, in which, of course, terms like "tabletop" and "brain" as well as "neuronal" and "electromagnetic quanta" do not appear and have been replaced by existentially quantified variables of appropriate type.

The realist view that I have defended under the name "structural realism" is, of course, similar to the "critical realism" of a group of somewhat earlier American philosophers. A representative, Roy Wood Sellars, is still an active contributor to the philosophical literature. A similar view is also held by Stephen Pepper.[23] It is certainly also closely related to the much maligned *representative realism* of Locke, and a comparison is interesting and instructive. We have already discussed the two stock objections to representative realism, the one from strict inductivism and the other from concept empiricism. The first vanishes when the untenability of strict inductivism is acknowledged and hypothetico-inferential confirmation is adopted. And the second one disappears when it is seen that no new concepts—no concepts other than logical ones and ones corresponding to items of direct experience—are needed to formulate propositions that express our (purely structural) knowledge (by description) about the (unobservable) external world. This is accomplished by the use of descriptions—more specifically by the use of the Ramsey sentence. When these features are added to Locke's view, it becomes identical with structural realism. Both the "primary" and the "secondary" properties of physical entities are structural properties, and, just as Locke insisted, the primary properties are those which resemble the corresponding (structural) sensory properties, while with the secondary properties there is no such resemblance. Even after strict inductivist holdovers are firmly rejected, this claim requires some comments. This is the first time that exemplifications of structural properties in private experience have been mentioned, but clearly they are abundantly present. When I observe that there are *three* black discs in my visual field, I observe a structural (second order) property of the group of discs. (Number or *cardinality* is perhaps the simplest structural property.) Consider now an approximately circular color expanse in the visual field. Such a circle will have, or approximate to, all of the formal (and therefore structural) properties of any circle on a Euclidean plane, all of which, of course, may be referred to by arithmetical and, thus, logical terms (using, for example, analytic geometry). We may now formulate a (Ramseyfied) hypothesis asserting the existence of an entity, call it a *circle in physical space* that has structural properties very similar to those of the circle in the visual field. The hypothesis would also describe the causal chains proceeding from positions on the physical circle to corresponding positions on the circle in the visual field. It should be emphasized that there are no purely logical or purely conceptual reasons that there be structural similarities between objects in the external world and items in our experience. Whether or not such similarities exist is entirely a contingent matter; if we have reason to believe that they do, it is because we have well-confirmed theories that assert this to be the case. But if such similarities were fewer or, even, virtually nonexistent, knowledge of the physical realm would be more difficult to come by but not necessarily impossible.

Not only are there numerous important structural properties common to items in our experience and physical objects; these same structural properties are, in general, also shared by the (nonexistent) naïve realist objects of common sense. This explains why our common-sense perceptual judgments, though strictly speaking false, nevertheless stand us in such excellent stead. They are close enough to the truth for most purposes because they attribute structural properties to ostensibly external objects that are identical with or very similar to the structural properties of the actual, external, physical objects. Moreover, there are structure-preserving causal chains between the parts of these physical objects and the corresponding parts of the patterns in our private experience that are mistakenly identified with the external objects that cause them.

There are other objections to this view, but these seem to me to have no cogency at all. I shall mention them only because I know students and colleagues whom I respect who take them seriously. Ryle[24] has claimed that the fact that words like "see" and "perceive" are not *process* words or *event* words—as shown by there not being any straightforward common use for their present participles—proves that there are no processes of "perceiving" or "seeing." So, since perception is not a process then, a fortiori, it is not a physiological process; nor is it an event caused by such a process. I am not so sure about the premises of this argument but, leaving that aside, I cannot see how they have much relevance for the conclusion. It may be true that in common-room usage we would seldom or never hear expressions such as "I am now *perceiving* [or *seeing*] a table." But surely we might hear or say, "I have been *watching* [or *looking at* or *keeping in sight*] that goldfinch continuously for ten minutes," or "I have been *hearing* a faint wailing in the other room for several minutes," or "Please don't interrupt; I am *watching* a wrestling match on TV." And even if it were true that there is no ordinary use of the present participle of "perceive," what does this imply for the traditional philosophical problems of perception? For surely philosophers have quite legitimately broadened the use of "perceive," and "perception" far beyond everyday colloquial usage. It may be true that words like "see" are "success" words, although it may be questioned whether this is always true. But even if "I see a table," as ordinarily used, entails "I observe directly an external, mind-independent object" (or perhaps something similarly but less offensively formulated), all that would follow, in my view, is that "I see a table" is always false. However, we do have visual experiences that we commonsensically (but mistakenly) identify with seeing tables, and we can use "see" in a new, similar but somewhat different sense to talk about seeing items in our (private) visual fields.

This brings us to the nub of this family of objections: it is supposed to be wicked, indeed impossible, to talk about private experience, much less *items* thereof. I believe that this objection is a simple, blatant example of what I called earlier the *fallacy of epistemologism,* the mistake of confusing questions about what is meant with questions about procedures for confirmation or verification. As might be expected, strict inductivism or strict confirmationism is often involved in instances of this fallacy. Let us review, then, the "argument against private languages," since it is supposed to preclude formulation of statements about private experiences. Suppose I want to label with a term, say "α," instances of what I claim to be a novel kind of experiences which occur from time to time. And suppose after a number of these alleged instances I say, "Oh yes, I am experiencing α again." But, I am told by the opponents of private languages that I cannot successfully do so. For, they say, I have no "criteria" for the *correct* application of "α." *I have no way of knowing* whether it is the same kind of experience that I last labeled "α" or not. I am not allowed to say that I distinctly and vividly *remember* the last instance and that it was virtually qualitatively identical with this instance, for, alas, I have no criteria for the correctness of this memory. Now get this: since there would be no way of *knowing* whether I merely think it is the same or whether I know it is the same, *it follows that,* so the argument goes, there would *be* no difference between merely thinking it the same and knowing it to be the same. Seldom do we see such an ingenuous and patent instance of the inference from what is or can be *known* to what *is* or *can be*. It is then supposed to follow that attempts like this, as well as all other attempts to give meaning to terms to refer to private experience, must fail. Why am I not allowed to contend that there *is* a difference between *merely thinking* it to be the same kind of experience and *knowing* it to be, namely that in the latter case it must have *been* the same, whereas in the former case it may not have been? I cannot see any cogent objection to this contention once the fallacious inference from lack of knowledge to lack of existence is rejected, unless "know" means "know with complete *logical* (as opposed to *psychological*) certainty." I am not clear as to what logical certainty about a contingent proposition could be, but even under strict confirmationism or, even, strict inductivism, it is apparent that no contingent proposition can be certain except perhaps those that report direct observations at the moment of observation. The requirement that all knowledge be certain knowledge would preclude any kind of confirmation, even strict inductivism. Are the proponents of "the argument against private languages" really prepared to reject all general knowledge claims and restrict their knowledge to singular statements about what they are observing at the moment?

But perhaps there are ways for them to escape this absurdity. Let us see. Suppose that they grant that there would indeed be a difference between knowing and merely thinking that can be drawn as I did above. But, they say, you would have no way of knowing that there is such a difference, even in a fairly liberal sense of "know." You would have no way to confirm your contention that you know it is α as opposed to merely thinking so, they continue; therefore, you would have no way of confirming your claim that you have succeeded in giving "α" any meaning. Suppose that I claim that it *is* highly confirmed that this occurrence of α is of the same kind as the previous one because my memory is, in general, reliable, especially when the instance of memory is as clear and vivid as this one. I believe that we now reach the true parting of the ways. There *is* no way to confirm the hypothesis that memory is generally reliable if confirmation is limited to strict inductivist or, even, strict confirmationist resources. For in addition to this hypothesis, there will always be an indefinitely large number of other hypotheses which, insofar as logical considerations are concerned, account for the "facts" equally well. If one does not know that the same is true for the hypotheses of science and everyday life and, thus, is ignorant of the total impotence of all varieties of strict confirmationism, one may try to uphold the argument on these strict confirmationist grounds.

One more move is usually made by the opponents of private languages, knowledge of private experiences, etc. Suppose, they continue, that it be granted that memory is generally reliable. Still, the case of a private experience is different. With other instances of memory, additional reasons can usually be given in their favor. We can check the alleged memory either by making further observations of our own or by calling upon the testimony of others; neither is the case for an ostensible memory of a private experience. My first reaction is: and so? Further testing by means of observations and testimony may indeed be desirable and may increase the degree of confirmation, but to hold that the possibility of such must exist in order for a knowledge claim even to have meaning seems to me another *epistemologistic* dogma and a grossly false one at that. Surely the possibility of both is not required. If I sincerely judge "This is red," the judgment may be bolstered by others' making the same perceptual judgment; but what other observations, beyond continuing the one in question, could reasonably be required of *me?*

Let us turn, then, to the requirement of testimony or, as it is sometimes called, "intersubjective agreement" or "public observability." The argument against the possibility of private languages is, of course, the same thing as the argument for the necessity of public observability. But when a sufficiently unrestrictive meaning of "know" is adopted and when strict confirmationism is abandoned so as to admit the general reliability of memory, it seems necessary to introduce the requirement of public observability as a *premise* if the argument is to retain any force. In view of this patent circularity, let us see if there are other grounds for the requirement of public observability. The *desirability* of corroboration by testimony is not in question. We are only asking whether testimony is *necessary* in order that terms *have* or *be given* meaning and whether, indeed, public observability is necessary in order for a *kind* of testimony to be operative. It must be admitted that the ordinary or common-sense account of how terms get meaning or of how a language is learned is simpler in form than the structural realist account. But it has already been noted that the growth of knowledge sometimes forces a simpler theory to give way to a more complex one. Since independent consideration seems to urge structural realism upon us, let us see whether it can yield a coherent theory of meaning. To summarize the common-sense account roughly, we learn the meaning or use of terms by hearing them used by others in the presence of the appropriate publicly observable entities. According to the structural realist, there *are* no public observables in the strict sense, because we do not directly observe anything in the external environment. There are, however, in this environment, entities that have what Russell has called "quasi-publicity." These produce similar perceptual effects in each of us so that, for example, under appropriate, easily arranged circumstances both you and I will be caused to have a red, circular color expanse in our visual field. If you can by any means cause me to focus attention and interest upon the expanse in my field and then say, "This is red," or, perhaps, just "Red," and I come to associate the sound you caused me to hear, "redd," with the colored expanse, then I will gradually learn the meaning of color words.

Let us take another example. Suppose, at some moment, you have reason to believe the hypothesis that pain is occurring in the private experience of your child because of evidence such as crying, grimacing, or because you saw him touch something hot or get a small scratch on his arm. (The hypothesis is hypothetico-inferentially supported by such evidence. We give it a much higher prior probability than alternative hypotheses that would be supported by the same evidence because we know by acquaintance that similar hypotheses are indeed the true explanation of similar behavior of our own and that pain *does* result when we touch something hot, etc. The common objections to this somewhat misnamed "argument from analogy" are all due to strict inductivism and other varieties of the fallacy of epistemologism. Once these fallacies are rejected, the grounds for belief in the existence of "other minds"—of the existence in others of thoughts and feelings very similar to our own—become clear and unproblematic. Such beliefs are the hypothetico-inferentially best confirmed explanations of the relevant observable evidence, which, in its turn, is causally produced in our own sense experience by the others' behavior and by other relevant events.) Believing, then, the hypothesis that your child is in pain, you say, "I'll bet it hurts, doesn't it," or, perhaps, just "Hurts!" or "Hurts?" etc. Again, if his interest is appropriately focused, he will learn to associate "hurt" with appropriate events in his private experience as well as to use it in hypotheses about the occurrence of similar events in others. Thus, although more complex, the structural realist account avoids the absurdities of the extreme *philosophical behaviorism* that the received account can avoid only at the expense of inconsistency. It goes without saying that we do not usually first consider the behavioral and other evidence and say to ourselves, "In the light of the evidence, the hypothesis that he is in pain seems the best confirmed among the other possible explanations." We immediately, automatically, and usually *correctly* judge that he is in pain, just as we immediately, automatically, but *incorrectly* make naïve realist judgments as a result of most of our sense experience. I have no doubt that we are constitutionally "wired" to perceive and conceive the world as naïve realists. Fortunately, this does not matter for most purposes, as was explained earlier, and it is undeniably simpler than the more correct interpretation provided by structural realism.

Even for most scientific purposes, it does little or no harm to interpret theoretical terms with the help of images and analogies with common-sense material objects, because, as already explained, the structural properties will usually be correctly ascribed. However, when considering some of the fundamental philosophical and scientific problems, we have seen that it is necessary to get a little closer to the truth and use the structural realist account. By so doing and by recognizing the untenability of strict confirmationism and the fallacy of epistemologism, we are able to remove the obstacles to a thoroughgoing realism regarding the "external world," the "unobservables" of scientific theories, and the private experience of ourselves and others. We thereby remove any puzzles about "other minds," and thus avoid the absurdities of both philosophical and methodological behaviorisms. Due to the very impotence of strict confirmationism, neither the strange philosophical alternatives to structural realism nor the familiar one of direct or naïve realism of common sense can be refuted, but recognition of this impotence also makes it clear that there is little or no reason for holding any of them. It is true that our common-sense prejudices are outraged by the proposition that we do not know what are the first order properties (the "intrinsic nature") of the external world. But this discomfort should be somewhat assuaged by recognizing that the second order or structural properties are the important ones for most of our purposes. Moreover, it turns out that admission of our ignorance of the "intrinsic nature" of the physical realm and the resulting necessity for changing our concepts and beliefs about the *physical* have surprising implications for that granddaddy of all philosophical problems, the mind-body problem. Our general ignorance about what first order properties are exemplified by the physical removes all of the usual conceptual and intuitive obstacles to claiming that we are, after all, acquainted with a small subset of first order, physical properties, namely those that are exemplified in our private experience, in other words, that these mental events are also physical events in the brain. This kind of mind-body monism avoids the counter-intuitive aspects of traditional (and contemporary) materialisms. Unlike them, it makes no attempt to ignore genuine mental events or to sweep them under the rug. While claiming that all mental events are physical events, it recognizes, even insists, that some physical events are mental events.[25]

In closing, let us take a brief look at the implications of these considerations for empiricism, in general, and for empiricist meaning criteria, in particular. We saw earlier that the most widespread kind of contemporary empiricism, judgment empiricism, turns out to be identical with strict confirmationism and must, therefore, be abandoned, as must be all meaning criteria based on confirmation. These are now seen to be instances of the fallacy of epistemologism. Surprisingly enough, however, it turns out that contemporary scientific knowledge affords strong support for a modified version of a rather old-fashioned theory of meaning, sometimes called "concept empiricism," which, in turn, entails a different kind of meaning criterion. In my exposition of it, I assumed logical terms to be unproblematic and used Russell's *principle of acquaintance* as a "meaning criterion" both for propositions and for descriptive (nonlogical) terms.[26] This is seen to be surprisingly unrestrictive both ontologically and epistemically once strict inductivism is abandoned and the powerful referential power of definite and indefinite descriptions, as explicated by Russell and Ramsey,[27] is recognized.

The alternative to judgment empiricism that I have proposed in no way diminishes the importance of testing our theories by observation and experiment. But certainly those who, like myself, are imbued with the spirit behind the philosophical motivation of most empiricists cannot but find it disquieting; for it abandons as logically impossible the goal of grounding our knowledge in logic and observational evidence alone. Although the probabilities or degrees of confirmation it is able to assign hypotheses are objectively existing relative frequencies, in our attempts to compute them, we are always forced to rely in part on *estimates* of prior probabilities in which hunches or intuition or other subjective elements are necessarily present. We *can* take some comfort in the well-known principle that if estimates of prior probabilities are not too near the extremes, their effect diminishes rapidly as more and more evidence accumulates. But since, I believe, the effect is surely there and the danger of an extreme but erroneous estimate is always present, surely it is better to recognize this and not pretend to a *complete* objectivity or a *totally* "disinterested search for knowledge," which cannot, *logically cannot*, exist.[28] Hopefully, if this pretense is viewed with appropriate suspicion, a certain amount of healthy humility regarding our claims to knowledge may result, and perhaps there will emerge a healthy reluctance to delegate completely to scientific, military, and political "expertise," supposedly based on such nonexistent complete objectivity, the responsibility for making policy decisions that shape the destiny of ourselves and our children.

NOTES

1. The thematic quotes by Russell are from *My Philosophical Development* (New York: Simon & Schuster, 1959) and *Human Knowledge: Its Scope and Limits* (New York: Simon & Schuster, 1948).
2. *My Philosophical Development*, p. 15.
3. See especially ibid.; B. Russell, *Human Knowledge;* and B. Russell, *The Analysis of Matter* (London: Allen and Unwin, 1927).
4. The terms "inductivism," "inductivist," etc., to the best of my knowledge, were first used by Sir Karl Popper. See, e.g., his *Conjectures and Refutations* (New York: Basic Books, 1962), p. 154. I am indebted to Popper and in general agree with his critique of induction, although I do believe it plays a role in reasoning as it is actually done but *not on a strict inductivist basis*.
5. "Hypothetico-inferential" is a better term, for sometimes inferences from hypotheses to evidence may be nondeductive, e.g., statistical.
6. See my "Corroboration Without Demarcation," in *The Philosophy of Karl Popper*, ed. P. S. Schilpp (LaSalle, Ill.: Open Court, forthcoming).
7. The arguments for the necessity of ranking and subsequently selecting among competing hypotheses on the basis of estimates of something like their prior probability seem to me overwhelming. I have discussed this in detail elsewhere (see my "Corroboration Without Demarcation"), and, since the matter is not crucial for the purposes of this paper, I shall not do so again here. The confirmation theory I defend there has points in common with contemporary "personalist" or "Bayesian" theories, but there are also important differences. A finite frequency theory of probability is used (see, e.g., Russell, *Human Knowledge*, pp. 350–62). The probability (or degree of confirmation) of the hypothesis in question is a "single case" probability, the *reference class* consisting of hypotheses similar to the one in question in certain relevant respects. (I am indebted to Wesley Salmon for this and certain other ideas employed. See his *Foundations of Scientific Inference* [Pittsburgh: University of Pittsburgh Press, 1967].) For every hypothesis being considered, after relevant background knowledge has been exhausted, it is always necessary to make estimates of prior probabilities on the basis of guessing, or intuition, or other nonevidential means. Our estimates may be good or bad, but the actual values of the prior probabilities that we are trying to estimate are objective (and contingent) relative frequencies that are independent of our knowledge or beliefs about them. The fact that we may have to resort to guessing or to intuition in order to estimate them no more makes them subjective than does my guess that it will snow tomorrow make the snow, when it falls, subjective (cf. Salmon, *Scientific Inference*). These are the most important differences between my view and those of the personalists. The theory uses hypothetico-inferential reasoning but differs from strict confirmationism most crucially in that contingent *but unconfirmed* estimates of prior probabilities are always involved in any confirmation.
8. P. A. Schilpp, ed., *The Philosophy of Rudolf Carnap* (La Salle, Ill.: Open Court, 1963), p. 868.
9. Ibid., p. 45.
10. See my "The Ontological Status of Theoretical Entities," in *Minnesota Studies in the Philosophy of Science*, III, eds. H. Feigl and G. Maxwell (Minneapolis: University of Minnesota Press, 1962).

11. See ibid., and "Structural Realism and the Meaning of Theoretical Terms," in *Minnesota Studies in the Philosophy of Science*, IV, eds. S. Winokur and M. Radner (Minneapolis: University of Minnesota Press, 1970).

12. Direct realism entails naïve realism but not conversely.

13. See, e.g., J. Hospers, *An Introduction to Philosophical Analysis*, 2d ed. (Englewood Cliffs, N.J.: Prentice-Hall, 1967), pp. 102 ff.

14. I have discussed this matter, including its extension to theories by using Ramsey sentences in more detail in my "Structural Realism and the Meaning of Theoretical Terms." Due to neglect of Russell's important later work, the nature of the principle of acquaintance is widely misunderstood, and sometimes phenomenalists, direct realists, and strict inductivists consider their views bolstered by it. However, Russell makes it clear that the principle has no ontological implications; it is a purely epistemic one and a surprisingly unrestrictive one at that.

15. Professor Herbert Feigl has been telling me for years in conversation that, in spite of the rebukes on the matter that he used to get from other members of the Vienna Circle, he still feels that realism's superior explanatory power tends to lend it confirmation by experiential evidence. As is obvious, I now entirely agree that it does. I hope that he will not feel that rejection of strict confirmationism is too high a price to pay for further support for his view.

16. I have not discussed Popper's views on testing hypotheses here because I do not believe that they would pose any serious obstacles for the position here defended, and I have discussed them in my "Corroboration Without Demarcation." I should think that he might not even disagree with my theory of confirmation, since it is a contingent theory and does not purport to be a *logic* of confirmation. At this stage of the discussion however, there is one important difference between us. He holds, I believe, that realism, while important, meaningful, and contingent, is, nevertheless, metaphysical and nonscientific because it is not falsifiable. But, if I am correct, we have seen that the same is true for almost any theory in the advanced sciences. His definition of metaphysics, therefore, seems unfortunate since just about every important general contingent statement would be, in this sense, metaphysical, which would not matter, but also, according to his classification, nonscientific, which *would* matter.

17. See, e.g., Russell's *Human Knowledge;* J. Beloff, *The Existence of Mind* (New York: Citadel Press, 1964); M. Mandelbaum, *Philosophy, Science, and Sense Perception* (Baltimore: Johns Hopkins Press, 1964); G. Maxwell, "Scientific Methodology and the Causal Theory of Perception" and "Reply" [to Professors W. V. Quine, Karl Popper, A. J. Ayer, and William Kneale], both in *Problems in the Philosophy of Science*, eds. I. Lakatos and A. Musgrave (Amsterdam: North Holland Publishing Co., 1968); G. Maxwell, "Philosophy and the Causal Theory of Perception," *The Graduate Review of Philosophy*, 5, no. 3 (1964); G. Maxwell, Review of *Knowledge, Mind, and Nature* by B. Aune, *The Philosophical Review*, 58 (1969), pp. 392–97.

18. If mind-body monism is true, the private experience is *identical* with certain components of the neuronal activity.

19. I am using "color," as well as color words like "red" in their primary, *occurrent* sense, that is, to refer to the colored expanses that are exemplified in our visual experiences. A *redefinition* such as: "to be red" means to look red under standard lighting conditions, etc., is absurd since it requires that "red" in the *definiens* have a different meaning from "red" in the *definiendum*—indeed that it have the primary, occurrent meaning. If this defect is repaired and a viable causal or dispositional redefinition of color words is produced so that they may be properly predicated of physical entities, then "colors" in this new sense will no longer be

first order properties, but rather, structural ones. Moreover, we will still need color words that have the primary, occurrent sense to refer to the first order properties that are exemplified in our direct experience. These cannot be eradicated by redefining words. I am sorry to take up space with such obvious matters, but sad experience has indicated that it is often necessary.

20. B. Russell, *The Autobiography of Bertrand Russell*, III (1944–1969) (New York: Simon & Schuster, 1969), p. 179.

21. See, e.g., John Hospers, *An Introduction to Philosophical Analysis*, pp. 502–05.

22. The term "naïve realism" is used here for historical reasons and is not quite apt, since I use it here to include any view that holds that *any* (as opposed to all) perceivable, first order property can be truly attributed to physical objects. To deny naïve realism, in this sense, is to claim that no perceivable, first order property is ever a property of a physical object.

23. Stephen Pepper, *Concept and Quality* (La Salle, Ill.: Open Court, 1967).

24. G. Ryle, *Dilemmas* (Cambridge: At the University Press, 1954).

25. A rough-and-ready definition of "mental" and "physical" may remove any apparent inconsistency here. We can define "mental" as "know or knowable by acquaintance" (or "occurring in private experience") and "physical" as "occurring in the spatio-temporal, causal order." Using an event ontology and "constructing" physical space-time out of temporal and causal relations among events removes any difficulty about locating mental events in physical space-time.

26. It is interesting that Carnap's latest formulation of a meaning criterion is primarily a criterion for *terms* and only derivatively one for propositions (R. Carnap, "The Methodological Character of Theoretical Concepts," in *Minnesota Studies in the Philosophy of Science*, I, eds. H. Feigl and M. Scriven (Minneapolis: University of Minnesota Press, 1956). It is thus much more like a concept empiricist criterion than a judgment empiricist or confirmationist criterion. I have discussed this at more length in "Criteria of Meaning and of Demarcation," in *Mind, Matter, and Method; Essays in Philosophy and Science in Honor of Herbert Feigl*, eds. P. Feyerabend and G. Maxwell (Minneapolis: University of Minnesota Press, 1966).

27. The Ramsey sentence has become such a household word among some philosophers that the original reference is seldom given; it is: Frank Ramsey, *The Foundations of Mathematics (and Other Essays)* (New York: Humanities Press, 1956).

28. The belief (or hope) that some of the theories that we are able to produce have a reasonably good prior probability and that our estimates of the prior probabilities are not too far off the mark is, I believe, a reasonable belief (or hope). I have attempted to give it a (weak) *vindication* in something like Feigl's sense of "vindication." (See his "De Principiis Non Disputandum . . . ? On the Meaning and the Limits of Justification," in *Philosophical Analysis*, ed. Max Black (Ithaca, N.Y.: Cornell University Press, 1950), pp. 119–56, and my "Corroboration Without Demarcation."

MARY HESSE

University of Cambridge

Is There an Independent Observation Language?

> Of all of the men of the century Faraday had the greatest
> power of drawing ideas out of his experiments and making
> his physical apparatus do his thinking, so that experimen-
> tation and inference were not two proceedings, but one.
> —C. S. Peirce
> *Values in a Universe of Chance*

I. Observation Predicates

RAPIDITY OF PROGRESS, or at least change, in the analysis of scientific theory structure is indicated by the fact that only a few years ago the natural question to ask would have been, "Is there an independent theoretical language?" The assumption would have been that theoretical language in science is parasitic upon observation language, and probably ought to be eliminated from scientific discourse by disinterpretation and formalization, or by explicit definition in or reduction to observation language. Now, however, several radical and fashionable views place the onus on believers in an observation language to show that such a concept has any sense in the absence of a theory. It is time to pause and ask what motivated the distinction between a so-called theoretical language and an observation language in the first place, and whether its retention is not now more confusing than enlightening.

In the light of the importance of the distinction in the literature, it is surprisingly difficult to find any clear statement of what the two lan-

Part of this paper was delivered as a lecture at the University of Pittsburgh on 17 October 1966. I am happy to acknowledge my indebtedness to many colleagues with whom I have discussed its contents, especially Professors H. Feigl, M. Brodbeck, G. Maxwell, and P. E. Meehl at the Minnesota Center for the Philosophy of Science, Professor T. S. Kuhn, and my colleagues in Cambridge, Professors R. B. Braithwaite and G. Buchdahl, and Dr. D. H. Mellor.

guages are supposed to consist of. In the classic works of twentieth-century philosophy of science, most accounts of the observation language were dependent on circular definitions of observability and its cognates, and the theoretical language was generally defined negatively as consisting of those scientific terms which are not observational. We find quasi definitions of the following kind: " 'Observation-statement' designates a statement which records an actual or possible observation"; "Experience, observation, and cognate terms will be used in the widest sense to cover observed facts about material objects or events in them as well as directly known facts about the contents or objects of immediate experience"; "The observation language uses terms designating observable properties and relations for the description of observable things or events"; "*observables*, i.e., . . . things or events which are ascertainable by direct observation."[1] Even Nagel, who gives the most thorough account of the alleged distinction between theoretical and observation terms, seems to presuppose that there is nothing problematic about the "direct experimental evidence" for observation statements, or the "experimentally identifiable instances" of observation terms.[2]

In contrast with the allegedly clear and distinct character of the observation terms, the meanings of theoretical terms, such as "electron," "electromagnetic wave," and "wave function,"[3] were held to be obscure. Philosophers have dealt with theoretical terms by various methods, based on the assumption that they have to be explained by means of the observation terms as given. None of the suggested methods has, however, been shown to leave theoretical discourse uncrippled in some area of its use in science. What suggests itself, therefore, is that the presuppositions of all these methods themselves are false, namely,

(a) that the meanings of the observation terms are unproblematic,
(b) that the theoretical terms have to be understood by means of the observation terms, and
(c) that there is, in any important sense, a distinction between two *languages* here, rather than different kinds of uses within the the same language.

In other words, the fact that we somehow understand, learn, and use observation terms does not in the least imply that the way in which we understand, learn, and use them is either different from or irrelevant to the way we understand, learn, and use theoretical terms. Let us then subject the observation language to the same scrutiny which the theoretical language has received.

Rather than attacking directly the dual language view and its underlying empiricist assumptions, my strategy will be first to attempt to construct a different account of meaning and confirmation in the observation language. This project is not the ambitious one of a general theory of meaning, nor of the learning of language, but rather the modest one of finding conditions for understanding and use of terms in science—some specification, that is to say, in a limited area of discourse, of the "rules of usage" which distinguish meaningful discourse from mere vocal reflexes. In developing this alternative account I shall rely on ideas which have become familiar particularly in connection with Quine's discussions of language and meaning and the replies of his critics, whose significances for the logic of science seem not yet to have been exploited nor even fully understood.[4]

I shall consider, in particular, the predicate terms of the so-called observation language. But first something must be said to justify considering the problem as one of "words" and not of "sentences." It has often been argued that it is sentences that we learn, produce, understand, and respond to, rather than words, that is, that in theoretical discussion of language, sentences should be taken as units. There are, however, several reasons why this thesis, whether true or false, is irrelevant to the present problem, at least in its preliminary stages. The observation language of science is only a segment of the natural language in which it is expressed, and we may for the moment assume that rules of sentence formation and grammatical connectives are already given when we come to consider the use of observation predicates. Furthermore, since we are interested in alleged distinctions between the observation and theoretical languages, we are likely to find these distinctions in the characteristics of their respective predicates, not in the connectives which we may assume that they share. Finally, and most importantly, the present enterprise does not have the general positive aim of describing the entire structure of a language. It has rather the negative aim of showing that there are no terms in the observation language which are sufficiently accounted for by "direct observation," "experimentally identifiable instances," and the like. This can best be done by examining the hardest cases, that is, predicates which do appear to have direct empirical reference. No one would seriously put forward the direct-observation account of grammatical connectives; and if predicates are shown not to satisfy the account, it is likely that the same arguments will suffice to show that sentences do not satisfy it either.

So much for preliminaries. The thesis I am going to put forward can be briefly stated in two parts:

i) All descriptive predicates, including observation and theoretical predicates, must be introduced, learned, understood, and used, either by means of direct empirical associations in some physical situations, or by means of sentences containing other descriptive predicates which have already been so introduced, learned, understood, and used, or by means of both together. (Introduction, learning, understanding, and use of a word in a language will sometimes be summarized in what follows as the *function* of that word in the language.)

ii) No predicates, not even those of the observation language, can function by means of direct empirical associations alone.

The process of functioning in the language can be spelled out in more detail:

A. Some predicates are initially learned in empirical situations in which an association is established between some aspects of the situation and a certain word. Given that any word with extralinguistic reference is ever learned, this is a necessary statement and does not presuppose any particular theory about what an association is or how it is established. This question is one for psychology or linguistics rather than philosophy. Two necessary remarks can, however, be made about such learning:

1) Since every physical situation is indefinitely complex, the fact that the particular aspect to be associated with the word is identified out of a multiplicity of other aspects implies that degrees of physical similarity and difference can be recognized between different situations.

2) Since every situation is in detail different from every other, the fact that the word can be correctly reused in a situation in which it was not learned has the same implication.

These remarks would seem to be necessarily implied in the premise that some words with reference are learned by empirical associations. They have not gone unchallenged, however, and it is possible to distinguish two sorts of objections to them. First, some writers, following Wittgenstein, have appeared to deny that physical similarity is necessary to the functioning of *any* word with extralinguistic reference. That similarity is not *sufficient*, I am about to argue, and I also agree that not all referring words need to be introduced in this way, but if *none* were, I am unable to conceive how an intersubjective descriptive language could ever get under way. The onus appears to rest upon those who reject similarity to show in what other way descriptive language is possible.[5] The other sort of objection is made by Popper, who argues that the notion of repetition of instances which is implied by (1) and (2) is essentially vacuous, because similarity is always similarity *in certain respects,* and "with a little ingenuity" we could always find similarities in *some* same respects between all members of any finite set of situations. That is to say, "anything can be said to be a repetition of anything else, if only we adopt the appropriate point of view." [6] But if this were true, it would make the learning process in empirical situations impossible. It would mean that however finitely large the number of presentations of a given situation-aspect, that aspect could never be identified as the desired one out of the indefinite number of other respects in which the presented situations are all similar. It would, of course, be possible to eliminate some other similarities by presenting further situations similar in the desired respect but not in others, but it would then be possible to find other respects in which all the situations, new and old, are similar— and so on without end.

However, Popper's admission that "a little ingenuity" may be required allows a less extreme interpretation of his argument, namely, that the physics and physiology of situations already give us some "point of view" with respect to which some pairs of situations are similar in more obvious respects than others, and one situation is more similar in some respect to another than it is in the same respect to a third. This is all that is required by the assertions (1) and (2). Popper has needlessly obscured the importance of these implications of the learning process by speaking as though, before any repetition can be recognized, we have to take thought, and *explicitly* adopt a point of view. If this were so, a regressive problem would arise about how we ever learn to apply the predicates in which we explicitly express that point of view. An immediate consequence of this is that there must be a stock of predicates in any descriptive language for which it is impossible to *specify* necessary and sufficient conditions of correct application. For if any such specification could be given for a particular predicate, it would introduce further predicates requiring to be learned in empirical situations for which there was no specification. Indeed such unspecified predicates would be expected to be in the majority, for those for which necessary and sufficient conditions can be given are dispensable except as a shorthand and, hence, essentially uninteresting. We must, therefore, conclude that the primary process of recognition of similarities and differences is necessarily *unverbalizable.* The emphasis here is of course on *primary,* because it may be perfectly possible to give empirical descriptions of the conditions, both psychological and physical, under which similarities are recognized, but such descriptions will themselves depend on further undescribable primary recognitions.

B. It may be thought that the primary process of classifying objects according to recognizable similarities and differences will provide us with exactly the independent observation predicates required by the traditional view. This, however, is to overlook a logical feature of relations of similarity and difference, namely, that they are not *transitive*. Two objects *a* and *b* may be judged to be similar to some degree in respect to predicate *P*, and may be placed in the class of objects to which *P* is applicable. But object *c* which is judged similar to *b* to the same degree may not be similar to *a* to the same or indeed to any degree. Think of judgments of similarity of three shades of color. This leads to the conception of some objects as being more "central" to the *P*-class than others, and also implies that the process of classifying objects by recognition of similarities and differences is necessarily accompanied by some loss of (unverbalizable) information. For if *P* is a predicate whose conditions of applicability are dependent on the process just described, it is impossible to *specify* the degree to which an object satisfies *P* without introducing more predicates about which the same story would have to be told. Somewhere this potential regress must be stopped by some predicates whose application involves loss of information which is present to recognition but not verbalizable. However, as we shall see shortly, the primary recognition process, though necessary, is not sufficient for classification of objects as *P*, and the loss of information involved in classifying leaves room for changes in classification to take place under some circumstances. Hence primary recognitions do not provide a stable and independent list of primitive observation predicates.

C. It is likely that the examples that sprang to mind during the reading of the last section were such predicates as "red," "ball," and "teddy bear." But notice that nothing that has been said rules out the possibility of giving the same account of apparently much more complex words. "Chair," "dinner," and "mama" are early learned by this method, and it is not inconceivable that it could also be employed in first introducing "situation," "rule," "game," "stomachache," and even "heartache." This is not to say, of course, that complete fluency in using these words could be obtained by this method alone; indeed, I am now going to argue that complete fluency cannot be obtained in the use of *any* descriptive predicate by this method alone. It should only be noticed here that it is possible for any word in natural language having some extralinguistic reference to be introduced in suitable circumstances in some such way as described in section A.

D. As learning of the language proceeds, it is found that some of these predicates enter into general statements which are accepted as true and which we will call *laws:* "Balls are round"; "In summer leaves are green"; "Eating unripe apples leads to stomachache." It matters little whether some of these are what we would later come to call analytic statements; some, perhaps most, are synthetic. It is not necessary, either, that every such law should be *in fact* true, only that it is for the time being accepted as true by the language community. As we shall see later, any one of these laws may be *false* (although not all could be false at once). Making explicit these general laws is only a continuation and extension of the process already described as identifying and reidentifying proper occasions for the use of a predicate by means of physical similarity. For knowledge of the laws will now enable the language user to apply descriptions correctly in situations other than those in which he learned them, and even in situations where nobody could have learned them in the absence of the laws, for example, "stomachache" of an absent individual known to have consumed a basketful of unripe apples, or even "composed of diatomic molecules" of the oxygen in the atmosphere. In other words, the laws enable generally correct inferences and predictions to be made about distant ("unobservable") states of affairs.

E. At this point the system of predicates and their relations in laws has become sufficiently complex to allow for the possibility of internal misfits and even contradictions. This possibility arises in various ways. It may happen that some of the applications of a word in situations turn out not to satisfy the laws which are true of other applications of the word. In such a case, since degrees of physical similarity are not transitive, a reclassification may take place in which a particular law is preserved in a subclass more closely related by similarity, at the expense of the full range of situations of application which are relatively less similar. An example of this would be the application of the word "element" to water, which becomes incorrect in order to preserve the truth of a system of laws regarding "element," namely, that elements cannot be chemically dissociated into parts which are themselves elements, that elements always enter as a whole into compounds, that every substance is constituted by one or more elements, and so on. On the other hand, the range of applications may be widened in conformity with a law, so that a previously incorrect application becomes correct. For example, "mammal" is correctly applied to whales, whereas it was previously thought that "Mammals live only on land" was a well-entrenched law providing criteria for correct use of "mammal." In such a case it is not adequate to counter with the suggestion that the correct use of "mammal" is *defined* in terms of animals which suckle their young, for it is conceivable that if other empirical facts had been different, the classification in terms of habitat would have been more useful and comprehensive than that in terms of milk production. And in regard to the first example, it cannot be maintained that it is the *defining* characteristics of "element" that are preserved at the expense of its application to water, because of the conditions mentioned it is not clear that any particular one of them is, or ever has been, taken as *the* defining characteristic; and since the various characteristics are logically independent, it is empirically possible that some might be satisfied and not others. *Which* is preserved will always depend on what system of laws is most convenient, most coherent, and most comprehensive. But the most telling objection to the suggestion that correct application is decided by definition is of course the general point made at the end of section A that there is always a large number of predicates for which *no* definition in terms of necessary and sufficient conditions of application can be given. For these predicates it is possible that the primary recognition of, for example, a whale as being sufficiently similar to some fish to justify its inclusion in the class of fish may be explicitly overridden in the interests of preserving a particular set of laws.

Properly understood, the point developed in the last paragraph should lead to a far-reaching reappraisal of orthodoxy regarding the theory-observation distinction. To summarize, it entails that no feature in the total landscape of functioning of a descriptive predicate is exempt from modification under pressure from its surroundings. That any empirical law may be abandoned in the face of counterexamples is trite, but it becomes less trite when the functioning of every predicate is found to depend essentially on some laws or other and when it is also the case that any "correct" situation of application—*even that in terms of which the term was originally introduced*—may become incorrect in order to preserve a system of laws and other applications. It is in this sense that I shall understand the "theory dependence" or "theory-ladenness" of all descriptive predicates.

One possible objection to this account is easily anticipated. It is not a *conventionalist* account, if by that we mean that any law can be assured of truth by sufficiently meddling with the meanings of its predicates.

Such a view does not take seriously the systematic character of laws, for it contemplates preservation of the truth of a given law irrespective of its coherence with the rest of the system, that is, the preservation of simplicity and other desirable internal characteristics of the system. Nor does it take account of the fact that not all primary recognitions of empirical similarity can be overridden in the interest of preserving a given law, for it is upon the existence of some such recognitions that the whole possibility of language with empirical reference rests. The present account on the other hand demands both that laws shall remain connected in an economical and convenient system and that at least most of its predicates shall remain applicable, that is, that they shall continue to depend for applicability upon the primary recognitions of similarity and difference in terms of which they were learned. That it is possible to have such a system with a given set of laws and predicates is not a convention but a fact of the empirical world. And although this account allows that *any* of the situations of correct application may change, it cannot allow that *all* should change, at least not all at once. Perhaps it would even be true to say that only a small proportion of them can change at any one time, although it is conceivable that over long periods of time most or all of them might come to change piecemeal. It is likely that almost all the terms used by the alchemists that are still in use have now changed their situations of correct use quite radically, even though at any one time chemists were preserving most of them while modifying others.

II. Entrenchment

It is now necessary to attack explicitly the most important and controversial question in this area, namely, the question whether the account of predicates that has been given really applies to all descriptive predicates whatsoever, or whether there are after all some which are immune to modification in the light of further knowledge and which might provide candidates for a basic and independent observation language. The example mentioned at the end of the last paragraph immediately prompts the suggestion that it would be possible at any time for both alchemists and chemists to "withdraw" to a more basic observation language than that used in classifying substances and that this language would be truly primitive and theory-independent. The suspicion that this may be so is not even incompatible with most of the foregoing account, for it may be accepted that we often do make words function without reflecting upon more basic predicates to which we could withdraw if challenged. Thus, it may not be disputed that we learn, understand, and use words like "earth," "water," "air," and "fire" in empirical situations and that their subsequent functioning depends essentially upon acceptance of some laws; and yet it may still be maintained that there are some more basic predicates for which cash value is payable in terms of empirical situations alone. Let us, therefore, consider this argument at its strongest point and take the case of the putative observation predicate "red." Is this predicate subject to changes of correct application in the light of laws in the way that has been described? The defense of our account at this point comes in two stages. First, it must be shown that *no* predicate of an observation language can function by mere empirical situations alone, independently of any laws. Second, it must be shown that there is no set of observation predicates whose interrelating laws are absolutely invariant to changes in the rest of the network of laws.

When a predicate such as "red" is claimed to be "directly" descriptive this claim is usually made in virtue of its use as a predicate of immediate experience—a sensation of a red postage stamp, a red spectral line, a red afterimage. It is unnecessary here to enter into the much discussed questions of whether there are any such "things" as sensations for "red" to be a predicate of, whether such predicates of sensations could be ingredients of either a public or a private language, and whether there is indeed any sense in the notion of a private language. The scientific observation language at least is not private but must be intersubjective; and whether some of its predicates are predicates of sensations or not, it is still possible to raise the further question: in *any* intersubjective language can the functioning of the predicates be independent of accepted laws? That the answer is negative can be seen by considering the original account of the empirical situations given in section I.A and by adopting one generally acceptable assumption. The assumption is that in using a public language, the correctness of any application of a predicate in a given situation must in principle be capable of intersubjective test.[7] Now if my careful response of "red" to each of a set of situations were all that were involved in my correct use of "red," this response would not be sufficient to ensure intersubjectivity. It is possible, in spite of my care, that I have responded mistakenly, in which case the laws relating "red" to other predicates can be appealed to in order to correct me (I can even correct myself by this method): "It can't have been red, because it was a sodium flame, and sodium flames are not red." If my response "red" is intended to be an ingredient of a public observation language, it carries at least the implication that disagreements can be publicly resolved, and this presupposes laws conditioning the function of "red." If this implication is absent, responses are mere verbal reflexes having no intersubjective significance (unless of course they are part of a physiological-psychological experiment, but then I am subject, not observer). This argument does not, I repeat, purport to show that there could not be a sense-datum language functioning as the observation language of science —only that if this were so, its predicates would share the double aspect of empirical situation and dependence on laws which belongs to all putative observation predicates.

Now consider the second stage of defense of our account. The suggestion to be countered here is that even if there are peripheral uses of "red" which might be subject to change in the event of further information about laws, there is nevertheless a central core of function of "red," with at least some laws which ensure its intersubjectivity, which remains stable throughout all extensions and modifications of the rest of the network of accepted laws. To illustrate the contrast between "periphery" and "core," take the following examples: we might come to realize that when "red" is applied to a portion of the rainbow, it is not a predicate of an object, as in the paradigm cases of "red," or that the ruddy hue of a distant star is not the color of the star, but an effect of its recession. But, it will be said, in regard to cherries, red lips, and the color of a strontium compound in a Bunsen flame, "red" is used entirely independently of the truth of or knowledge of the great majority of laws in our network. We might, of course, be mistaken in application of "red" to situations of this central kind, for we may mistake color in a bad light or from defects of vision; but there are enough laws whose truth cannot be in doubt to enable us to correct mistakes of this kind, and by appealing to them we are always able to come to agreement about correct applications. There is no sense, it will be argued, in supposing that in cases like this we could all be mistaken all the time or that we might, in any but the trivial sense of deciding to use another word equivalent to "red," come to change our usage in these central situations.

One possible reply [8] is to point out that the admission that there are *some* situations in which we might change our use even of a predicate like "red" is already a significant one, especially in the examples given above. For the admission that the "red" of a rainbow or a receding star is not the color of an object is the admission that in these cases at least it is a *relational* predicate, where the relata, which may be quite complex, are spelled out by the laws of physics. Now no doubt it does not *follow* that "red" ascribed to the book cover now before me is also a relational predicate, unless we take physics to provide the real truth about everyday objects as well as those that are more remote. The schizophrenia induced by not taking physics seriously in this way raises problems of its own which we cannot pursue here. But suppose our critic accepts the realist implication that "red" is, on all occasions of its use as a predicate of objects, in fact a relational predicate, and then goes on to discount this admission by holding that such a relatively subtle logical point is irrelevant to the ordinary function of "red" in the public language. Here we come near the heart of what is true in the critic's view. The truth might be put like this: Tom, Dick, and Mary do indeed use the word "red" with general indifference to logical distinctions between properties and relations. Even logicians and physicists continue to use it in such a way that in ordinary conversation it need never become apparent to others, or even to themselves, that they "really believe" that color predicates are relational. And more significantly for the ultimate purpose of this essay, the conversation of a Newtonian optician about sticks and stones and rolls of bread need never reveal a difference of function of "red" from the conversation of a postrelativity physicist.

Such a concession to the critic with regard to invariance of function in limited domains of discourse is an important one, but it should be noticed that its force depends not upon fixed stipulations regarding the use of "red" in particular empirical situations, but rather upon empirical facts about the way the world is. Physically possible situations can easily be envisaged in which even this central core of applicability of "red" would be broken. Suppose an isolated tribe all suffered a congenital color blindness which resulted in light green being indistinguishable from red and dark green from black. Communication with the outside world, or even the learning of physics without any such communication, might very well lead them to revise the function of "red" and "black" even in paradigm cases.

A more realistic and telling example is provided by the abandonment of Newtonian time simultaneity. This is an especially striking case, because time concepts are among the most stable in most languages and particularly in a physics which has persistently regarded spatial and temporal qualities as primary and as providing the indispensable framework of a mechanistic science. As late as 1920 N. R. Campbell, usually a perceptive analyst of physical concepts, wrote: "Is it possible to find any judgement of sensation concerning which all sentient beings whose opinion can be ascertained are always and absolutely in agreement? . . . I believe that it is possible to obtain absolutely universal agreement for judgements such as, the event A happened at the same time as B, or A happened between B and C." [9] Special relativity had already in 1905 shown this assumption to be false. This means that at any time before 1905 the assumption was one from which it was certainly possible to withdraw; it was in fact "theory-laden," although it had not occurred to anybody that this was the case. Now let us cast Einstein in the role of the "operationist" physicist who, wiser than his contemporaries, has detected the theory-ladenness and wishes to withdraw from it to a "level of direct observation," where there are no theoretical implications, or at least where these are at a minimum.[10] What can he do? He can try to set up an operational definition of time simultaneity. When observers are at a distance from each other (they are always at *some* distance), and when they are perhaps also moving relatively to each other, he cannot assume that they will agree on judgments of simultaneity. He will assume only that a given observer can judge events that are simultaneous in his own field of vision, provided they occur close together in that field. The rest of Einstein's operational definition in terms of light signals between observers at different points is well known. But notice that this definition does not carry out the program just proposed for an operationist physicist. For far from withdrawal to a level of direct observation where theoretical implications are absent or at a minimum, the definition requires us to assume, indeed to postulate, that the velocity of light *in vacuo* is the same in all directions and invariant to the motions of source and receiver. This is a postulate which is logically prior in special relativity to any experimental measurement of the velocity of light, because it is used in the very definition of the time scale at distant points. But from the point of view of the operationist physicist before 1905, the suggestion of withdrawing from the assumption of distant absolute time simultaneity to this assumption about the velocity of light could not have appeared to be a withdrawal to more direct observation having fewer theoretical implications, but rather the reverse. This example illustrates well the impossibility of even talking sensibly about "levels of more direct observation" and "degrees of theory-ladenness" *except in the context of some framework of accepted laws.* That such talk is dependent on this context is enough to refute the thesis that the contrast between "direct observation" and "theory-ladenness" is itself theory-independent. The example also illustrates the fact that at any given stage of science it is never possible to know *which* of the currently entrenched predicates and laws may have to give way in the future.

The operationist has a possible comeback to this example. He may suggest that the process of withdrawal to the directly observed is not a process of constructing another theory, as Einstein did, but properly stops short at the point where we admitted that at least one assumption of Newtonian physics is true and must be retained, namely, that "a given observer can judge events that are simultaneous in his own field of vision, provided they occur close together in that field"—call this assumption (S). This, it may be said, is a genuine withdrawal to a less theory-laden position, and all that the rest of the example shows is that there is in fact no possibility of advance again to a more general conception of time simultaneity without multiplying insecure theoretical assumptions. Now, of course, the game of isolating some features of an example as paradigms of "direct observation," and issuing a challenge to show how *these* could ever be overthrown, is one that can go on regressively without obvious profit to either side. But such a regress ought to stop if either of the following two conditions are met:

(a) that it is logically possible for the alleged paradigm to be overthrown and that its overthrow involves a *widening* circle of theoretical implications or

(b) that the paradigm becomes less and less suitable as an observation statement, because it ceases to have the required intersubjective character.

The time simultaneity example made its point by illustrating condition (a). The assumption (S) to which it is now suggested we withdraw can be impaled on a dilemma between (a) and (b). Suppose it were shown that an observer's judgment of simultaneity in his field of sensation were quite strongly dependent on the strength of the gravitational field in his neighborhood, although this dependence had not yet been shown up in the fairly uniform conditions of observation on the surface of the earth. Such a discovery, which is certainly conceivable, would satisfy condition (a). As long as the notion of simultaneity is so interpreted as to allow for intersubjective checking and agreement, there is always an indefinite number of such possible empirical situations whose variation might render the assumption (S) untenable. The only way to escape this horn of the dilemma is to interpret (S) as referring to the direct experience of simultaneity of a single observer, and this is intersubjectively and hence scientifically useless, and impales us on the horn of condition (b).

The comparative stability of function of the so-called observation predicates is logically speaking an accident of the way the world is. But it may now be suggested that since the way the world is is not likely to alter radically during the lifetime of any extant language, we might define an observation language to be just that part of language which the facts allow to remain stable. This, however, is to take less than seriously the effects of scientific knowledge on our ways of talking about the world and also to underestimate the tasks that ordinary language might be called upon to perform as the corpus of scientific knowledge changes. One might as well hold that the ordinary language of Homer, which identifies life with the breath in the body and fortuitous events with interventions of divine personages, and was no doubt adequate to discourse before the walls of Troy, should have remained stable in spite of all subsequent changes in physics, physiology, psychology, and theology. Our ordinary language rules for the use of "same time," which presuppose that this concept is independent of the distance and relative motion of the spatial points at which time is measured, are not only contradicted by relativity theory, but would possibly need fundamental modification if we all were to take habitually to space travel. Another point to notice here is that the comparatively stable area within which it is proposed to define an observation language itself is partly known to us because its stability is explained by the theories we now accept. It is certainly not sufficiently defined by investigating what observation statements have in fact remained stable during long periods of time, for this stability might be due to accident, prejudice, or false beliefs. Thus any attempted definition itself would rely upon current theories and, hence, not be a definition of an observation language which is theory-independent. Indeed, it might justly be concluded that we shall know what the most adequate observation language is only when, if possible, we have true and complete theories, including theories of physiology and physics which tell us what it is that is most "directly observed." Only then shall we be in a position to make the empirical distinctions that seem to be presupposed by attempts to discriminate theoretical and observation predicates.

The upshot of all this may be summarized by saying that although there is a nucleus of truth in the thesis of invariance of the observation language and, hence, of the theory-observation distinction among predicates, this truth has often been located in the wrong place and used to justify the wrong inferences. The invariance of observation predicates has been expressed in various ways, not all equivalent to one another and not all equally valid. Let us summarize the discussion so far by examining some of these expressions:

i) "There are some predicates which are *better entrenched* than others, for instance, 'red' than 'ultra-violet,' 'lead' than 'π-meson.' "

If by "better entrenched" is meant less subject to change of function in ordinary discourse and, therefore, less revelatory of the speaker's commitments to a system of laws or of his relative ignorance of such systems, then (i) is true. But this is a *factual* truth about the relative invariance of some empirical laws to increasing empirical information, not about the a priori features of a peculiar set of predicates, and it does not entail that any predicate is *absolutely* entrenched, nor that any subsystems of predicates and the laws relating them are immune to mutual modification under pressure from the rest of the system.

ii) "There are some predicates which refer to aspects of situations more *directly observable* than others."

If this means that their function is more obviously related to empirical situations than to laws, (ii) is true, but its truth does not imply that a line can be drawn between theoretical and observation predicates in the place it is usually desired to draw it. For it is not at all clear that highly complex and even theoretical predicates may not sometimes be directly applicable in appropriate situations. Some examples were given in section I.C.; other examples are thinkable where highly theoretical descriptions would be given directly: "particle-pair annihilation" in a cloud chamber, "glaciation" of a certain landscape formation, "heart condition" of a man seen walking along the street. To the immediate rejoinder that these examples leave open the possibility of withdrawal to less "theory-laden" descriptions, a reply will be given in (v) below. Meanwhile it should be noticed that this sense of "observable" is certainly not coextensive with that of (i).

iii) "There are some predicates which are learnable and applicable in a *pragmatically* simpler and quicker manner than others."

This is true, but does not necessarily single out the same set of predicates in all language communities. Moreover, it does not necessarily single out all or only the predicates which are "observable" in senses (i) and (ii).

iv) "There are some predicates in terms of which others are *anchored to the empirical facts*."

This may be true in particular formulations of a theory, where the set of anchor predicates is understood as in (i), (ii), or (iii), but little more needs to be said to justify the conclusion that such a formulation and its set of anchoring predicates would not be unique. In principle it is conceivable that any predicate could be used as a member of the set. Thus, the commonly held stronger version of this assumption is certainly false, namely, that the anchor predicates have unique properties which allow them to endow theoretical predicates with empirical meaning which these latter would not otherwise possess.

v) The most important assumption about the theory-observation distinction, and the one which is apparently most damaging to the present account, can be put in a weaker and a stronger form:

(a) "There are some predicates to which we could always *withdraw* if challenged in our application of others."

(b) "These form a unique subset in terms of which 'pure descriptions' free from 'theory-loading' can be given."

Assumption (a) must be accepted to just the extent that we have accepted the assumption that there are degrees of entrenchment of predicates, and for the same reasons. It is indeed sometimes possible to withdraw from the implications of some ascriptions of predicates by using others better entrenched in the network of laws. To use some of the examples already mentioned, we may withdraw from "particle-pair annihilation" to "two white streaks meeting and terminating at an angle"; from "heart condition" to a carefully detailed report of complexion, facial structure, walking habits, and the like; and from "epileptic fit" to a description of teeth-clenching, falling, writhing on the floor, and so on. So far, these examples show only that some of the lawlike implications that are in mind when the first members of each of these pairs of descriptions are used can be withdrawn from and replaced by descriptions which do not have *these* implications. They do not show that the second members of each pair are free from lawlike implications of their own, nor even that it is possible to execute a series of withdrawals in such a way that each successive description contains fewer implications than the description preceding it. Far less do they show that there is a unique set of descriptions which have *no* implications; indeed the arguments already put forward should be enough to show that this assumption, assumption (b), must be rejected. As in the case of entrenchment, it is in principle possible for any particular lawlike implication to be withdrawn from, although not all can be withdrawn from at once. Furthermore, although in any given state of the language some descriptive predicates are more entrenched than others, it is not clear that withdrawal to those that are better entrenched is withdrawal to predicates which have *fewer* lawlike implications. Indeed, it is likely that better entrenched predicates have in fact far more implications. The reason why these implications do not usually seem doubtful or objectionable to the observational purist is that they have for so long proved to be true, or been believed to be true, in their relevant domains that their essentially inductive character has been forgotten. It follows that when well-entrenched predicates and their implications are from time to time abandoned under pressure from the rest of the network, the effects of such abandonment will be more far-reaching, disturbing, and shocking than when less well entrenched predicates are modified.

III. The Network Model

The foregoing account of theories, which has been presented as more adequate than the deductive two-language model, may be dubbed the *network model* of theories. It is an account that was first explicit in Duhem and more recently reinforced by Quine. Neither in Duhem nor Quine, however, is it quite clear that the netlike interrelations between more directly observable predicates and their laws are in principle just as subject to modifications from the rest of the network as are those that are relatively theoretical. Duhem seems sometimes to imply that although there is a network of relatively phenomenological representations of facts, once established this network remains stable with respect to the changing explanations. This is indeed one reason why he rejects the view that science aims at explanation in terms of unobservable entities and restricts theorizing to the articulation of mathematical representations which merely systematize but do not explain the facts. At the same time, however, his analysis of the facts is far subtler than that presupposed by later deductivists and instrumentalists. He sees that what is primarily significant for science is not the precise nature of what we directly observe, which in the end is a *causal* process, itself susceptible of scientific analysis. What is significant is the interpretative expression we give to what is observed, what he calls the *theoretical facts*, as opposed to the "raw data" represented by *practical facts*. This distinction may best be explained by means of his own example. Consider the theoretical fact "The temperature is distributed in a certain manner over a certain body." [11] This, says Duhem, is susceptible of precise mathematical formulation with regard to the geometry of the body and the numerical specification of the temperature distribution. Contrast the practical fact. Here geometrical description is at best an idealization of a more or less rigid body with a more or less indefinite surface. The temperature at a given point cannot be exactly fixed, but is only given as an average value over vaguely defined small volumes. The theoretical fact is an imperfect translation, or interpretation, of the practical fact. Moreover, the relation between them is not one-to-one, but rather many-to-many, for an infinity of idealizations may be made to more or less fit the practical fact, and an infinity of practical facts may be expressed by means of one theoretical fact.

Duhem is not careful in his exposition to distinguish *facts* from *linguistic expressions of facts*. Sometimes both practical and theoretical facts seem to be intended as linguistic statements (for instance, where the metaphor of "translation" is said to be appropriate). But even if this is his intention, it is clear that he does not wish to follow traditional empiricism into a search for forms of expression of practical facts which will constitute the basis of science. Practical facts are not the appropriate place to look for such a basis—they are imprecise, ambiguous, corrigible, and on their own ultimately meaningless. Moreover, there is a sense in which they are literally inexpressible. The absence of distinction between fact and linguistic expression here is not accidental. As soon as we begin to try to capture a practical fact in language, we are committed to some theoretical interpretation. Even to say of the solid body that "its points are more or less worn down and blunt" is to commit ourselves to the categories of an ideal geometry.

What, then, is the "basis" of scientific knowledge for Duhem? If we are to use this conception at all, we must say that the basis of science is the set of theoretical facts in terms of which experience is interpreted. But we have just seen that theoretical facts have only a more or less loose and ambiguous relation with experience. How can we be sure that they provide a firm empirical foundation? The answer must be that we cannot be sure. There is no such foundation. Duhem himself is not consistent on this point, for he sometimes speaks of the persistence of the network of theoretical facts as if this, once established, takes on the privileged character ascribed to observation statements in classical positivism. But this is not the view that emerges from his more careful discussion of examples. For he is quite clear, as in the case of the correction of the "observational" laws of Kepler by Newton's theory, that more comprehensive mathematical representations may show particular theoretical facts to be false.

However, we certainly seem to have a problem here, because if it is admitted that subsets of the theoretical facts may be removed from the corpus of science, and if we yet want to retain some form of empiricism, the decision to remove them can be made only by reference to *other* theoretical facts, whose status is in principle equally insecure. In the traditional language of epistemology some element of correspondence with experience, though loose and corrigible, must be retained but also be supplemented by a theory of the coherence of a network. Duhem's account of this coherence has been much discussed but not always in the context of his complete account of theoretical and practical facts, with the result that it has often been trivialized. Theoretical facts do not stand on their own but are bound together in a network of laws which constitutes the total mathematical representation of experience. The putative theoretical fact that was Kepler's third law of planetary motion, for example, does not fit the network of laws established by Newton's theory. It is, therefore, modified, and this modification is possible without violating experience because of the many-to-one relation between the theoretical fact and that practical fact understood as the ultimately inexpressible situation which obtains in regard to the orbits of planets.

It would seem to follow from this (although Duhem never explicitly draws the conclusion) that there is no theoretical fact or lawlike relation whose truth or falsity can be determined in isolation from the rest of the network. Moreover, many conflicting networks may more or less fit the same facts, and which one is adopted must depend on criteria other than the facts: criteria involving simplicity, coherence with other parts of science, and so on. Quine, as is well known, has drawn this conclusion explicitly in the strong sense of claiming that any statement can be maintained true in the fact of any evidence: "Any statement can be held true come what may, if we make drastic enough adjustments elsewhere in the system. . . . Conversely, by the same token, no statement is immune to revision." [12] In a later work, however, he does refer to "the philosophical doctrine of infallibility of observation sentences" as being sustained in his theory. Defining the stimulus meaning of a sentence as the class of sensory stimulations that would prompt assent to the sentence, he regards observation sentences as those sentences whose stimulus meanings remain invariant to changes in the rest of the network and for which "the stimulus meanings may without fear of contradiction be said to do full justice to their meanings." [13] This seems far too conservative a conclusion to draw from the rest of the analysis, for in the light of the arguments and examples I have presented, it appears very dubious whether there are such invariant sentences if a long enough historical perspective is taken.

There are other occasions on which Quine seems to obscure unnecessarily the radical character of his own position by conceding too much to more traditional accounts. He compares his own description of theories to those of Braithwaite, Carnap, and Hempel in respect of the "contextual definition" of theoretical terms. But his own account of these terms as deriving their meaning from an essentially *linguistic* network has little in common with the formalist notion of "implicit definition" which these deductivists borrow from mathematical postulate systems in which the terms need not be interpreted empirically. In this sense the implicit definition of "point" in a system of Riemannian geometry is entirely specified by the formal postulates of the geometry and does not depend at all on what would count empirically as a realization of such a geometry.[14] Again, Quine refers particularly to a net analogy which Hempel adopts in describing theoretical predicates as the knots in the net, related by definitions and theorems represented by threads. But Hempel goes on to assert that the whole "floats . . . above the plane of observation" to which it is anchored by *threads of a different kind,* called "rules of interpretation," *which are not part of the network itself.*[15] The contrast between this orthodox deductivism and Quine's account could hardly be more clear. For Quine, and in the account I have given here, there is indeed a network of predicates and their lawlike relations, but it is not floating above the domain of observation; it is attached to it at some of its knots. *Which* knots will depend on the historical state of the theory and its language and also on the way in which it is formulated, and the knots are not immune to change as science develops. It follows, of course, that "rules of interpretation" disappear from this picture: *all* relations become laws in the sense defined above, which, it must be remembered, includes near analytic definitions and conventions as well as empirical laws.

IV. Theoretical Predicates

So far it has been argued that it is a mistake to regard the distinction between theoretical and observational predicates either as providing a unique partition of descriptive predicates into two sets or as providing a simple ordering such that it is always possible to say of two predicates that one is under all circumstances more observational than or equally observational with the other. Various relative and noncoincident distinctions between theoretical and observational have been made, none of which is consistent with the belief that there is a unique and privileged set of observation predicates in terms of which theories are related to the empirical world. So far in the network model it has been assumed that any predicate may be more or less directly ascribed to the world in some circumstances or other, and that none is able to function in the language by means of such direct ascription alone. The second of these assumptions has been sufficiently argued; it is now necessary to say more about the first. Are there any descriptive predicates in science which could not under any circumstances be directly ascribed to objects? If there are, they will not fit the network model as so far described, for there will be nothing corresponding to the process of classification by empirical associations, even when this process is admitted to be fallible and subject to correction by laws, and they will not be connected to other predicates by laws, since a law presupposes that the predicates it connects have all been observed to co-occur in some situation or other.

First, it is necessary to make a distinction between theoretical *predicates* and theoretical *entities,* a distinction which has not been sufficiently considered in the deductivist literature. Theoretical entities have sometimes been taken to be equivalent to unobservable entities. What does this mean? If an entity is unobservable in the sense that it never appears as the subject of observation reports, and is not in any other way related to the entities which do appear in such reports, then it has no place in science. This cannot be what is meant by "theoretical" when it is applied to such entities as electrons, mesons, genes, and the like. Such applications of the terms "theoretical" and "unobservable" seem rather to imply that the entities do not have predicates ascribed to them in observation statements, but only in theoretical statements. Suppose the planet Neptune had turned out to be wholly transparent to all electromagnetic radiation and, therefore, invisible. It might still have entered planetary theory as a theoretical entity in virtue of the postulated force relations between it and other planets. Furthermore, the monadic predicate "mass" could have been inferred of it, although mass was never ascribed to it in an observation statement. Similarly, protons, photons, and mesons have monadic and relational predicates ascribed to them in theoretical but not in observation statements, at least not in prescientific language. But this distinction, like others between the theoretical and the observational domains, is relative; for once a theory is accepted and further experimental evidence obtained for it, predicates may well be ascribed directly to previously unobservable entities, as when genes are identified with DNA molecules visible in micrographs or when the ratio of mass to charge of an elementary particle is "read off" the geometry of its tracks in a magnetic field.

In contrasting theoretical with observable entities, I shall consider that theoretical entities are sufficiently specified as being those to which monadic predicates are not ascribed in relatively observational statements. It follows from this specification that relational predicates cannot be ascribed to them in observation statements either, for in order to recognize that a relation holds between two or more objects, it is necessary to recognize the objects by means of at least some monadic properties. ("The tree is to the left of x" is not an observation statement; "the tree is to the left of x and x is nine stories high" may be.) A theoretical entity must, however, have some postulated relation with an observable entity in order to enter scientific theory at all, and both monadic and relational predicates may be postulated of it in the context of a theoretical network. It must be emphasized that this specification is not intended as a close analysis of what deductivists have meant by "theoretical entity" (which is in any case far from clear), but rather as an explication of this notion in terms of the network account of theories. At least it can be said that the typical problems that have seemed to arise about the existence of and reference to theoretical entities have arisen only in so far as these entities are not subjects of monadic predicates in observation statements. If a monadic predicate were ascribed to some entity in an observation statement it would be difficult to understand what would be meant by calling such an entity "unobservable" or by questioning its "existence." The suggested explication of "theoretical entity" is, therefore, not far from the apparent intentions of those who have used this term, and it does discriminate electrons, mesons, and genes on the one hand from sticks and stones on the other.

When considering the relatively direct or indirect ascription of predicates to objects, it has already been argued that the circumstances of use must be attended to before the term "unobservable" is applied. In particular it is now clear that a predicate may be observable of some kinds of entities and not of others. "Spherical" is observable of baseballs (entrenched and directly and pragmatically observable), but not of protons; "charged" is observable in at least some of these senses of pith balls but not of ions, and so on. No monadic predicate is observable of a theoretical entity; some predicates may be observable of some observable entities but not of others; for example, "spherical" is not directly or pragmatically observable of the earth. The question whether there are absolutely theoretical *predicates* can now be seen to be independent of the question of theoretical entities; if there are none, this does not imply that there are no theoretical entities, nor that predicates ascribed to them may not also be ascribed to observable entities.

How is a predicate ascribed to theoretical entities or to observable entities of which it is not itself observable? If it is a predicate which has already been ascribed directly to some observable entity, it may be inferred of another entity by analogical argument. For example, stones released near the surface of Jupiter will fall toward it because Jupiter is in other relevant respects like the earth. In the case of a theoretical entity, the analogical argument will have to involve relational predicates: high energy radiation arrives from a certain direction; it is inferred from other instances of observed radiation transmission between pairs of objects that there is a body at a certain point of space having a certain structure, temperature, gravitational field, and so on.

But it is certain that some predicates have been introduced into science which do not appear in the relatively entrenched observation language. How are they predicated of objects? Consistently with the network model, there seem to be just two ways of introducing such newly minted predicates. First, they may be introduced as new observation predicates by assigning them to recognizable empirical situations where descriptions have not been required in prescientific language. Fairly clear examples are "bacteria" when first observed in microscopes and "sonic booms" first observed when aircraft "broke the sound barrier." Such introductions of novel terms will of course share the characteristic of all observation predicates of being dependent for their function on observed associations or laws as well as direct empirical recognitions. In some cases it may be difficult to distinguish them from predicates introduced by *definition* in terms of previously familiar observation predicates. Fairly clear examples of this are "molecule," defined as a small particle with certain physical and chemical properties such as mass, size, geometrical structure, and combinations and dissociations with other molecules, which are expressible in available predicates (most *names* of theoretical entities seem to be introduced this way); or "entropy," defined quantitatively and operationally in terms of change of heat content divided by absolute temperature. In intermediate cases, such as "virus," "quasar," and "oedipus complex," it may be difficult to decide whether the function of these predicates is exhausted by logical equivalence with certain complex observation predicates or whether they can be said to have an independent function in some empirical situations where they are relatively directly observed. Such ambiguities are to be expected, because in the network model, laws which are strongly entrenched may sometimes be taken to be definitional, and laws introduced as definitions may later be regarded as being falsifiable empirical associations.

Notice that in this account the view of the function of predicates in theories that has been presupposed is explicitly nonformalist. The account is in fact closely akin to the view that all theories require to be interpreted in some relatively observable model, for in such a model their predicates are ascribed in observation statements. It has been assumed that when familiar predicates such as "charge," "mass," and "position" are used of theoretical entities, these predicates are the "same" as the typographically similar predicates used in observation statements. But it may be objected that when, say, elementary particles are described in terms of such predicates, the predicates are not used in their usual sense, for if they were, irrelevant models and analogies would be imported into the theoretical descriptions. It is important to be clear what this objection amounts to. If it is the assertion that a predicate such as "charge" used of a theoretical entity has a sense related to that of "charge" used of an observable entity only through the apparatus of formal deductive system plus correspondence rules, then the assertion is equivalent to a formal construal of theories, and it is not clear why the word "charge" should be used at all. It would be less conducive to ambiguity to replace it with an uninterpreted sign related merely by the theoretical postulates and correspondence rules to observation predicates. If, however, the claim that it is used of theoretical entities in a different sense implies only that charged elementary particles are different kinds of entities from charged pith balls, this claim can easily be admitted and can be expressed by saying that the predicate co-occurs and is co-absent with different predicates in the two cases. The fact that use of the predicate has different lawlike implications in relatively theoretical contexts from those in observation contexts is better represented in the network model than in most other accounts of theories, for it has already been noticed that in this model the conditions of correct application of a predicate depend partly on the other predicates with which it is observed to occur. This seems sufficiently to capture what is in mind when it is asserted that "charge" "means" something different when applied to elementary particles and pith balls, or "mass" when used in Newtonian and relativistic mechanics.

Since formalism has been rejected, we shall regard predicates such as those just described as retaining their identity (and hence their logical substitutivity) whether used of observable or theoretical entities, though they do not generally retain the same empirical situations of direct application. But the formalist account, even if rejected as it stands, does suggest another possibility for the introduction of new theoretical predicates, related to observation neither by assignment in recognizable empirical situations nor by explicit definition in terms of old predicates. Can the network model not incorporate new predicates whose relations with each other and with observation predicates are "implicit." not in the sense intended by formalists, but rather as a new predicate might be coined in myth or in poetry, and understood in terms of its context, that is to say, of its asserted relations with both new and familiar predicates? This suggestion is perhaps nearer the intentions of some deductivists than is pure formalism, from which it is insufficiently discriminated.[16]

It is not difficult to see how such a suggestion could be incorporated into the network model. Suppose instead of relating predicates by known laws, we *invent a myth* in which we describe entities in terms of some predicates already in the language, but in which we introduce other predicates in terms of some mythical situations and mythical laws. In other words we build up the network of predicates and laws partly imaginatively, but not in such a way as to contradict known laws, as in a good piece of science fiction.[17] It is, moreover, perfectly possible that such a system might turn out to have true and useful implications in the empirical domain of the original predicates, and in this way the mythical predicates and laws may come to have empirical reference and truth. This is not merely to repeat the formalist account of theoretical predicates as having meaning only in virtue of their place within a postulate system, because it is not necessary for such a formal system to have any interpretation, whereas here there is an interpretation, albeit an imaginary one. Neither are the predicates introduced here by any mysterious "implicit definition" by a postulate system; they are introduced by the same two routes as are all other predicates, except that the laws and the empirical situations involved are imaginary.

Whether any such introduction of new predicates by mythmaking has ever occurred in science may be regarded as an open question. The opinion may be hazarded that no convincing examples have yet been identified. All theory construction, of course, involves an element of mythmaking, because it makes use of *familiar* predicates related in new ways by postulated laws not yet accepted as true. Bohr's atom, for example, was postulated to behave as no physical system had ever been known to behave; however, the entities involved were all described in terms of predicates already available in the language. There is, moreover, a reason why the mythical method of introducing new predicates is not likely to be very widespread in science. The reason is that use of known predicates which already contain some accepted lawlike implications allow inductive and analogical inference to further as yet unknown laws, which mythical predicates do not allow. There could be no prior inductive confidence in the implications of predicates and laws which were wholly mythical, as there can be in the implications of predicates at least some of whose laws are accepted. How important such inductive confidence is, however, is a controversial question which cannot be pursued here. But it is sufficient to notice that the network model does not demand that theories should be restricted to use of predicates already current in the language or observable in some domain of entities.

V. Theories

Under the guise of an examination of observational and theoretical predicates, I have in fact described a full-fledged account of theories, observation, and the relation of the one to the other. This is, of course, to be expected, because the present account amounts to a denial that there is a fundamental distinction between theoretical and observation predicates and statements and implies that the distinction commonly made is both obscure and misleading. It should not, therefore, be necessary to say much more about the place of theories in this account. I have so far tried to avoid the term "theory," except when describing alternative views, and have talked instead about laws and lawlike implications. But a theory *is* just such a complex of laws and implications, some of which are well entrenched, others less so, and others again hardly more than suggestions with as yet little empirical backing. A given theory may in principle be formulated in various ways, and some such formulations will identify various of the laws with postulates; others with explicit definitions; others with theorems, correspondence rules, or experimental laws. But the upshot of the whole argument is that these distinctions of function in a theory are relative not only to the particular formulation of the "same" theory (as with various axiomatizations of mechanics or quantum theory), but also to the theory itself, so that what appears in one theory as an experimental law relating "observables" may in another be a high-level theoretical postulate (think of the chameleonlike character of the law of inertia, or the conservation of energy). It is one of the more misleading results of the deductive account that the notion of "levels," which has proper application to proofs in a formal postulate system in terms of the order of deducibility of theorems, has become transferred to an ordering in terms of more and less "theory-laden" constituents of the theory. It should be clear from what has already been said that these two notions of "level" are by no means co-extensive.

So much is merely the immediate application to theories of the general thesis here presented about descriptive predicates. But to drive the argument home, it will be as well to consider explicitly some of the problems which the theory-observation relation has traditionally been felt to raise and how they fare in the present account.

A. *The Circularity Objection*   An objection is sometimes expressed as follows: if the use of all observation predicates carry theoretical implications, how can they be used in descriptions which are claimed to be evidence for these same theories? At least it must be possible to find terms in which to express the evidence which are not laden with the theory for which they express the evidence.

This is at best a half-truth. If by "theory-laden" is meant that the terms used in the observation report presuppose the *truth* of the very theory under test, then, indeed, this observation report cannot contribute evidence for this theory. If, for example, "motion in a straight line with uniform speed" is *defined* (perhaps in a complex and disguised fashion) to be equivalent to "motion under no forces," this definition implies the truth of the law of inertia, and an observation report to the effect that a body moving under no forces moves in a straight line with uniform speed does not constitute evidence for this law. The logic of this can be expressed as follows:

Definition. $P(x) \equiv_{df} Q(x)$
Theory. $(x)[P(x) \supset Q(x)]$
Observation. $P(a)\&Q(a)$

Clearly neither theory nor observation report states anything empirical about the relation of P and Q.

Contrast this with the situation where the "theory-loading" of $P(a)$ is interpreted to mean "Application of P to an object a implies acceptance of the truth of some laws into which P enters, and these laws are part of the theory under test," or, colloquially, "The meaning of P presupposes the truth of some laws in the theory under test." In the inertia example the judgment that a is a body moving in a straight line with uniform speed depends on the truth of laws relating measuring rods and clocks, the concept of "rigid body," and ultimately on the physical truth of the postulates of Euclidean geometry, and possibly of classical optics. All these are part of the theory of Newtonian dynamics and are confirmed by the very same kinds of observation as those which partially justify the assertion $P(a)\&Q(a)$. The notion of an observation report in this account is by no means simple. It may include a great deal of other evidence besides the report that $P(a)\&Q(a)$, namely, the truth of other implications of correct application of P to a and even the truth of universal laws of a high degree of abstractness. It is, of course, a standard objection to accounts such as the present, which have an element of "coherence" in their criteria of truth, that nothing can be known to be true until everything is known. But although an adequate confirmation theory for our account would not be straightforward, it would be perfectly possible to develop one in which the correct applicability of predicates, even in observation reports, is strongly influenced by the truth of some laws into which they enter, and only vanishingly influenced by others. The notion of degrees of entrenchment relative to given theories would be essential to expressing the total evidence in such a confirmation theory.[18]

The reply to the circularity objection as it has been stated is, then, that although the "meaning" of observation reports is "theory-laden," the truth of particular theoretical statements depends on the coherence of the network of theory and its empirical input. The objection can be put in another way, however: if the meaning of the terms in a given observation report is even partially determined by a theory for which this report is evidence, how can the same report be used to decide between two theories, as in the classic situation of a crucial experiment? For if this account is correct the same report cannot have the same meaning as evidence for two different theories.

This objection can be countered by remembering what it is for the "meaning" of the observation report to be "determined by the theory." This entails that ascription of predicates in the observation report implies acceptance of various other laws relating predicates of the theory, and we have already agreed that there may be a hard core of such laws which are more significant for determining correct use than others. Now it is quite possible that two theories which differ very radically in most of their implications still contain some hard-core predicates and laws which they both share. Thus, Newtonian and Einsteinian dynamics differ radically in the laws into which the predicate "inertial motion" enter, but they share such hard-core predicates as "acceleration of falling bodies near the earth's surface," "velocity of light transmitted from the sun to the earth," and so on, and they share some of the laws into which these predicates enter. It is this area of *intersection* of laws that must determine the application of predicates in the report of a crucial experiment. The situation of crucial test between theories is not correctly described in terms of "withdrawal to a neutral observation language," because, as has already been argued, there is no such thing as an absolutely neutral or non-theory-laden language. It should rather be described as exploitation of the area of intersection of predicates and laws between the theories; this is, of course, entirely relative to the theories in question.

An example originally due to Feyerabend [19] may be developed to illustrate this last point. Anaximenes and Aristotle are devising a crucial experiment to decide between their respective theories of free fall. Anaximenes holds that the earth is disc-shaped and suspended in a nonisotropic universe in which there is a preferred direction of fall, namely the parallel lines perpendicular to and directed toward the surface of the disc on which Greece is situated. Aristotle, on the other hand, holds that the earth is a large sphere, much larger than the surface area of Greece, and that it is situated at the center of a universe organized in a series of concentric shells, whose radii directed toward the center determine the direction of fall at each point. Now clearly the word "fall" as used by each of them is, in a sense, loaded with his own theory. For Anaximenes it refers to a preferred direction uniform throughout space; for Aristotle it refers to radii meeting at the center of the earth. But equally clearly, while they both remain in Greece and converse on nonphilosophical topics, they will use the word "fall" without danger of mutual misunderstanding. For each of them the word will be correlated with the direction from his head to his feet when standing up, and with the direction from a calculable point of the heavens toward the Acropolis at Athens. Also, both of them will share most of these lawlike implications of "fall" with ordinary Greek speakers, although the latter probably do not have any expectations about a preferred direction throughout universal space. This is not to say, of course, that the ordinary Greek speaker uses the word with *fewer* implications, for he may associate it with the passage from truth to falsehood, good to evil, heaven to hell—implications which the philosophers have abandoned.

Now suppose Anaximenes and Aristotle agree on a crucial experiment. They are blindfolded and carried on a Persian carpet to the other side of the earth. That it is the other side might be agreed upon by them, for example, in terms of star positions—this would be part of the intersection of their two theories. They now prepare to let go of a stone they have brought with them. Anaximenes accepts that this will be a test of his theory of fall and predicts, "The stone will fall." Aristotle accepts that this will be a test of his theory of fall and predicts, "The stone will fall." Their Persian pilot performs the experiment. Aristotle is delighted and cries, "It falls! My theory is confirmed." Anaximenes is crestfallen and mutters, "It rises; my theory must be wrong." Aristotle now notices that there is something strange about the way in which they have expressed their respective predictions and observation reports, and they embark upon an absorbing analysis of the nature of observation predicates and theory-ladenness.

The moral of this tale is simply that confirmation and refutation of competing theories does not depend on all observers using their language with the same "meaning," nor upon the existence of any neutral language. In this case, of course, they could, if they thought of it, agree to make their predictions in terms of "moves from head to foot," instead of "falls," but *this* would have presupposed that men naturally stand with their feet on the ground at the Antipodes, and this is as much an uncertain empirical prediction as the original one. Even "moves perpendicularly to the earth" presupposes that the Antipodes is not a series of steeply sloping enclosed caves and tunnels in which it is impossible to know whether the stars occasionally glimpsed are reflected in lakes or seen through gaps in thick clouds. In Anaximenes's universe, almost anything might happen. But we are not trying to show that in any particular example there are *no* intersections of theories, only that the intersection does not constitute an independent observation language, and that some predicates of the observation reports need not even lie in the intersection in order for testing and mutual understanding to be possible. The final analysis undertaken by Anaximenes and Aristotle will doubtless include the learning of each other's theories and corresponding predicates or the devising of a set of observation reports in the intersection of the two theories, or, more probably, the carrying out of both together.

As a corollary of this account of the intersections of theories, it should be noted that there is no a priori guarantee that two persons brought up in the same language community will use their words with the same meanings in all situations, even when each of them is conforming to standard logic within his own theory. If they discourse only about events which lie in the intersection of their theories, that they may have different theories will never be behaviorally detected. But such behavioral criteria for "same meaning" may break down if their theories are in fact different and if they are faced with new situations falling outside the intersection. Misunderstanding and logical incoherence cannot be logically guarded against in the empirical use of language. The novelty of the present approach, however, lies not in that comparatively trivial remark, but in demonstrating that rational communication can take place in intersections, even when words are being used with "different meanings," that is, with different implications in areas remote from the intersection.

B. The Two-Language Account and Correspondence Rules   Many writers have seen the status of the so-called rules of interpretation, or correspondence rules, as the key to the proper understanding of the problem of theory and observation. The concept of correspondence rules presupposes the theory-observation distinction, which is bridged by the rules and, therefore, seems to have been bypassed in the present account. But there are cases where it seems so obvious that correspondence rules are both required and easily identifiable, that it is necessary to give some attention to them, in case some features of the theory-observation relation have been overlooked.

These cases arise most persuasively where it seems to be possible to give two descriptions of a given situation, one in theoretical and one in observation terminology, and where the relation between these two descriptions is provided by the set of correspondence rules. Take, for example, the ordinary-language description of the table as hard, solid, and blue, and the physicist's description of the same table in terms of atoms, forces, light waves, and so on—or the familiar translation from talk of the pressure, volume, and temperature of a gas to talk of the energy and momentum of random motions of molecules. It seems clear that in such examples there is a distinction between theoretical and observational descriptions and also that there are correspondence rules which determine the relations between them. How does the situation look in the network account?

It must be accepted at once that there is something more "direct" about describing a table as hard, solid, and blue, than as a configuration of atoms exerting forces. "Direct" is to be understood on our account in terms of the better entrenchment of the predicates "hard," "solid," and "blue" and the laws which relate them, and in terms of the practical ease of learning and applying these predicates in the domain of tables, compared with the predicates of the physical description. This does not imply, however, that "atom," "force," and "light wave" function in a distinct theoretical language nor that they require to be connected with observation predicates by extraneous and problematic correspondence rules. Consider as a specific example, usually regarded as a correspondence rule: " 'This exerts strong repulsive forces' implies 'This is hard.' " Abbreviate this as "Repulsion implies hardness," and call it (C). What is the status of (C)? Various suggestions have been made, which I shall now examine.[20]

a) It is an analytic definition. This is an uninteresting possibility, and we shall assume it to be false, because "repulsion" and "hardness" are not synonymous in ordinary language. They are introduced in terms of different kinds of situation and generally enter into different sets of laws. Furthermore, in this domain of entities "hardness" has the pragmatic characteristics of an observation predicate, and "repulsion" of a theoretical predicate, and, hence, they cannot be synonymous here. Therefore, (C) is a synthetic statement.

How, then, do "hard" and "repulsion" function in the language? Consistently with our general account we should have to say something like this: meaning is given to "hard" by a complex process of learning to associate the sound with certain experiences and also by accepting certain empirical correlations between occurrences reported as "This is hard," "This exerts pressure" (as of a spring or balloon), "This is an area of strong repulsive force" (as of iron in the neighborhood of a magnet), "This is solid, impenetrable, undeformable . . . ," "This bounces, is elastic. . . ." Similarly, "repulsion" is introduced in a set of instances including some of those just mentioned and also by means of Newton's second law and all its empirical instances. Granted that this is how we *understand* the terms of (C), what kind of synthetic statement is it?

b) It may be suggested that it is a theorem of the deductive system representing the physical theory. This possibility has to be rejected by two-language philosophers because for them "hard" does not occur in the language of the theory and, hence, cannot appear in any theorem of the theory. But for us the possibility is open, because both terms of (C) occur in the same language, and it is perfectly possible that having never touched tables, but knowing all that physics can tell us about the forces exerted by atoms and knowing also analogous situations in which repulsive forces are in fact correlated with the property of hardness (springs, balloons, etc.), we may be able to deduce (C) as a theorem in this complex of laws.

c) More simply, (C) may be not so much a deductive inference from a system of laws as an inductive or analogical inference from other accepted empirical correlations of repulsive force and hardness. This is a possibility two-language philosophers are prone to overlook, because they are wedded to the notion that "repulsion" is a theoretical term in the context of tables and, therefore, not a candidate for directly observed empirical correlations. But it does not follow that it is not comparatively observable in other domains—springs, magnets, and the like. Observation predicates are, as we have remarked, relative to a domain of entities.

d) Unable to accept (a), (b), or (c), the two-language philosopher is almost forced to adopt yet another alternative in his account of correspondence rules, namely, that (C) is an independent empirical postulate,[21] which is added to the postulates of the theory to make possible the deduction of observable consequences from that theory. There is no need to deny that this possibility may sometimes be exemplified. It should only be remarked that if all correspondence rules are logically bound to have this status, as a two-language philosopher seems forced to hold, some very strange and undesirable consequences follow. If there are no deductive, inductive, or analogical reasons other than the physicist's fiat why particular theoretical terms should be correlated with particular observation terms, how is it possible for the "floating" theory ever to be refuted? It would seem that we could always deal with an apparent refutation at the observation level by arbitrarily modifying the correspondence rules, for since on this view these rules are logically and empirically quite independent of the theory proper, they can always be modified without any disturbance to the theory itself. It may be replied that considerations of simplicity would prevent such arbitrary salvaging of a theory. But this objection can be put in a stronger form: it has very often been the case that a well-confirmed theory has enabled predictions to be made in the domain of observation, where the deduction involved one or more *new* correspondence rules, relating theoretical to observation terms in a new way. If these correspondence rules were postulates introduced for no reason intrinsic to the theory, it is impossible to understand how such predictions could be made with confidence.

On the present account, then, it need not be denied that there is sometimes a useful distinction to be made between comparatively theoretical and comparatively observational descriptions, nor that there are some expressions with the special function of relating these descriptions. But this does not mean that the distinction is more than pragmatically convenient, nor that the correspondence rules form a logically distinct class of statements with unique status. Statements commonly regarded as correspondence rules may in different circumstances function as independent theoretical postulates, as theorems, as inductive inferences, as empirical laws, or even in uninteresting cases as analytic definitions. There is no one method of bridging a logical gap between theory and observation. There is no such logical gap.

C. Replaceability   Granted that there is a relative distinction between a set of less entrenched (relatively theoretical) predicates and
better entrenched observation predicates, and that correspondence rules
do not form a special class of statements relating these two kinds of
predicates, there still remains the question, What is the relation between
two descriptions of the same subject matter, one referring to theoretical
entities and the other observation entities?

First of all, it follows from the present account that the two descriptions are not equivalent or freely interchangeable. To describe a table as
a configuration of atoms exerting forces is to use predicates which enter
into a system of laws having implications far beyond the domain of
tables. The description of the table in ordinary language as hard and
solid also has implications, which may not be fewer in number but are
certainly different. One contrast between the two descriptions which
should be noted is that the lawlike implications of the theoretical descriptions are much more explicit and unambiguous [22] than those of
ordinary-language predicates like "hard" and "solid." Because of this
comparative imprecision, it is possible to hold various views about the
status of an observational description. It is sometimes argued that an
observational description is straightforwardly *false,* because it carries
implications contradicted by the theoretical description, which are probably derived from out-of-date science. Thus, it is held that to say a table
is hard and solid implies that it is a continuum of material substance
with no "holes" and that to touch it is to come into immediate contact
with its substance. According to current physics these implications are
false. Therefore, it is claimed, in all honesty we must in principle replace
all our talk in observation predicates by talk in theoretical predicates in
which we can tell the truth.

This view has a superficial attraction, but as it stands, it has the very
odd consequence that most of the descriptions we ever give of the world
are not only false but known to be false in known respects. While
retaining the spirit of the replaceability thesis, this consequence can be
avoided in two ways. First, we can make use of the notion of intersection
of theories to remark that there will be a domain of discourse in which
there is practical equivalence between some implications of observational description and some implications of theoretical description. In
this domain the observation language user is telling the truth so long as
he is not tempted to make inferences outside the domain. This is also the
domain in which pragmatic observation reports provide the original

evidence for the theory. Within this domain ordinary conversation can go on for a long time without its becoming apparent that an observation language user, an ordinary linguist, and a theoretician are "speaking different languages" in the sense of being committed to different implications outside the domain. It may even be the case that an ordinary linguist is not committed to *any* implications outside the domain which conflict with those of the theoretician. For example, and in spite of much argument to the contrary, it is not at all clear that the user of the ordinary English word "solid" *is*, or ever was, committed to holding that a table is, in the mathematically infinitesimal, a continuum of substance. The question probably never occurred to him, either in the seventeenth century or the twentieth, unless he had been exposed to some physics. Secondly, the network account of predicates makes room for change in the function of predicates with changing knowledge of laws. In this case it may very well be that use of "hard" and "solid" in the observation language comes to have whatever implications are correct in the light of the laws of physics or else to have built-in limitations on their applicability, for example, to the domain of the very small.

These suggestions help to put in the proper light the thesis that it is in principle possible to replace the observation language by the theoretical, and even to teach the theoretical language initially to children without going through the medium of the observation language. Such teaching may indeed be *in principle* possible, but consider what would happen if we assume that the children are being brought up in normal surroundings without special experiences devised by physicists. They will then learn the language in the intersection of physics and ordinary language; and though they may be taught to mouth such predicates as "area of strong repulsive force" where other children are taught "hard," they will give essentially the same descriptions in this intersection as the ordinary linguist, except that every observation predicate will be replaced by a string of theoretical predicates. Doubtless they will be better off when they come to learn physics, much of which they will have learned implicitly already; and if they were brought up from the start in a highly unnatural environment, say in a spaceship, even ordinary discourse might well be more conveniently handled in theoretical language. But all these possibilities do not seem to raise any special problems or paradoxes.

D. Explanation   It has been presupposed in the previous section that when an observational description and a theoretical description of the same situation are given, both have reference to the same entities, and that they can be said to contradict or to agree with one another. Furthermore, it has been suggested that there are circumstances in which the two descriptions may be equivalent, namely, when both descriptions are restricted to a certain intersection of theoretical implications and when the implications of the observation predicates have been modified in the light of laws constituting the theory. Sometimes the objection is made to this account of the relation of theoretical and observational descriptions that, far from being potentially equivalent descriptions of the same entities, the theory is intended to *explain* the observations; and explanation, it is held, must be given in terms which are different from what is to be explained. And, it is sometimes added, explanation must be a description of *causes* which are distinct from their observable effects.

It should first be noticed that this argument cannot be used in defense of the two-language view. That explanations are supposed to refer to entities different from those referred to in the explananda does not imply that these sets of entities have to be described in different languages. Explanation of an accident, a good crop, or an economic crisis will generally be given in the same language as that of the explananda.

It does seem, however, that when we give the theoretical description of a table as a configuration of atoms exerting repulsive forces, we are saying something which *explains* the fact that the table is hard and states the *causes* of that hardness. How then can this description be in any sense *equivalent* to the observational description of the table as hard? It does of course follow from the present account that they are not equivalent in the sense of being *synonymous*. That much is implied by the different function of the theoretical and observation predicates. Rather, the descriptions are equivalent in the sense of having the same reference, as the morning star is an alternative description of the evening star and also of the planet Venus. It is possible for a redescription in this sense to be explanatory, for the redescription of the table in theoretical terms serves to place the table in the context of all the laws and implications of the theoretical system. It is not its reference to the *table* that makes it explanatory of the observation statements, which also have reference to the table. It is rather explanatory because it says of the table that in being "hard" ("exerting repulsive force") it is *like* other objects which are known to exert repulsive force and to feel hard as the table feels hard, and that the table is, therefore, an instance of general laws relating dynamical properties with sensations. And in regard to the *causal* aspects of explanation, notice that the repulsive forces are not properly said to be the causes of the table having the property "hard," for they *are* the property "hard"; but rather repulsive forces are causes of the table *feeling* hard, where "hard" is not a description of the table but of a sensation. Thus the cause is not the same as the effect, because the referents of the two descriptions are different; and the explanans is not the same as the explanandum, because although their referents are the same, the theoretical description explains by relating the explanadum to other like entities and to a system of laws, just in virtue of its use of relatively theoretical predicates. There is some truth in the orthodox deductive account of explanation as deducibility in a theoretical system, but there is also truth in the contention that explanation involves stating as well what the explanandum *really* is and, hence, relating it to other systems which are then seen to be essentially similar to it. Initial misdescription of the function of descriptive predicates precludes the deductive account from doing justice to these latter aspects of explanation, whereas in the present account they are already implied in the fact that redescription in theoretical predicates carries with it lawlike relations between the explanandum and other essentially similar systems.

VI. Conclusion

In this essay I have outlined a network model of theoretical science and argued that it represents the structure of science better than the traditional deductivist account, with its accompanying distinction between the theoretical and the observational.

First, I investigated some consequences of treating the theoretical and the observational aspects of science as equally problematic from the point of view of truth conditions and meaning. I described the application of observation terms in empirical situations as a classificatory process, in which unverbalized empirical information is lost. Consequently, reclassification may in principle take place in any part of the observational domain, depending on what internal constraints are imposed by the theoretical network relating the observations. At any given stage of science there are *relatively* entrenched observation statements, but any of these may later be rejected to maintain the economy and coherence of the total system.

This view has some similarity with other nondeductivist accounts in which observations are held to be "theory-laden," but two familiar objections to views of this kind can be answered more directly in the network account. First, it is not a conventionalist account in the sense that any theory can be imposed upon any facts regardless of coherence conditions. Secondly, there is no vicious circularity of truth and meaning, for at any given time *some* observation statements result from correctly applying observation terms to empirical situations according to learned precedents and independently of theories, although the relation of observation and theory is a self-correcting process in which it is not possible to know at the time which of the set of observation statements are to be retained as correct in this sense, because subsequent observations may result in rejection of some of them.

Turning to the relatively theoretical aspects of science, I have argued that a distinction should be made between theoretical *entities* and theoretical *predicates*. I have suggested that if by theoretical predicates is meant those which are never applied in observational situations to any objects, and if the open-ended character of even observation predicates is kept in mind, there are no occasions on which theoretical predicates are used in science, although of course there are many theoretical entities to which predicates observable in other situations are applied. It follows that there is no distinction in kind between a theoretical and an observation language. Finally, I have returned to those aspects of scientific theories which are analyzed in the deductive view in terms of the alleged theory-observation distinction and shown how they can be reinterpreted in the network model. Correspondence rules become empirical relations between relatively theoretical and relatively observational parts of the network; replaceability of observational descriptions by theoretical descriptions becomes redescription in more general terms in which the "deep" theoretical similarities between observationally diverse systems are revealed; and theoretical explanation is understood similarly as redescription and not as causal relationship between distinct theoretical and observable domains of entities mysteriously inhabiting the same space-time region. Eddington's two tables are one table.

NOTES

1. A. J. Ayer, *Language, Truth, and Logic*, 2d ed. (London: Gollancz, 1946), p. 11; R. B. Braithwaite, *Scientific Explanation* (New York: Cambridge University Press, 1953), p. 8; R. Carnap, "The Methodological Character of Theoretical Concepts," in *Minnesota Studies in the Philosophy of Science*, I, ed. H. Feigl and M. Scriven (Minneapolis: University of Minnesota Press, 1956), p. 38; C. G. Hempel, "The Theoretician's Dilemma," in *Minnesota Studies in the Philosophy of Science*, II, ed. H. Feigl, M. Scriven, and G. Maxwell (Minneapolis: University of Minnesota Press, 1958), p. 41.

2. E. Nagel, *The Structure of Science* (New York: Harcourt, Brace & World, 1961).
3. It would be possible to give examples from sciences other than physics: "adaptation," "function," "intention," "behavior," "unconscious mind"; but the question whether these are theoretical terms in the sense here distinguished from observation terms is controversial, so is the question whether, if they are, they are eliminable from their respective sciences. These questions would take us too far afield.
4. The account which follows is by no means original; in fact versions of it have been lying about in the literature for so long that it may even sound trite. It may be useful to bring together here references to those discussions which I have found particularly helpful, but no claim is made for exhaustiveness, especially in regard to work published after 1966. My own more recent developments of some of the ideas of the present essay are to be found in my chapter, "Duhem, Quine, and a New Empiricism," in *Knowledge and Necessity*, Royal Institute of Philosophy Lectures, Vol. 3, 1968–69 (London: Macmillan, 1970), p. 191, and in the references in note 18 below.

Among general works in analytic philosophy which contain clues for a revised analysis of the observation language, the following should be specially mentioned: L. Wittgenstein, *Philosophical Investigations*, trans. G. E. M. Anscombe (London and New York: Macmillan, 1953); F. Waismann, "Verifiability," in *Logic and Language*, 1st ser., ed. A. G. N. Flew (Oxford: Blackwell, 1952), p. 117; F. Waismann, *Principles of Linguistic Philosophy*, ed. R. Harré (London: Macmillan; New York: St. Martin's Press, 1965); Peter Geach, *Mental Acts* (London: Routledge & Kegan Paul, 1957); D. W. Hamlyn, *The Psychology of Perception* (London: Routledge & Kegan Paul, 1957); P. F. Strawson, *Individuals, an Essay in Descriptive Metaphysics* (London: Methuen, 1959); Renford Bambrough, "Universals and Family Resemblances," *Proceedings of the Aristotelian Society*, 62 (1961), pp. 207–22.

Early discussions of observation and experiment which distinguish the observational from the theoretical, without taking the meaning of observation terms for granted, are to be found in P. Duhem, *The Aim and Structure of Physical Theory*, 2d ed., trans. P. Wiener (Princeton: Princeton University Press, 1954) pt. II, chaps. 4–6; N. R. Campbell, *Physics, the Elements* (Cambridge University Press, 1920; reprint ed., *Foundations of Science*, New York: Dover, 1957), chap. 2. In both these works experimental laws are analyzed in terms of the mutually supporting parts of a network of relationships among observation terms. The question of the reference of descriptive terms and the analogy of the network is pursued in W. V. O. Quine, "Two Dogmas of Empiricism," *Philosophical Review*, 60 (1951), p. 20, reprinted in *From a Logical Point of View* (Cambridge, Mass.: Harvard University Press, 1953), p. 10, and in his *Word and Object* (Cambridge, Mass.: MIT Press, 1960), chaps. 1–3. Not dissimilar theses with regard to the interrelations of meaning and inference are stated in W. Sellars, "Some Reflections on Language-Games," *Philosophy of Science*, 21 (1954), p. 204. Also see A. Kaplan, "Definition and Specification of Meaning," *Journal of Philosophy*, 43 (1946), p. 281, and A. Kaplan and H. F. Schott, "A Calculus for Empirical Classes," *Methodos*, 3 (1951), p. 165. (I owe the last two of these references to Dr. N. Jardine.)

A radical reinterpretation of the positivist understanding of the observation language is implied in Karl R. Popper, *The Logic of Scientific Discovery* (London: Hutchinson, 1959), where all observation terms are analyzed as "dispositional," i.e., lose their alleged "direct reference." In Popper's "The Aim of Science," *Ratio*, 1 (1957), p. 24, the mutual adjustments and corrections of theoretical and observational laws are illustrated. Both aspects of Popper's work are taken further by P. K. Feyerabend, "An Attempt at a Realistic Interpretation of Experience," *Proceedings of the Aristotelian Society,* 58 (1957–58), p. 143, and "Explanation, Reduction, and Empiricism," in *Minnesota Studies in the Philosophy of Science,* III, ed. H. Feigl and G. Maxwell (Minneapolis: University of Minnesota Press, 1962), p. 28, and subsequent writings where he attacks the alleged "stability" and theory independence of the observation language and the alleged deductive relation between theory and observation. The notion of theory independence of the observation language is also attacked from various points of view in N. R. Hanson, *Patterns of Discovery* (New York: Cambridge University Press, 1958); M. Scriven, "Definitions, Explanations and Theories," in *Minnesota Studies in the Philosophy of Science,* II, ed. H. Feigl, M. Scriven, and G. Maxwell, p. 99; W. Sellars, "The Language of Theories," in *Current Issues in the Philosophy of Science,* ed. H. Feigl and G. Maxwell (New York: Holt, Rinehart and Winston, 1961), p. 57; and H. Putnam, "What Theories Are Not," in *Logic, Methodology and Philosophy of Science,* ed. E. Nagel, P. Suppes, and A. Tarski (Stanford, Calif.: Stanford University Press, 1962), p. 240, and his "The Analytic and the Synthetic," in *Minnesota Studies in the Philosophy of Science,* III, ed. H. Feigl and G. Maxwell, p. 358. There are further discussions of these issues in May Brodbeck, "Explanation, Prediction, and 'Imperfect' Knowledge," in *Minnesota Studies in the Philosophy of Science,* III, ed. H. Feigl and G. Maxwell, p. 231; Mary Hesse, "Theories, Dictionaries and Observation," *British Journal of the Philosophy of Science,* 9 (1958), pp. 12, 128, and her "Gilbert and the Historians," *British Journal of the Philosophy of Science,* 11 (1960), pp. 1, 130; P. Alexander, "Theory-Construction and Theory-Testing," *British Journal of the Philosophy of Science,* 9 (1958), p. 29, and his *Sensationalism and Scientific Explanation* (London: Routledge & Kegan Paul, 1963); Dudley Shapere, "Space, Time, and Language—An Examination of Some Problems and Methods of the Philosophy of Science," in *Philosophy of Science,* The Delaware Seminar, 2, ed. Bernard Baumrin (New York: Interscience, 1963), p. 139, and his "Meaning and Scientific Change," in *Mind and Cosmos,* ed. R. G. Colodny (Pittsburgh: University of Pittsburgh Press, 1966), p. 41.

Interpretations of the history of science in terms of successive "conceptual frameworks" or "paradigms" may be found in Stephen Toulmin, *Foresight and Understanding* (London: Hutchinson, 1961); T. S. Kuhn, *The Structure of Scientific Revolutions* (Chicago: University of Chicago Press, 1962); R. Harré, *Matter and Method* (London: Macmillan, 1964); Mary Hesse, *Forces and Fields* (London: Nelson, 1961; and Totowa, N.J.: Littlefield, 1965). Related to this view of the primacy of theoretical models in the development of science are various general arguments on the role of physical models and metaphors in the structure of theories. See, for example, N. R. Campbell, *Physics, the Elements,* chap. 6; G. Buchdahl, "Theory Construction: The Work of N. R. Campbell," *Isis,* 55 (1964), p. 151; E. Hutten, On Semantics and Physics," *Proceedings of the Aristotelian Society,* 49 (1948–49), p. 115, and his "The Role of Models in Physics," *British Journal of the Philosophy of Science,* 4 (1953), p. 284; Mary Hesse, "Models in Physics," *British Journal of the Philosophy of Science,* 4 (1953), p. 198, and her *Models and Analogies in Science* (London: Sheed and Ward, 1963; and Notre Dame, Ind.: University of Notre Dame Press, 1966); R. B. Braithwaite, *Scientific Explanation,* chap. 4, and his "Models in the Empirical Sciences," in *Logic, Methodology and Philosophy of Science,* ed. E. Nagel, P. Suppes, and A. Tarski, p. 224; M. Black, *Models and Metaphors* (Ithaca, N.Y.: Cornell University Press, 1962), chaps. 3, 13; E. Nagel, *The Structure of Science,* chap. 6; R. Harré, *Theories and Things* (London: Sheed and Ward, 1961); C. M. Turbayne, *The Myth of Metaphor* (New Haven and London: Yale University Press, 1962); D. Schon, *The Displacement of Concepts* (London: Tavistock Publications, 1963); P. Achinstein, "Theoretical Terms and Partial Interpretation," *British Journal of the Philosophy of Science*, 14 (1963), p. 89, and his "Theoretical Models," ibid., 16 (1965), p. 102; Marshall Spector, "Theory and Observation," ibid., 17 (1966), pp. 1, 89; E. McMullin, "What Do Physical Models Tell Us?" in *Logic, Methodology and Philosophy of Science*, ed. B. van Rootselaar and J. F. Stahl (Amsterdam: North Holland Publishing Co., 1968), p. 385.

In *Concepts of Science* (Baltimore: Johns Hopkins Press, 1968), Peter Achinstein develops an analysis of meaning of the observation and theory languages, and a critique of the observation-theory distinction, from a point of view similar to that presented here.

5. See, for example, Alan Gauld, "Could a Machine Perceive?" *British Journal of the Philosophy of Science*, 17 (1964), p. 44, and especially p. 53.

The a priori character of this account of descriptive language has been challenged by D. Davidson and by N. Chomsky. Davidson, in "Theories of Meaning and Learnable Languages," *Logic, Methodology and Philosophy of Science*, ed. Y. Bar-Hillel (Amsterdam: North Holland Publishing Co., 1965), p. 383, claims that there is no need for a descriptive predicate to be learned in the presence of the object to which it is properly applied, since, for example, it might be learned in "a skilfully faked environment" (p. 386). This possibility does not, however, constitute an objection to the thesis that it must be learned in *some* empirical situation and that this situation must have some similarity with those situations in which the predicate is properly used. Chomsky, on the other hand ("Quine's Empirical Assumptions," *Synthese*, 19 [1968], p. 53) attacks what he regards as Quine's "Humean theory" of language acquisition by stimulus and conditioned response. But the necessity of the *similarity* condition for language learning does not depend on the particular empirical mechanism of learning. Learning by patterning the environment in terms of a set of "innate ideas" would depend equally upon subsequent application of the same pattern to similar features of the environment. Moreover, "similar" cannot just be *defined as* "properly ascribed the same descriptive predicate in the same language community," for, as is argued below, similarity is a matter of degree and is a nontransitive relation, whereas "properly ascribed the same descriptive predicate" is not. The two terms cannot, therefore, be synonymous. See also Quine's reply to Chomsky, ibid., p. 274.

6. Karl R. Popper, *The Logic of Scientific Discovery*, Appendix x, p. 422.
7. See, for example, L. Wittgenstein, *Philosophical Investigations*, sec. 258 ff.; A. J. Ayer, *The Concept of a Person* (London: Macmillan, 1953), p. 39 ff.; Karl R. Popper, *The Logic of Scientific Discovery*, p. 44–45.
8. Cf. P. K. Feyerabend, "An Attempt at a Realistic Interpretation of Experience," *Proceedings of the Aristotelian Society*, 58 (1957–58), p. 143, 160.
9. *Physics, the Elements*, p. 29.
10. That this way of putting it is a gross distortion of Einstein's actual thought processes is irrelevant here.
11. P. Duhem, *The Aim and Structure of Physical Theory*, p. 133.
12. W. V. O. Quine, *From a Logical Point of View*, p. 43.
13. *Word and Object*, p. 42.
14. Ibid., p. 11. For an early and devastating investigation of the notion of "implicit definition" in a formal system, see G. Frege, "On the Foundations of Geometry,"

trans. M. E. Szabo, *Philosophical Review*, 69 (1960), p. 3 (first published 1903), and in specific relation to the deductive account of theories, see C. G. Hempel, "Fundamentals of Concept Formation in Empirical Science," *International Encyclopedia of Unified Science*, II, no. 7 (Chicago: University of Chicago Press, 1952), p. 81.

15. C. G. Hempel, "Fundamentals of Concept Formation," p. 36.
16. It certainly represents what Quine seems to have *understood* some deductive accounts to be (cf. pp. 54–55).
17. We build up the network in somewhat the same way that Black (*Models and Metaphors*, p. 43) suggests a poet builds up a web of imagined associations within the poem itself in order to make new metaphors intelligible. He might, indeed, in this way actually coin and give currency to wholly new words.
18. That no simple formal examples can be given of the present account is perhaps one reason why it has not long ago superseded the deductive account. I have made some preliminary suggestions toward a confirmation theory for the network account in my chapter on "Positivism and the Logic of Scientific Theories," in *The Legacy of Logical Positivism for the Philosophy of Science*, ed. P. Achinstein and S. Barker (Baltimore: Johns Hopkins Press, 1969), p. 85, and in "A Self-correcting Observation Language," in *Logic, Methodology and Philosophy of Science*, ed. B. van Rootselaar and J. F. Stahl, p. 297.
19. P. K. Feyerabend, "Explanation, Reduction and Empiricism," p. 85.
20. I owe several of these suggestions to a private communication from Paul E. Meehl. See also E. Nagel, *The Structure of Science*, p. 354 ff.
21. Cf. Carnap's "meaning postulates," *Philosophical Foundations of Physics* (New York and London: Basic Books, 1966), chap. 27.
22. This is not to say, as formalists are prone to do, that theoretical terms are completely unambiguous, precise, or exact, like the terms of a formal system. If they were, this whole account of the functioning of predicates would be mistaken. The question of precision deserves more extended discussion. It has recently been investigated by Stephen Korner, *Experience and Theory* (London: Routledge & Kegan Paul, 1966); D. H. Mellor, "Experimental Error and Deducibility," *Philosophy of Science*, 32 (1965), p. 105, and his "Inexactness and Explanation," *Philosophy of Science*, 33 (1966), p. 345.

ABNER SHIMONY
Boston University

Scientific Inference

> There's a divinity that shapes our ends,
> Rough-hew them how we will.
> —Shakespeare, *Hamlet*

I. Introduction

A. MY EPISTEMOLOGICAL POSITION can be described as naturalistic, or, with obvious latitude, as Copernican.[1] It recognizes that a human being is a minute part of a universe which existed long before his birth and will survive long after his death; it considers human experience to be the result of complex interactions of human sensory apparatus with entities having careers independent of their being perceived; and it acknowledges the probability that the fundamental principles governing the natural order will seem extremely strange from the standpoint of ordinary human conceptions. In order to proclaim oneself a Copernican at the

This essay is an expanded version of a paper presented to a philosophy of science workshop at the University of Pittsburgh in May, 1965. The comments of the other participants in the workshop were very helpful in revising the original paper. Conversations with Marx Wartofsky, Joseph Agassi, and Imre Lakatos were also very stimulating. For many years Richard Jeffrey has patiently listened to my conjectures concerning induction and has given expert criticism. Howard Stein read several drafts of the essay with great care and made a large number of valuable suggestions on matters of principle, on the structure of the argument, and on style; furthermore, conversations with him over many years have strongly influenced my philosophical outlook. I am grateful to the National Science Foundation for a Senior Postdoctoral Fellowship, during the tenure of which most of this essay was written. I also wish to thank the Harvard Physics Department and the M.I.T. Humanities Department for providing secretarial assistance in preparing the typescript.

My former teacher, Rudolf Carnap, was always very generous and tolerant in giving encouragement and technical advice, even though my approach to induction has diverged from his. I would like to dedicate this essay to the memory of this great scholar, who passed away on September 14, 1970.

present stage in history, one admittedly does not have to be radical and nonconformist. Copernicanism is part of the generally accepted scientific world view, and it has been the doctrine of a number of philosophical schools, such as the American naturalists, the eighteenth-century materialists, the Lockean empiricists, and (anachronistically) the Greek atomists. Nevertheless, this familiar point of view is worthy of restatement, partly because it provides a perspective which can prevent narrowness in the technical investigations of philosophy of science and partly because its implications have by no means been exhaustively explored.

From the Copernican point of view it is natural to see two grand philosophical problems, which may be described by borrowing the vivid phrases of Heraclitus. The first problem is "the way up": to show how it is that human beings, whose experience is extremely restricted in space and time and is conditioned by physiological and psychological peculiarities, can obtain an understanding of the universe beyond themselves. The second is "the way down": to understand in terms of fundamental principles how there can be entities such as ourselves in nature, capable of the kind of experience and endowed with the kind of faculties which make natural knowledge possible. "The way up" is essentially the subject matter of methodology and epistemology, whereas "the way down" is roughly the concern of philosophical psychology, though it overlaps the domains of many sciences. A complete solution to either of these two grand problems will surely depend upon a complete solution of the other, and even partial solutions can be expected to be mutually dependent. As a result, the Copernican point of view involves some circularity, although, I believe, the circularity need not be vicious.[2] In this essay I shall not examine the question of circularity in full generality, nor give detailed criticisms of epistemological theories, like those of Descartes and Russell, which purport to escape circularity by developing the structure of knowledge along an architectural plan of successive stories resting upon a firm foundation. I shall, however, discuss circularity in the theory of inductive inference, where it is particularly acute, as Hume pointed out, and I hope that the proposals made in this context will be capable of generalization to a wider range of philosophical problems.

B. I shall not be concerned with all varieties of inductive inference but only with scientific inference. If "induction" is construed broadly as meaning any valid nondeductive reasoning, then all but the logical and mathematical components of scientific inference are inductive. However, there are problems of induction which are not subsumed under scientific inference, especially problems of making practical decisions in a rational manner under conditions of uncertainty or partial knowledge. My treatment of scientific inference is instrumentalistic, in the sense that the justification for certain prescriptions is their conduciveness to achieving certain ends, but the ends admitted are only the theoretical ones of learning the truth about various aspects of the world; accordingly, these prescriptions are characterized by open-mindedness, caution, and a long-range point of view.[3] In making practical decisions there are other ends than pure knowledge, and a short-range point of view is often unavoidable. Although I believe that my treatment of scientific inference can be embedded in a more general theory of induction, in which the interrelations of theory and practice are properly taken into account, I shall not undertake to formulate such a theory.

The treatment of scientific inference in the following sections will be divided into two stages. In the first stage (sections II, III, and IV) the examination will proceed as far as possible by methodological considerations. Scientific inference will be formulated in terms of probability, and a conception of probability (essentially due to Harold Jeffreys) which I call "tempered personalism" will be presented. It will be argued that scientific inference, thus formulated, is a method of reasoning which is sensitive to the truth, whatever the actual constitution of the universe may be. In this stage very little will be assumed about the place of human beings in nature, but the result of such austerity is that scientific method is only minimally justified, as a kind of counsel of desperation.[4] In the second stage (section V) the justification is enlarged and made more optimistic by noting the biological and psychological characteristics of human beings which permit them to use induction as a discriminating and efficient instrument. Thus the second stage considers scientific methodology from the standpoint of "the way down," and it evidently presupposes a considerable body of scientific knowledge about men and their natural environment, thereby raising the issue of circularity.

The outline of this treatment of scientific inference is due to Peirce. The first stage derives from Peirce's conception of the scientific method as a deliberate and systematic "surrendering" to the facts, whatever they may be (for example, Peirce 5.581, 7.78). The second stage derives from Peirce's proposition that all men, and the great discoverers preeminently, possess "an inward power, not sufficient to reach the truth by itself, but yet supplying an essential factor to the influences carrying their minds to the truth" (1.80). He was the first philosopher to understand fully that a study of instinct and intelligence in the light of evolutionary biology could provide a surrogate for classical rationalism, thereby reconciling a

Copernican skepticism about the reliability of intuition with a recognition of the role of innate ideas in human knowledge.

There are two important respects, both in the first stage, in which my treatment of scientific inference departs from the main line of Peirce's argument. First, my treatment makes essential use of a concept of probability which is entirely distinct from the only one considered by Peirce to be legitimate, namely, the frequency concept (for example, 2.673–85). Second, Peirce usually claims that in the infinitely long run it is certain that scientific inference will asymptotically approach the truth, whereas the claims I put forth are qualified and conditional, and concern the results of investigations in finite intervals of time. However, there are some interesting passages in Peirce's later writings, which will be cited in section IV, in which he appears to depart from his usual position in the direction that I am proposing.

In a broad sense the structure [5] of the theory of scientific inference which I am proposing, and indeed of my entire program of naturalistic epistemology, is Platonic. I try above all to retain the characteristics of Plato's dialectic of beginning *in medias res* with hints and clues rather than with certainties, and of regarding knowledge of fundamental principles as the terminus of inquiry. The interplay of considerations of "the way up" and "the way down," by invoking some of the propositions of natural science at the same time that the validity of scientific procedures is being examined, fits into the dialectic pattern, even if the wording is not typically Platonic. Even Plato's doctrine of recollection is found in modern form in the conjecture of Peirce, which is incorporated into my theory, that the phylogenetic heritage of ideas and dispositions which result from evolutionary adaptation are of great heuristic value for scientific investigations of the nature of things.

C. I hope that in time the themes mentioned above will be developed in a comprehensive naturalistic epistemology and applied in detailed studies of many special problems. There are, however, great difficulties in trying to carry out this program. One is the obvious need for a critical examination of the relationship between philosophy and the natural sciences. A naturalistic epistemology is supposed to be based upon scientific knowledge, but the term "based upon" is elusive. After all, there are few if any modern philosophers, even among those whom I would classify as non-Copernican (such as Berkeley, Hume, Kant, the phenomenologists, the logical positivists, and the ordinary language analysts) who do not admit the validity, on an appropriate level, of the scientific propositions to which a Copernican appeals. Since philosophical systems have "many degrees of freedom," differing in their treatments of experience, of evidence, of confirmation, of existence, etc., they can accommodate the results of natural science in various ways (cf. McKeon 1951). Nevertheless, it may be possible to go beyond a relativism of systems by carefully applying criteria of philosophical adequacy which are recognized (to be sure, with varying interpretations) from all points of view—for example, coherence, fineness of reasoning, comprehensiveness, openness to evidence, and richness of content. Furthermore, the application of the last three criteria requires serious and detailed attention to the results of natural science. In subtle ways, then, the natural sciences (and, to the extent that these sciences are veridical, the facts of the world) would exercise a guiding and perhaps even a controlling role in philosophical investigations. However, this methodological supposition, which is Platonic in character, can be made convincing only by a detailed examination of the relevance of scientific material to philosophical controversies.[6] This indicates a second difficulty in carrying out the program of a naturalistic epistemology: it requires extensive knowledge of the sciences, particularly of psychology, far exceeding the amount that philosophers have usually deemed necessary for their purposes. (A case in point is the problem of the relation between observational and theoretical terms, which has often been treated with considerable logical ingenuity, but seldom with more than armchair information about empirical psychology.) A third difficulty is the most formidable but also the most interesting: the currently accepted body of scientific propositions does not provide all the information which a naturalistic epistemologist requires; nor are accepted propositions immune to critical probing by philosophers. For example, a naturalistic epistemology can offer no more than speculative proposals concerning some problems of perception, because the mind-body relationship is a vast desert of ignorance in the sciences of psychology and physiology. As a result, a naturalistic epistemology should not merely be an appendage to natural science, but rather the relation should be one of mutual criticism, stimulation, and illumination. If the demarcation between philosophy and natural science is thereby blurred, this will be regarded as healthy by anyone who appreciates the seventeenth-century sense of "natural philosophy."

In view of these difficulties a program of naturalistic epistemology must be carried out cumulatively and with contributions from many workers. I find it encouraging that in recent years a number of epistemologists have been working, although diversely and with little coordination, along the lines that I envisage.[7] The present essay is intended to be a contribution to a comprehensive program of naturalistic epistemology in two ways: first, as a treatment of one important problem from a naturalistic standpoint, and second, as an illustration of a set of themes, particularly concerning the structure of knowledge, which may be capable of incorporation into a more general theory. I feel strongly that my proposals are steps in the right direction, but I am fully aware that at many points they need to be made more precise, augmented, and corrected.

II. Bayesian Formulations of Scientific Inference

A. The general theses of Copernicanism in epistemology are too broad to determine the character of scientific inference in any detail, but they do impose some demands. Above all, they require that scientific inference not be apodictic, since Copernicanism doubts the existence of human powers (such as those claimed in Aristotle's theory of intuitive induction) for attaining certainty with regard to propositions which are not entailed by the evidence. Also, they require that the kind of "inverse" reasoning which is exhibited in the hypothetico-deductive method be given a central place in scientific methodology, since Copernicanism recognizes that important scientific truths may be remote from direct experience and, therefore, unattainable by methods (like those of Bacon) which test only straightforward empirical generalizations.

A natural way to satisfy these demands is to formulate scientific inference in "Bayesian" terms. This means (a) using as the central concept in all nondeductive reasoning a probability function $P(h|e)$ (read "the probability of the hypothesis h upon evidence e"); (b) ensuring that P satisfies the standard axioms of probability,[8] namely,

(i) $0 \leq p(h|e) \leq 1$,
(ii) $P(h|e) = 1$ if e entails h,
(iii) $P(h \vee h'|e) = P(h|e) + P(h'|e)$, if e entails $\sim(h\&h')$,
(iv) $P(h\&h'|e) = P(h|e) \cdot P(h'|h\&e) = P(h'|e) \cdot P(h|e\&h')$;

and (c) allowing as admissible values for h not only statistical hypotheses, referring to the composition of very large reference classes, but arbitrary hypotheses, including such radically nonstatistical ones as singular propositions and scientific theories. The name "Bayesian" is applied to a method of inference which satisfies (a), (b), and (c) because it permits the use of Bayes's theorem

$$P(h|e\&a) = \frac{P(h|a) \cdot P(e|h\&a)}{P(e|a)}$$

(an immediate consequence of axiom [iv]), in order to evaluate the *posterior probability* $P(h|e\&a)$ in terms of the *prior probability* $P(h|a)$ and the *likelihood* $P(e|h\&a)$. Other treatments of inductive inference, such as the "orthodox" statistical theories of Fisher and of Neyman and Pearson, disallow the application of Bayes's theorem to arbitrary hypotheses, on the grounds that prior probability is in general not well defined.

A Bayesian is able to construe the hypothetico-deductive method as an application of Bayes's theorem in special circumstances. Suppose that there is a class of admissible hypotheses $h_1, h_2, \ldots$, a body of auxiliary information and assumptions a, and a body of evidence e such that for every integer i either e or $\sim e$ is entailed by $h_i\&a$. (This is a great idealization, since in realistic situations an unequivocal relationship of entailment or inconsistency between the hypothesis, together with the auxiliary assumptions, and the evidence is very rare. Also, there is a deep methodological problem, related to the central problem of section III, in demarcating the class of admissible hypotheses; but this will be set aside in the present formal analysis of hypothetico-deductive inference.) Suppose further that $h_1\&a$ entails e, and that if h_j is any other hypothesis which shares this property with h_1, then $P(h_1|a) >> P(h_j|a)$. It follows from Bayes's theorem that the posterior probability $P(h_1|a\&e)$ is much greater than the posterior probability of any other member of the admissible class, and this conclusion can be presented by a Bayesian as the essential content of the hypothetico-deductive confirmation of h_1. (A Bayesian may wish to construe the hypothetico-deductive confirmation of h_1 more strongly, for example, as asserting that $P(h_1|a\&e) > 1-\epsilon$, where ϵ is some preassigned small positive number; but to do this the suppositions regarding the prior probabilities must be strengthened.) That the usual formulations of the hypothetico-deductive method seldom compare the prior probabilities of those hypotheses which share with h_1 the property of "saving the appearances" can be regarded by a Bayesian as merely a deficiency in explicitness of analysis. Furthermore, a Bayesian can correctly claim that in many cases the hypothetico-deductive method cannot be used without drastically idealizing and distorting the conditions of investigation, whereas other inferences using Bayes's theorem are valid (cf. Salmon 1966: pp. 250–51).

There exist non-Bayesian treatments of scientific inference which attempt to satisfy the demands imposed by a Copernican point of view, the most important perhaps being that of Popper, who construes the hypothetico-deductive method as a procedure of elimination or refutation. In this paper, however, I shall not try to evaluate such treatments,[9] since my purpose is the constructive one of exhibiting a specific Bayesian formulation of scientific inference which is both thoroughly Copernican in spirit and defensible against the usual objections to the application of Bayes's theorem. It is generally granted, even by opponents, that the Bayesian approach is systematic, treating a great variety of special modes of inference, from the hypothetico-deductive method to statistical decision procedures, as cases of a few general principles (in contrast to orthodox statistical theory, which proceeds in a much more piecemeal fashion).[10] But the opponents maintain that this systematization is specious, since it is based upon the concept of prior probability, which they claim is either empty or arbitrary or otherwise useless for scientific inference, depending upon its interpretation. Their criticism is partially valid, I believe, when it is directed against the two conceptions of probability which most Bayesians have espoused, namely, the "logical concept" and the "personalist concept." Again, since my purpose is constructive rather than critical, I shall only briefly discuss these two concepts, in order to provide the motivation for a version of Bayesianism which is based upon the "tempered personalist" concept of probability. It will be seen that this is essentially the concept of probability which Harold Jeffreys develops in his treatise, *Theory of Probability*, his verbal allegiance to the logical concept notwithstanding.

B. The leading exponents of the logical concept of probability are Keynes and Carnap, but the concepts which they present are different because of the divergence of their views of logic.

For Keynes, logic is concerned with relations between propositions, which are characterized as "the objects of knowledge and belief" (1921: p. 12) and evidently are conceived by him to be nonlinguistic entities. Probability theory is part of logic because "probability-relations" are a kind of relation between propositions: "if a knowledge of h justifies a rational belief in a of degree α, we say that there is a *probability-relation* of degree α between a and h" (ibid.: p. 4). (The "degrees" are not assumed to be ordered as the real numbers or even to constitute a simply ordered set [p. 34].) These relations are asserted to be "real" and "objective" (p. 5), independent of what human beings happen to believe (p. 4), and incapable of analysis in terms of simpler ideas (p. 8). Human beings are attributed the power to have direct knowledge of "secondary propositions" that probability relations hold between various pairs of propositions (p. 16), but the extent and accuracy of this power varies from person to person.[11] "The perception of some relations of probability may be outside the powers of some or all of us" (p. 18). Keynes thus conceives of probability relations in much the same way as a Platonist in mathematics, such as Gödel, conceives of the propositions of set theory —as true in virtue of internal relations among sets, which are real entities independent of human thought, language, or axiom systems.

It is just this comparison between mathematics and the logical theory of probability which exhibits the weakness of Keynes's position. For a Platonist in mathematics can point to a rich set of principles which have very strong intuitive appeal and which are sufficient for the "construction" (in the set theoretical sense rather than in the sense of recursive function theory) of all the structures ordinarily investigated by mathematicians. Although mathematical intuition becomes "astigmatic" (Gödel 1946: pp. 150–52) concerning esoteric matters, such as the continuum hypothesis, the edifice of set theory is sufficiently impressive to provide good cases both for the ontological independence of mathematical entities and for the excellence of human intuition concerning them. By contrast, the intuitive knowledge which Keynes cites about probability is slight, the only systematic components being: (a) a set of axioms regarding comparisons of degrees of probability (later strengthened by Koopman [1940]) and (b) a weak form of the principle of indifference, which in spite of the caution of its statement is nevertheless too strong to be defensible as it stands (cf. Russell 1948: p. 375). With trivial exceptions (for example, when e entails h or $\sim h$), the axioms only permit the derivation of probability propositions from other probability propositions, and these may be compared with those elementary principles of set theory which assert that the results of Boolean operations upon sets are sets—a very small part of set theory! Furthermore, the axioms of probability can be justified without any assumption about objective probability relations (cf. sections II.C and III.D). An extremely weak version of the principle of indifference, asserting only that $P(h_i|e) = P(h_j|e)$ if h_i and h_j are symmetrical in all respects relative to e, is indeed intuitively irresistible, but this version is of no use in any actual scientific inference without supplementary judgments to the effect that the symmetry-breaking features of the alternatives (which are always present in actual cases) are probabilistically irrelevant. Even though I am sympathetic with Platonism in mathematics, I find it hard to withstand Ramsey's argument

that there really do not seem to be any such things as the probability relations he describes. He supposes that, at any rate in certain cases, they can be perceived; but speaking for myself I feel confident that this is not true. I do not perceive them, and if I am to be persuaded that they exist it must be by argument; moreover I shrewdly suspect that others do not perceive them either, because they are able to come to so very little agreement as to which of them relates any two given propositions. All we appear to know about them are certain general propositions, the laws of addition and multiplication; . . . and I find it hard to imagine how so large a body of general knowledge can be combined with so slender a stock of particular facts. It is true that about some particular cases there is agreement, but these somehow paradoxically are always immensely complicated; we all agree that the probability of a coin coming down heads is 1/2, but we can none of us say exactly what is the evidence which forms the other term for the probability relation about which we are then judging. If, on the other hand, we take the simplest possible pairs of propositions such as 'This is red' and 'That is blue' or 'This is red' and 'That is red', whose logical relations should surely be the easiest to see, no one, I think, pretends to be sure what is the probability relation which connects them. (1931: pp. 161–62) [12]

For Carnap a sentence is logically true if its truth is a consequence of the syntactical and semantical rules of the language in which it occurs. Statements of "probability" in the sense of "degree of confirmation" (as contrasted with the frequency sense) are metalinguistic statements of the form "$c(h,e) = r$," where h and e are sentences in some object language and r is a real number; and the true sentences of this form are consequences of metalinguistic rules governing the function c. (In an unpublished work [1968] Carnap continues to refer to some language or class of languages in discussing inductive logic, but his probability statements are now of the form "$C(H|E) = r$," where "H" and "E" denote propositions which are nonlinguistic entities.) In order to guide the selection of a c-function, Carnap lays down a number of conditions of adequacy (1963: pp. 974–76) (which are converted into axioms for the function C [1968]). These have been criticized both on the ground that some of them have too much content to be imposed a priori (for example, Nagel 1963: pp. 797–99 and Salmon 1961: p. 249) and on the ground that they are so weak that they admit too large a range of functions (for example, Lenz 1956). Carnap has also been criticized for saying that a c-function is an instrument (specifically, "an instrument for the task of constructing a picture of the world on the basis of observational data and especially of forming expectations of future events as a guidance for practical conduct" [1963: p. 55]) and, therefore, that the choice of a c-function may properly be evaluated by its performance, "that is, the values it supplies and their relation to later estimates" (ibid.); according to critics, it is only future performance which is relevant, and any assertion about that must be based upon induction, the rules of which are in question when deliberating upon the choice of c (for example, Lenz 1956). On these disputed points, however, I believe that the general direction of Carnap's thought is more correct and more suggestive than his critics have allowed. There may well be a place in the theory of induction for principles which are not justifiable a priori, but which are entrenched deeply though without an unconditional commitment (cf. section V.D). It may very well be that any reasonable set of general rules will fall short of determining numerically the probabilities of most hypotheses of interest upon most bodies of evidence, and yet the need to rely upon subjective decisions to supplement such rules may preclude neither eventual consensus among reasonable men nor sensitivity to the truth (cf. section III). And the need to appeal to induction in order to evaluate the performance of an inductive method may be both an inescapable consequence of the human situation and, with proper qualifications, an opportunity to use a legitimate and nonvicious kind of circular reasoning (cf. section V.E). (Carnap [1968: sec. 4] makes an analogy to geometry and distinguishes "pure" from "applied" inductive logic, placing principles which are justifiably a priori in the first class and those which are not in the second. The accuracy of this analogy is questionable, since inductive logic has a kind of normative character which is alien to geometry. Nevertheless, the general suggestion of dichotomizing inductive principles may be fruitful.)

In my opinion, the crucial error in Carnap's published works on induction is his taking the choice of a c-function, which is tantamount to one sweeping decision regarding the evaluation of all probabilities, as a desideratum. There are two major objections to this procedure: (1) the rules governing a c-function are purely syntactical and semantical and take no account of whether a given hypothesis h has been considered worthy of serious consideration by any one. But treating on the same footing those hypotheses which have been seriously proposed and those which have not has the effect of drastically decreasing the value set upon intelligent guessing and insight into the subject matter of investigation. In other words, undiscriminating impartiality toward hypotheses which no one has seriously suggested is a veiled kind of skepticism, since it would leave very little prior probability to be assigned to any specific proposal. Putnam presents in detail an objection along these lines in his contribution to the volume on Carnap in *The Library of Living Philosophers* (1963: pp. 770–74). Carnap's reply is remarkably receptive: "it is worthwhile to consider the possibility of preserving an interesting suggestion which Putnam offers, namely to make inductive results dependent not only on the evidence but on the class of actually proposed laws" (1963: p. 916). This reply can be made consistent with a purely logical definition of a c-function by increasing the number of arguments of c from two to three, that is, by defining a function $c(h,e,l)$, which is the degree of confirmation of the hypothesis h on the basis of evidence e, if the set of actually proposed laws is l. If this modification were worked out in detail, it would provide an answer to objection (1). I can also see an attractive alternative modification, which has the advantage of not increasing the number of arguments of c: the serious proposal of a hypothesis h by some one could be recognized as a relevant piece of evidence, expressed in a sufficiently rich language by the sentence s; and if e expresses all other evidence, there is no reason to suppose that $c(h,e{\cdot}s)$ is even approximately equal to $c(h,e)$. It would be difficult to work out this answer in detail, however, for according to the multiplication axiom (axiom iv),

$$c(h,e \cdot s) = \frac{c(h \cdot s,e)}{c(s,e)} \, ,$$

and the technical work so far presented by Carnap gives no help in evaluating the ratio on the right-hand side of this equation.

(2) Even if there is a "best" c-function, which is supposed somehow to be chosen, it would by no means be clear how to use it in practice. According to Carnap's "requirement of total evidence," a person applying inductive logic in a given situation "must use as evidence e his total observation knowledge" (1963: p. 972). But how can this total knowledge be marshaled and expressed in a sentence suitable for substitution in the function $c(h,e)$, when by far the larger part of it is only dimly remembered or has been transformed into beliefs, habits, and adjustments? In order to conform to the requirement of total evidence, an inductive method is needed which effectively deploys the sharp part of our observational knowledge, such as might be recorded in an experimenter's notebook, but also somehow takes account of the massive background of experience.

The difficulties associated with a single sweeping decision regarding a c-function are largely avoided in Carnap's unpublished work. There he suggests that each separate kind of scientific investigation be taken as a separate domain of "applied inductive logic" (1968: secs. 4 and 5). He does not discuss the extent or the autonomy of the various domains. Nevertheless, his suggestion is at least partially in agreement with the principle of locality, recommended in section III below.

C. The personalist concept of probability was proposed independently by Ramsey and DeFinetti and has been elaborated recently by a number of Bayesian statisticians, notably Savage, Good, Raiffa, and Schlaiffer. They find it convenient to measure a person's credence (that is, the intensity of his belief) in h, conditional upon the truth of e, by finding the largest fraction of a unit of utility which he is willing to stake in order to receive one unit if h is true, the bet being cancelled if e turns out to be false. Personal probabilities are the intensities, thus measured, of a person's subjective beliefs, provided that these are examined so as to eliminate lapses of *coherence* (a reasonable extension of consistency, which will be explained in the next paragraph). Sometimes "subjective probability" is used as synonymous with "personal probability," but it seems preferable to distinguish these terms, applying the former to actual beliefs, which may violate the condition of coherence (Edwards, Lindman, and Savage 1963: p. 197). The personal probability of h, conditional on e, for X is denoted by "$P_X(h|e)$", where "X" refers to a person at a definite time (the temporal qualification being required because of fluctuations in a man's beliefs); the subscript X is often omitted if it is clear from context whose beliefs are being considered. It should be noted that e is not assumed to be the total body of evidence as in the similar notation for the logical concept of probability.

The only normative constraint upon X's personal probabilities, according to the personalists, is the condition of coherence: that no set of bets can be proposed which are acceptable to X according to his probability evaluations, but which cause him to suffer a net loss regardless of the truth values of the hypotheses bet upon (that is, no "book" can be made against X).[13] This is a nontrivial constraint, however, since, as DeFinetti has demonstrated, it implies that X's personal probabilities satisfy the standard axioms of probability (axioms [i] through [iv] in section II.A). And from these axioms follows Bayes's theorem, which asserts that in a certain sense X is constrained to learn from experience. For if the ratio $P_X(e|a\&h)/P_X(e|a)$ is either much greater or much less than 1, then the posterior probability differs sharply from the prior probability; and even though this ratio is a matter of subjective belief, once e is specified, the evidence e actually obtained is not under X's control, and to this extent the posterior probability is shaped by experience. The personalists claim that in this way the system of personal probabilities is adequate for guiding our practical expectations and for checking the theoretical hypotheses of natural science. They say, furthermore, that all attempts to evade the subjective basis of induction are deceptive, and they cite in particular the subjective character of choices of significance levels and of confidence intervals which are made by statisticians committed to an "objectivist" concept of probability (that is, probability in the sense of relative frequency).[14] The critics of personalism are parodied as saying: "We see that it is not secure to build on sand. Take away the sand, we shall build on the void" (paraphrase of DeFinetti in Edwards, Lindman, and Savage 1963: p. 208).

It is difficult to see, however, that the personalists have adequately explicated the concept of probability which is involved in scientific inference. A man's beliefs are intimately bound up with his emotional attachments, his subconscious associations, and the ideals of his culture, and it is likely that the belief systems of almost everyone in most cultures, and of the majority of men even in "advanced" cultures, are not conducive to performing good inductions on theoretical matters beyond the concerns of ordinary life. I do not find an adequate discussion of these complicating features of belief systems by members of the personalist school of probability. The only kind of irrationality which they explicitly discuss is incoherence, but there are surely other properties of belief systems which are characterizable as irrational and which would inhibit the progress of natural science if they were universal. In particular, the assignment of extremely small prior probabilities to unfamiliar hypotheses is compatible with coherence, but may be irrational in the sense of being the result of prejudice or of obeisance to authority or of narrowness in envisaging possibilities. But unless a hypothesis is assigned a nonnegligible prior probability, Bayes's theorem does not permit it to have a large posterior probability even upon the grounds of a very large body of favorable observational data, and thus the theorem does not function as a good instrument for learning from experience. Perhaps the personalists make an implicit assumption that those who employ their methods are reasonable in some broad sense; after all, anyone who is willing to use the machinery of probability theory in order to adjust his opinions is unlikely to be a hardened fanatic. It is interesting that one occasionally finds personalists stating obiter dicta which seem to qualify their official attitude toward subjective belief. Thus

a prior distribution which has a region of zero probability is therefore undesirable unless you really consider it impossible that the true parameter might fall in that region. Moral: Keep the mind open, or at least ajar. (Edwards, Lindman, and Savage 1963: p. 211)

And

more generally, two people with widely divergent prior opinions but reasonably open minds will be forced into arbitrarily close agreement about future observations by a sufficient amount of data. (Ibid.: p. 201)

Implicit in these passages is a normative principle supplementing coherence, which ought to be made explicit if inductive reasoning is to be understood.

D. Despite the differences between Carnap's logical concept of probability and the personalist concept, both seem to be attempts to explicate the same informal concept (in Carnap's terminology, the same explicandum), namely, "rational credibility function" (1963: p. 971). Furthermore, Carnap, like the personalists, imagines betting situations in order to treat belief quantitatively: "$P(h,e)$ is a fair betting quotient for X with respect to a bet on h" (ibid.: p. 967).[15] The essential difference between their conceptions is that the personalists admit no criterion for rationality or fairness other than coherence, whereas Carnap formulates a series of additional criteria and hopes to find yet more.

There are reasons for doubting that the concept of probability as it occurs in scientific inference can be properly explained in terms of betting quotients, or even that it can be equated with the concept of a rational credibility function. The source of the difficulties is the indispensability of general theories in natural science as it is now studied and, indeed, as it conceivably could be studied if the ideal of general insight into the nature of things is not to be abandoned. Although probability is a valuable instrument in reasoning about scientific theories, it does not seem to make sense to explain the probability of a theory as a fair betting quotient. As Putnam argues, if the bet concerns the truth or falsity of a scientic theory, then how should the outcome of the bet be decided? Even if we can agree to call the theory false in certain definite cases, under what circumstances would we decide that the theory is certainly true and that the stakes should be paid to the affirmative bettor (1963a: p. 3)? There is no doubt that bets with definite payoff conditions can be made *concerning* a theory, for instance, on whether strongly unfavorable evidence will be found by the year 2000 or on whether the theory will be accepted by working scientists and enshrined in textbooks by that date, but such bets are evidently not about the *truth* or *falsity* of the theory. Moreover, even if one does not explain "credence" in terms of "acceptable betting quotient" and, therefore, "rational credibility" in terms of "fair betting quotient," the identification of the probability of a theory with its rational credibility is problematic, at least in the case of theories which are both general and precise and, therefore, very strong in their assertions (as in fundamental physics).[16] For it can be argued that upon any body of supporting evidence the rational degree of belief that such a theory is literally true, without qualification in conditions far different from those in which experiments have been conducted (for example, at much higher energies, or in much smaller space-time regions, or at much higher or lower temperatures than in the experiments) and without exception throughout all space and time, is zero or extremely small. Most scientists who are aware of the history of scientific revolutions can be expected to have very small credence in the literal and unqualified truth of theories of great strength. But if probability evaluations are to be guides to the acceptance (even tentatively) of general theories, the smallness of the probabilities of theories would be troublesome. Although this argument does not appear to me to be as decisive as the one against reasoning about theories in terms of bets, since conditions of acceptance might be formulated in terms of ratios of posterior probabilities (provided these are not all zero) rather than in terms of absolute values, it nevertheless provides a motivation for reexamining the probability of theories and for seeking a different explicandum than rational credibility.

Some suggestions for an alternative explicandum can be gathered from the history of science. It has often happened that a scientific theory which was once successful, in the sense of being the only one among the commonly known rival theories to survive severe tests by independent experimenters, later had to be displaced. However, one of the reasons for believing that the scientific enterprise is progressive, despite the limitations of human powers which Copernicans emphasize, is that the displacement of once successful theories is seldom ignominious. (Evidently, this claim is cogent only if "successful" is used in the sense specified above, which drastically limits the reference class of theories, and not in the sense of "generally accepted.") Usually something, at least in the form of suggestions, is salvaged from an old theory. A select few (such as classical mechanics and Euclidean physical geometry) were displaced under such honorable circumstances that they may be regarded as promoted to the rank of "theory emeritus": they not only yielded the same observational predictions as the new theories over a wide domain, but the new theories were in a certain sense conceptual generalizations of them. This suggests that a person whose belief in the literal truth of a general proposition h, given evidence e, is extremely small may nevertheless have a nonnegligible credence that h is related to the truth in the following way: (i) within the domain of current experimentation h yields almost the same observational predictions as the true theory; (ii) the concepts of the true theory are generalizations or more complete realizations of those of h; (iii) among the currently formulated theories competing with h, there is none which better satisfies conditions (i) and (ii).[17] If a word is needed for this modality of belief, I suggest "commitment." If "rational degree of commitment," is identified as the explicandum of "probability" in the context of scientific inference, there will be no prima facie obstacle to finding that even very strong theories upon appropriate evidence have probabilities close to 1. Furthermore, the use of "probability" in this way would permit a formulation of scientific inference which fits the contours of the actual thinking of investigators, by focussing attention upon the progress of knowledge rather than upon the ultimate truth of theories. In the case of singular propositions and other special hypotheses, the compunctions stated above about a high credence are not applicable, and, therefore, credence and degree of commitment should be the same or very close. In these cases, then, "rational degree of commitment" virtually coincides with Carnap's explicandum of "probability," namely, "rational credibility function."

There is a difficulty concerning the concept of rational degree of commitment which arises from the occurrence of several vague expressions ("almost the same observational predictions" and "more complete realizations") in the first two of conditions (i) to (iii). This vagueness is troublesome in treating the "catchall" hypothesis h_n, which is the negation of the disjunction of all the specific hypotheses $h_1, \ldots, h_{n-1}$ considered in a scientific investigation. The ratio of the probabilities, upon given evidence, of h_i and h_j, where neither i nor j equals n, does not seem to depend upon the exact meanings of these expressions. On the other hand, the ratio of the probability of h_i $(i \neq n)$ to that of h_n is sensitive to their exact meanings: the more stringently "almost the same observational predictions" and "more complete realizations" are construed, the stronger is the proposition that h_i will be displaced only under conditions (i) to (iii), and hence the smaller is the ratio of its probability to that of h_n. I can see three possible ways of dealing with this difficulty:

a) Conditions (i) and (ii) might be reformulated so as to eliminate vagueness. In the case of condition (i) this reformulation would be relatively straightforward, consisting in a quantitative specification of "almost the same." In the case of condition (ii) I do not know a straightforward procedure of reformulation, since "more complete realization" does not lend itself in an obvious way to quantitative treatment. Furthermore, there is not merely a question of quantitative specification, since the character of the relationship between a moderately successful old theory and a more successful revolutionary theory cannot be known in advance of the scientific revolution.

b) It might be possible to formulate scientific inference in a way that avoids treating the catchall hypothesis on the same footing with the specific hypotheses $h_1, \ldots, h_{n-1}$. Numerical weights might be attached to the specific hypotheses but not to h_n, and the assignment of these weights might be governed by principles similar to the standard probability axioms. However, there would necessarily be some departure from Bayesianism, for h_n is equivalent to $\sim (h_1 \vee \ldots \vee h_{n-1})$, and hence the disallowance of a numerical weight to h_n implies that the domain of the weight function is not closed under Boolean operations. Because of the arguments already given in section II.A and also because of the justification of the axioms of probability to be presented in section III.C, I am very reluctant to abandon a Bayesian formulation of scientific inference. Nevertheless, solution (b) would be attractive if two conditions could be fulfilled: the development of a reasonable set of principles, comparable to the standard axioms of probability, governing the assignment of numerical weights; and the provision of a reasonably open-minded treatment of the possibility that none of the specific hypotheses $h_1, \ldots, h_{n-1}$ is a good approximation to the truth. I conjecture that all the methodological proposals of the remainder of this paper (the "localization" of problems, the role assigned to subjective judgment, the preference accorded to seriously proposed hypotheses, the rejection of a priori orderings of hypotheses according to simplicity, and the derivation of methodological guidelines from the tentative body of scientific knowledge) would be compatible with a partial modification of Bayesianism.

c) The procedure for evaluating the rational degree of commitment to a hypothesis h relative to evidence e may be such that the vagueness of the concept of commitment is methodologically innocuous. One thesis of section III is that a major component of the *rationality* of a rational degree of commitment is conduciveness to the progress of knowledge, and this criterion in turn leads to a prescription of open-mindedness toward seriously proposed hypotheses (that is, giving each of them a chance to be accepted into one's tentative body of knowledge if supported by sufficiently favorable evidence). But the application of this prescription in numerically weighing hypotheses is compatible with the vagueness of the concept of commitment. It will be argued in section III.C that persons who disagree sharply regarding prior probability evaluations and yet conform to the prescription of open-mindedness (which is there called "the tempering condition") are able to achieve rough consensus regarding posterior probabilities relative to a moderate amount of experimental data; and this argument can be adapted to show that rough consensus regarding posterior probabilities can be achieved even if quite different senses of "commitment" are used. Consequently, despite arbitrariness in the ratios of the prior probability of h_i ($i \neq n$) to that of h_n, it is possible to make numerical probability evaluations fruitfully without a prior clarification of the concept of commitment. In fact, the converse can be expected to occur: as knowledge progresses, and in particular as we learn how crude but partially successful theories are replaced by more refined theories, we may be able to clarify conditions (i) and (ii) in a nonarbitary manner. I shall tentatively accept solution (c) in this paper. It fits well into a theory of scientific inference in which an a priori logical structure is supported and enriched by a posteriori considerations. And it illustrates one of Plato's central ideas: that the clarification of concepts is inseparable from the progress of knowledge. [Note added, July 1969: The acceptance of solution (c) now appears to me to be based upon wishful thinking. Consequently, I now prefer solution (b) but have not worked it out in detail.]

III. The Tempered Personalist Concept of Probability

A. The discussion of section II has indicated that a rationalization of scientific inference should be sought in terms of a concept of probability which is endowed with a stronger normative character than that of the personalist concept, which derives this additional normative character from conditions for the progress of knowledge (rather than from internal relations among propositions as in the theory of Keynes), and which possesses greater openness to the contingencies of inquiry than Carnap's c-functions. A concept satisfying these requirements is to be found, a little beneath the surface, in the work of Harold Jeffreys. Although Jeffreys is generally recognized as one of the masters of Bayesian statistics (Edwards, Lindman, and Savage 1963: p. 194; Hacking 1965: pp. 201–02; Carnap 1950; pp. 245–46), he has not, in my opinion, been completely understood. The misunderstanding is largely due to the discrepancy between his philosophical statements on the foundations of probability theory and his treatment of probability evaluations in concrete contexts—the latter being, again in my opinion, by far the more profound. His characterization of probability resembles that of Keynes in having a Platonic ring, for he seems to regard the probability of *h* on *e* as objectively fixed by internal relations between *h* and *e*. Thus, he states,

probability theory . . . is in fact the system of thought of an ideal man that entered the world knowing nothing, and always worked out his inferences completely, just as pure mathematics is part of the system of thought of an ideal man who always gets his arithmetic right. (1961: p. 38)

Also

differences between individual assessments . . . can be admitted without reducing the importance of a unique standard of reference. (Ibid.: p. 37)

But it is by no means clear that these statements should be taken literally, especially in view of his recommendation that difficult problems of evaluating prior probabilities be settled by the decision of an "International Research Council" (ibid.: p. 37). After all, nothing concerned Plato more deeply than to refute the contention of the Sophists that norms are matters of convention! An implicit conception of probability as an instrument for extending knowledge, rather than logical considerations, underlies Jeffreys's actual proposals for evaluating prior probabilities. Thus, he says that

philosophers often argue that induction has so often failed in the past that Laplace's estimate of the probability of a general law is too high, whereas the main point of the present work is that scientific progress demands that it is far too low. (Ibid.: p. 132)

His fundamental rule for assigning prior probabilities to scientific hypotheses (a rule which he misleadingly calls "the simplicity postulate") is the following:

Any clearly stated law has a finite prior probability, and therefore an appreciable posterior probability until there is definite evidence against it. (Ibid.: p. 129)

This rule makes no pretense at providing an ideal prior ordering and weighing of all possible laws as did an earlier proposal of Wrinch and Jeffreys (1921) (for which the name "simplicity postulate" was appropriate), but is rather a methodological prescription of open-mindedness. Because of the vagueness of this prescription, there is much arbitrariness, which he tends to gloss over, in his actual evaluation of probabilities (for example, his assignment of probability 1/2 to the null hypothesis [1961: sec. 5.0]). However, he correctly points out that within wide limits the exact values of prior probabilities are often unimportant in comparing posterior probabilities, because the ratio of likelihoods becomes the dominant factor (ibid.: p. 194). My feeling at this point is that the basic insight of Jeffreys, which is the instrumental character of probability, would be better presented if all claims of approximating an objectively fixed value of the prior probability were dropped in favor of the acknowledgment that his prescription of open-mindedness leaves great latitude to subjective judgment.

A methodological principle of great importance, which permits probability to be a manageable instrument in scientific inference, seems to underlie tacitly much of Jeffreys's treatment of specific problems (especially in 1961: chap. VI, "Significance Tests: Various Complications"): it is that *the individual investigation delimits an area in which probabilities are calculated*. This principle in no way denies that investigations overlap and influence one another in intricate ways, but asserts only that the conditions of a single investigation establish a kind of "local" universe of discourse within which calculations strictly governed by the axioms of probability can be performed. The conditions of the investigation are (1) a set of hypotheses $h_1, \ldots, h_n$ (of which the last may be a "catchall" hypothesis equivalent to $\sim [h_1 \vee \ldots \vee h_{n-1}]$) which have been "suggested as worth investigating" (ibid.: pp. 268, 270); (2) a set of possible outcomes $e_1, \ldots, e_m$ of envisaged observations; and (3) the information i initially available. This information in actual circumstances is very heterogeneous, consisting partly of vague experience concerning the matter of interest, partly of experience which may be sharp but is of dubious relevance, partly of sharp evidence which does not seem to bear directly on the question at hand but is relevant to other questions in the same field, and partly of propositions which are regarded as established even though they go beyond the actual evidence and, therefore, have been accepted because of previous investigations.[18] According to the principle of letting the investigation delimit an area in which probabilities are calculated, $P(q|r)$ need not be taken as well defined in the present context unless the information i is included in (that is, entailed by) r; and, therefore, in particular, the prior probability of h can be taken to be $P(h|i)$. Without such a restriction on the domain of definition of P, one could formally apply Bayes's theorem to any decomposition of i into the conjunction of i_1 and i_2 to obtain $P(h|i) = P(h|i_2) \cdot P(i_1|h\&i_2)/P(i_1|i_2)$, thereby effectively assigning to probabilities upon i_2 the role of prior probabilities. The principle need not be taken as a *prohibition* against taking $P(q|i_2)$ to be well defined, and there may be instances of i and of decompositions of i into i_1 and i_2 in which it would be convenient to do so. But the *option* of relativizing prior probabilities to i should be preserved in order to permit the apparatus of probability theory to be applied *in medias res* (where actual investigations are always located), and also to avoid being driven to the ultimate decomposition in which i_1 is taken to be i itself and i_2 is taken to be the empty or tautological proposition t, so that the application of Bayes's theorem expresses $P(h|i)$ in terms of the "tabula rasa" [19] probabilities $P(h|t)$ and $P(i|t)$.

It should be noted that the methodological principle of letting the individual investigation delimit an area in which probabilities are calculated is nowhere stated explicitly by Jeffreys, but it is a natural complement to such dicta as the following:

> The best way of testing differences from a systematic rule is always to arrange our work so as to ask and answer one question at a time. Thus William of Ockham's rule, 'Entities are not to be multiplied without necessity' achieves for scientific purposes a precise and practically applicable form: *Variation is random until the contrary is shown; and new parameters in laws, when they are suggested, must be tested one at a time unless there is specific reason to the contrary.* (1961: p. 342)

It should also be noted that Jeffreys often apologizes when he follows this principle: for example, "there is a limit to the amount of calculation that can be undertaken at all—another imperfection of the human mind" (ibid.: p. 366). In other words, he seems to treat the principle merely as an expedient; this fact is not surprising, since the principle entails a "local" view of probability that conflicts with his espousal (however superficial) of a logical concept of probability. However, if I am correct that an instrumental approach to probability is the profounder part of Jeffreys's work, then the principle of letting the individual investigation delimit an area for probability calculations can be regarded as partially constitutive of his working concept of probability.

B. The tempered personalist concept of probability, which I propose as a suitable explication of "rational degree of commitment," is very close to Jeffreys's working sense of "probability," except for the role explicitly assigned to subjective judgment. Alternatively, it can be regarded, like the personalist concept of probability, as an idealization of subjective probability, except that it is governed not only by the axioms of probability but by a prescription of open-mindedness. A definition of the proposed term will now be given, and the remainder of section III will be devoted to comments and clarifications. The definition assumes that "the investigation" is schematized as in section III.A.

Suppose that at the beginning of an investigation a person X has a body i of information and assumptions, and suppose that S is a set of propositions closed under truth-functional operations and containing all seriously proposed hypotheses of which X is aware on the matter under investigation and all propositions regarding the possible outcomes of an envisaged set of observations. *Then a tempered personalist probability function over S for X with the body i of information and assumptions* is a map assigning to every ordered pair (q,r) of members of S (such that r is consistent with i) a real number which is approximately equal to X's subjective degree of commitment to q upon supposition that r is his total evidence supplementary to i, subject to two conditions: (1) the map satisfies the standard axioms of probability, and (2) the prior probability (that is, the number assigned when r is tautological) of each seriously proposed hypothesis must be sufficiently high to allow the possibility that it will be preferred to all rival, seriously proposed hypotheses as a result of the envisaged observations.

The crucial features of this concept of probability are its locality, its latitude toward subjective judgment, and its prescription of open-mindedness, the last being expressed by condition (2), which I call "the tempering condition." [20] I do not preclude that additional conditions may reasonably be imposed on methodological grounds in order to make the concept of probability a more sensitive instrument. One such possibility is the invariance rule of Jeffreys, discussed in section III.E. At present, however, I do not see that the incorporation of this rule would be an improvement.

It will be convenient to use a notation fitted to a somewhat idealized decomposition of i into a part a, which is both sharply formulable and prima facie relevant to the matter under investigation, and a residue b, which is either vague or of no apparent relevance and, therefore, classifiable as background information. If this decomposition is specified, then the tempered personalist probability assigned under the conditions given in the definition will be denoted by "$P_{X,b}(q|r\&a)$"; if it is clear from the context whose probability function is being considered and at what time (thereby implicitly fixing b as well as possible, since an explicit inventory of X's experience is out of the question), the abbreviated notation "$P(q|r\&a)$" can be used. There are two advantages in a notation which displays a explicitly: first, the logical relations between i and the propositions of S are essential to calculations which employ the axioms of probability, but only the part a of the total body i of information and assumptions has evident logical relations to the members of S. For example, assumptions about the design of instruments and about the physical laws governing their operation, in conjunction with a hypothesis about the matter at hand, often entail definite conclusions about what appearances may *normally* be anticipated (if there is no malfunctioning or mishandling of the equipment, no unexpected external disturbance, etc.), whereas most of the experience which is crystallized in a man's skill at using equipment is inaccessible for such sharp deductions, essential though it may be for his capacity to function as an experimentalist. Second, in the comparison of personal probability evaluations implicit in public discussions of the status of scientific hypotheses (which, a Bayesian must acknowledge, are often conducted in terms that obscure the probabilistic structure of scientific inference), it is important that all parties disclose to the others their information and assumptions, in order to determine whether discrepancies in probability evaluations may not be attributable at least in part to differences regarding these. But since it is hopeless to compare total bodies of experience, the process of diagnosing disagreement is expedited if the sharp and prima facie relevant part of each person's information and assumptions is set forth explicitly. In spite of these advantages the decomposition of i into a and b is admittedly a psychological and methodological oversimplification. Occasionally, in fact, a crucial step in an inquiry (particularly concerning singular hypotheses, which are, however, of less importance in the natural sciences than in law and history and in practical concerns) is to retrieve and weigh more carefully some element of experience which had previously been relegated to the background. In most cases, however, the decomposition of i corresponds well to what happens in effective scientific procedure, and the advantages far outweigh the disadvantages.

Tempered personalism recognizes the possibility that X and X', with background information respectively b and b', may agree with respect to a and yet have unequal $P_{X,b}(h|a)$ and $P_{X',b'}(h|a)$; but the latitude toward subjective judgment in the tempered personalist concept of probability permits the question of whether this inequality is due to deeplying differences between b and b' to be circumvented. The tempering condition, moreover, makes it possible for X and X' to move toward consensus without having to agree first upon a probability function and second upon the total body of evidence. For if X and X' are aware of the same set of seriously proposed hypotheses and are open-minded toward all of them, then their prior probabilities $P_{X,b}(h_i|a)$ and $P_{X',b'}(h_i|a)$ are not radically different in order of magnitude, and, consequently, a moderate amount of data e is capable of ensuring agreement as to which hypothesis is to be preferred. Thus, if for any two seriously proposed hypothesis h_i and h_j

$$P_{X,b}(h_i|a)/P_{X,b}(h_j|a) < 10^4 \quad \text{and} \quad P_{X',b'}(h_i|a)/P_{X',b'}(h_j|a) < 10^4$$

(thus allowing considerable free play to the subjective judgments of both X and X'), and if the likelihood of e upon h_1 is at least 10^6 times as great as upon h_i ($i \neq 1$) for both X *and* X' (not atypical with a moderate amount of data in sampling problems), then X and X' would agree that the posterior probability of h_1 upon e is at least one hundred times greater than that of any of its rivals. This is an example of what Jeffreys calls "swamping" the prior probabilities. If personal probabilities were used, unqualified by the tempering condition, then deep-seated prejudices on the part of either X or X' could produce immense (and possibly infinite) prior probability ratios; swamping would consequently be impossible with a moderate amount of data, and consensus would be severely delayed. In a sense, therefore, my commitment to the tempered personalist concept of probability in preference to personalism is a reflection of Peirce's "social theory of logic": "He who would not sacrifice his own soul to save the whole world, is, as it seems to me, illogical in all his inferences, collectively. Logic is rooted in the social principle" (2.654). Or less dramatically, "the progress of science cannot go far except by collaboration; or to speak more accurately, no mind can take one step without the aid of other minds" (2.220). The tempering condition, which prescribes that seriously proposed hypotheses be given an opportunity to show their virtues under scrutiny, incorporates Peirce's social principle into a concept of probability.[21]

C. One of the points in the characterization of tempered personalism which requires clarification is the status of the axioms of probability. That $P_{X,b}(h|r\&a)$ satisfies these axioms is built into the definition of a tempered personalist probability function, but it is questionable whether this ought to be done in an instrumentalist explication of "rational degree of commitment." Because of the considerations in section II.D against interpreting "rational degree of commitment" in terms of the coherence of a set of betting quotients, the method of justifying the axioms due to Ramsey and DeFinetti—and accepted by most Bayesians—is inapplicable. A further question is whether the axioms of probability are even consistent with the tempering condition, since the latter prescribes that a person who learns of a new seriously proposed hypothesis on the matter under investigation must give it a nonnegligible prior probability, and it is not clear how this can be done without redistributing the probabilities assigned to the previously known hypotheses in a manner which violates the axioms.[22]

The second question is easily answered. The tempered personalist concept of probability is "local," in the sense that its application is circumscribed by the conditions of an individual investigation. When the set of seriously proposed hypotheses is augmented, the conditions are changed, and prior probabilities are reevaluated. The axioms of probability can be used to calculate posterior probabilities once prior probabilities are specified, but cannot be used to infer the prior probabilities associated with the new conditions from those associated with the old. Indeed, if one attempts to make such an inference formally, he will see that some very weird probability evaluations are required, such as $P_{X,b}(s|a)$, where s is the proposition that X *will soon become aware that* h_{n+1} *is seriously proposed by someone*. (In section II.B it was pointed out that a similar peculiar probability evaluation may be required in order to reconcile a purely logical set of confirmation rules with Carnap's concession that the laws actually proposed by scientists should perhaps be taken into account.) It is most unlikely that s would belong to the set S, explained in the definition of "tempered probability function," which is associated with the investigation at hand. Furthermore, an intellectual contortion would be required in order to evaluate subjectively the probability that a certain hitherto neglected hypothesis will be seriously proposed by someone, since posing this question ipso facto singles out the hypothesis for special consideration.[23]

The first question, whether a concept which explicates "rational degree of commitment" can be justifiably assumed to satisfy the axioms of probability, is more difficult. A promising approach, which derives the axioms from a relatively weak set of assumptions, is to use with some supplementation a remarkable argument invented independently by a number of writers, the earliest to my knowledge being Cox (1946, 1961) and Good (1950). An extensive bibliography and a rigorous presentation of the purely mathematical aspects of the argument are given by Aczél (1966: pp. 319–24). I shall begin by stating and sketching the proof of a modification of a theorem in Aczél's book, and then I shall discuss the application of this theorem to the present question. (Throughout the following discussion the notation $P(c|e)$ will be used without a subscript "X" even when a tempered personalist interpretation of the formalism is intended.)

Theorem: Let $P(c|e)$ be a real-valued function of pairs of propositions (c,e) such that $c \epsilon S$ (a nonempty set of propositions closed under truth-functional operations) and $e \epsilon T$ (where $T \subseteq S$, no member of T is a contradiction, and for every $e, e' \epsilon T$ there exists a finite set of propositions $e_1, \ldots, e_n$ such that $e \& e_1 \epsilon T$, $e_i \& e_{i+1} \epsilon T$ $[i = 1, \ldots, n-1]$, and $e_n \& e' \epsilon T$). Suppose P satisfies the following conditions:

1) if c is logically equivalent to c' and e to e' then $P(c|e) = P(c'|e')$;
2) or any $e \epsilon T$ there is a real number u_0 (perhaps dependent on e) such that if o is a contradiction belonging to S then $u_0 = P(o|e) \leq P(c|e)$;
3) there is a real number u_1 such that for all $e, f \epsilon T$ $P(e|e) = P(f|f) = u_1 > u_0$;
4) $P(c \& e|e) = P(c|e)$;
5) for every $e \epsilon T$ there is a function F_e such that $P(c \& d|e) = F_e[P(c|d \& e), P(d|e)]$; and
6) for every $e \epsilon T$ there is a function G_e which is continuous and monotonically increasing in both variables such that if e entails $\sim (c \& d)$ then $P(c \vee d|e) = G_e[P(c|e), P(d|e)]$.

Then there exists a continuous and monotonically increasing function $h(t)$ such that $h(u_0) = 0$, $h(u_1) = 1$, and $P'(c|e) = h[P(c|e)]$ satisfies the standard axioms of probability.

This theorem can be proved by slightly modifying the demonstration on pages 321–24 of Aczél's book. Since P here is a function of propositions rather than of sets, as in the book, the propositional calculus together with conditions (1) through (6) must be used as follows in order to establish the functional equations which Aczél needs on page 322:

$$F_e(x, u_1) = F_e[P(c|e), P(e|e)] = F_e[P(c|e \& e), P(e|e)] = P(c \& e|e)$$
$$= P(c|e) = x,$$
$$F_e(u_1, x) = F_e[P(e|e), P(c|e)] = F_e[P(c \& e|c \& e), P(c|e)]$$
$$= P[(c \& e) \& c|e] = P(c \& e|e) = P(c|e) = x,$$
$$G_e(u_0, x) = G_e[P(o|e), P(c|e)] = P(o \vee c|e) = P(c|e) = x,$$
$$G_e[G_e(x, y), z] = P[(c \vee d) \vee f|e] = P[c \vee (d \vee f)|e] = G_e[x, G_e(y, z)],$$
$$F_e[G_{f \& e}(v, w), z] = F_e[P(c \vee d|f \& e), P(f|e)] = P[(c \vee d) \& f|e]$$
$$= P[(c \& f) \vee (d \& f)|e] = G_e[P(c \& f|e), P(d \& f|e)]$$
$$= G_e[F_e(v, z), F_e(w, z)],$$

where

$$x = P(c|e), \quad y = P(d|e), \quad z = P(f|e), \quad v = P(c|f \& e), \quad w = P(d|f \& e).$$

Aczél manipulates these equations and the premises of the theorem to show that $G_e(v, w) = G_{e \& f}(v, w)$, so that the last of these functional equations becomes a distributivity equation $F_e[G_e(x, y), z] =$

$G_e[F_e(x,z),F_e(y,z)]$. A previous theorem permits him to assert that the most general solution to this equation (with the assumptions regarding G_e) is

$$G_e(x,y) = H^{-1}[H(x) + H(y)],$$
$$F_e(x,y) = H^{-1}[H(x)C(y)],$$

where C is an arbitrary continuous function and H is an arbitrary monotonically increasing function such that $H(u_0) = 0$ and $H(u_1) = 1$. But from the second of the functional equations $x = F_e(u_1,x) = H^{-1}[H(u_1)C(x)] = H^{-1}[C(x)]$, so that $C(x) = H(x)$ for any x in the range of P. Hence the most general forms of G_e and F_e are

$$F_e(x,y) = H^{-1}[H(x)H(y)]$$

and

$$G_e(x,y) = H^{-1}[H(x) + H(y)].$$

If $P'(c|e)$ is defined as $H[P(c|e)]$, then substitution into conditions (1) through (6) yields the following:

    1') if c is logically equivalent to c' and e to e', then $P'(c|e) = P'(c'|e')$,
    2') $0 = P'(o|e) \leq P'(c|e)$,
    3') $P'(e|e) = P'(f|f) = 1$,
    4') $P'(c\&e|e) = P'(c|e)$,
    5') $P'(c\&d|e) = P'(c|d\&e) \cdot P'(d|e)$,
    6') if e entails $\sim (c\&d)$, then $P'(c \vee d|e) = P'(c|e) + P'(d|e)$.

Aczél's argument ends at this point, but axioms (i) through (iv) trivially follow from conditions (1') through (6'). If $e\epsilon T$ and e entails h, then $h\&e$ is logically equivalent to e, and therefore $1 = P'(e|e) = P'(h\&e|e) = P'(h|e)$, which is axiom (ii). Axiom (i) asserts that $0 \leq P'(c|e) \leq 1$; the first inequality is the same as condition (2'), while the second follows from $1 = P'(c \vee \sim c|e) = P'(c|e) + P'(\sim c|e)$ together with $0 \leq P'(\sim c|e)$. Axiom (iii) follows immediately from conditions (5') and (1').

The utility of this theorem for the question at hand obviously depends upon whether its premises are justifiable if $P(c|e)$ is taken as an explicatum of "rational degree of commitment," and especially if locality and instrumentalism (which are characteristics of the tempered personalist concept of probability) are maintained. The condition on S is evidently reasonable if one wishes to use the operations of deductive logic in the context of scientific inference. The first two conditions on T

are trivial: $T \subseteq S$ is required if one wishes $P(e|e)$ to be defined for all $e \epsilon T$, and a contradictory proposition cannot possibly serve as evidence. The third condition on T is a direct consequence of locality: for if e and f are admissible evidential propositions in the probability calculations associated with an investigation, then they are respectively of the form $e'\&i$ and $f'\&i$, where i is the initial body of information and assumptions. Conditions (1) through (4) are justifiable partly as straightforward rules for ensuring that inductive and deductive procedures should mesh and partly as conventions for ordering the rational degree of commitment to a contradiction relative to the rational degrees of commitment to other propositions. Conditions (5) and (6), by contrast, are nontrivial. The functional dependence in condition (5) seems very natural, since it merely asserts that one can evaluate the rational degree of commitment to a conjunction $c\&d$ relative to evidence e by proceeding in a stepwise fashion: by evaluating the rational degree of commitment to d on e, and to c on $d\&e$. Even more appealing intuitively is the functional dependence in condition (6), for it is hard to see what else $P(c \lor d|e)$ could depend upon than $P(c|e)$ and $P(d|e)$. However, there is an obscurity in calling upon intuition in this way, just because the concept of commitment is unclear. The naturalness of these assumptions may be derivative from the consideration that the addition and multiplication principles of probability—axioms (iii) and (iv)—are necessary conditions for the coherence of beliefs in a set of bettable propositions (that is, those for which definite payoff conditions exist). One may feel that in treating commitment, which is a modality of belief appropriate to nonbettable propositions, it is quite conservative to retain the bare assertion of functional dependence instead of the specific dependencies of axioms (iii) and (iv). But more would have to be known about the general character of commitment in order to make this reasoning forceful. Perhaps the best defense of the assumption of functional dependence in conditions (5) and (6) is just that they contribute to the clarification of the concept of commitment in a methodologically fruitful way. They permit the argument to proceed to the conclusion that the explicatum of "rational degree of commitment" satisfies the axioms of probability, thereby permitting a formulation of scientific inference in which there is a firm mathematical structure. On the other hand, the fact that these are quite weak assumptions makes it plausible that their denial will result in a loose formulation of scientific inference rather than in a formulation with a firm though non-Bayesian structure. In order to strengthen this plausibility argument, however, one should systematically investigate the alternatives to the functional dependencies of conditions (5) and (6) in the light of the fundamental purpose of learning about the world from experience. Once the existence of the function G_e of condition (6) is granted, its continuity is entirely natural. If I introspect about my subjective commitments, I certainly feel strongly that a very small change in my commitment to one of two exclusive propositions induces a very small change in my commitment to their disjunction, and I can see no reason why the normative concept of degree of rational commitment should differ from the concept of subjective commitment in this respect. (It is relevant to note that discontinuities in prior probability distributions are sometimes of great instrumental value, as will be argued in section III.E, since they permit the possibility of accepting a hypothesis asserting the exact value of a parameter which a priori can have any of a continuum of values; but the continuity of the function G_e is clearly compatible with this kind of discontinuity in the distribution function.)

The one premiss which prima facie seems unacceptable in view of locality is that $P(c|e)$ takes on all values between u_0 and u_1, for it is possible that the set of propositions S mentioned in the definition of "tempered personalist probability function" is denumerable or even finite, in which case the premiss would be false. However, if S is augmented by an appropriate set I of "ideal elements," which are propositions irrelevant to the investigation at hand but which are introduced in order to permit refined judgments of subjective commitment, then $P(c|e)$ will take on all values as a ranges over S', the closure of $S \cup I$ under truth-functional operations. A convenient choice of I is the set of all propositions c_r ($0 \leq r \leq 1$) to the effect that a needlepoint constrained to come to rest somewhere on a scale one meter long will, because of an unspecified physical mechanism, come to rest at a point less than r meters from the o-end. It is reasonable to suppose that the degree of subjective commitment to c_r (which in this case can surely be equated with belief) relative to information i is a continuous function of r, and because c_0 violates the assumed constraints while c_1 is required by them, the rational degree of commitment to c_0 and c_1 must respectively be u_0 and u_1. Then $P(c_r|i)$ takes on all values between u_0 and u_1. No other assumption need be made about the functional dependence of P upon r, and in particular there is no appeal to a principle of indifference regarding subintervals of [0,1] of equal length. When a man reflects upon a proposition belonging to S, a comparison with the propositions of a set like I is a very useful method for arriving at a numerical assessment of his subjective commitments. Indeed, unless the propositions of S are bettable, it is hard to see how numerical assessments could be arrived at except by systematic comparisons of this kind (or by a variant procedure like Koopman's use of a sequence of "n-scales" [1940: pp. 290–91] instead of the continuous set I); and if they are all bettable, then considerations of coherence ensure that P satisfies the axioms of probability so that the present inquiry is unnecessary. The "ideal elements" are, to be sure, not mentioned in the definition of "tempered personalist probability function," but it is unnecessary to do so, since there a scale for measuring subjective commitments is presupposed. However, the present inquiry concerns the justification for conjoining the various conditions in this definition, and, therefore, it is legitimate to pay attention here to the ideal elements which are auxiliary to numerical assessments of subjective degree of commitment.

The argument up to this point is that for any function $P(c|e)$ which is suitable as an explicatum of "rational degree of commitment," there exists a continuous monotonic function $H(x)$ such that $P'(c|e) = H[P(c|e)]$ satisfies the standard axioms of probability. Because of the monotonicity of H, P' is suitable as a measure of degree of rational commitment if P is, and the choice between them is claimed by Cox (1961: p. 16) and by Good (1950: p. 106) to be merely a matter of the convenient selection of a scale. An improvement can be made upon their conventionalistic reasoning by considering I', the closure under truth-functional operations of the set I discussed in the preceding paragraph. All the members of I' are bettable, and furthermore they are such that the degree of subjective commitment to each is equal to the credence in it. DeFinetti showed that if credence is measured by the maximum acceptable betting quotient and if coherence is recognized as a necessary condition for rationality, then a rational credibility function must satisfy the axioms of probability. Hence, if $c,d\epsilon I'$, then $P(c\&\sim c|i) = 0$, $P(c \vee \sim c|i) = 1$, and $P(c \vee d|i) = P(c|i) + P(d|i)$ if i entails $\sim (c\&d)$. It follows that $H(0) = 0$, $H(1) = 1$, and $H(x+y) = H(x) + H(y)$ (where $x = P(c|i)$ and $y = P(d|i)$). But by the construction of I' it is possible to choose c and d so as to let x and y be arbitrary real numbers in $[0,1]$ subject to the constraint $x + y \leq 1$. One easily sees then that $H(x) = x$ for all dyadic rationals (x equals the sum of a finite number of terms of the form 2^{-n}, where $n = 1,2,3, \ldots$), and hence by the continuity of H, $H(x) = x$ for all $x\epsilon[0,1]$. Consequently, P is identical with P', so that P itself satisfies axioms (i) through (iv) for all $c\epsilon S$ and $e\epsilon T$.

The foregoing argument admittedly rests upon several idealizations, but they are just those which one must make in order to use the real number system freely in assessing subjective commitments. It should be emphasized that this argument does not depend upon construing all commitments in terms of dispositions to bet, but rather the fact that some commitments can be so construed is used in order to complete an argument which is based primarily upon considerations of orderliness in scientific inference.

D. The phrase "seriously proposed hypothesis" in the definition of "tempered personalist probability function" requires clarification. Although there is often general agreement in the context of a specific investigation as to which hypotheses are to be considered seriously proposed, it is very difficult to state a reasonable set of conditions upon the intrinsic characteristics of hypotheses and upon the circumstances of their proposal which would permit one to distinguish unambiguously those which are seriously proposed from those which are not. Probably the request for a sharp set of conditions ought not to be honored because of the danger of arbitrariness and of diminishing the flexibility of the scientific method. On the other hand, if decisions about classifying hypotheses as seriously proposed or not are left entirely to the subjective judgment of individual investigators, then the advantages which I have claimed for tempered personalism—as a "social theory of inductive logic" —over unqualified personalism are in danger of being lost. In order to navigate between these two dangers, I shall try to formulate some methodologically sensible guidelines for decisions on this question, without pretending to eliminate entirely the subjective judgment of the investigator. I shall also argue that under social conditions which are generally favorable to theoretical inquiry, the prospect of scientific progress is not imperiled by this informality, while social conditions which are adverse to theoretical inquiry are not likely to be ameliorated merely by formalizing scientific methodology.

A part of the problem is to determine whose conjectures are to be regarded as seriously proposed. Again there are opposing dangers of being too strict and too loose. If the only persons who are to be so respected belong to some delimited group, such as a definite profession, then there is a danger of "blocking the road to inquiry," for such groups are subject to stagnation, parochialism, and obeisance to authority. On the other hand, if the conjectures of everyone are to be taken seriously, won't a conscientious investigator find himself overwhelmed by capricious and crankish hypotheses? The dilemma is not very painful if one recognizes that there are different kinds of situations. Sometimes the mode of presentation of a hypothesis to an investigator conforms to professional standards: the hypothesis is clearly stated, the motivation for proposing it is explained, and the explanation indicates understanding (though not necessarily complete acceptance) of the recognized body of propositions regarding the subject, and it is not an arbitrary choice from a family of hypotheses which answer to the same motivation. (See the following paragraphs on this last point.) Whether the proposer belongs to the scientific establishment or not can reasonably be regarded as irrelevant in such cases. On the other hand, there are cases in which the presentation of a hypothesis fails to meet professional standards in one or more ways. For example, the formulation may be obscure, or it may seem obscure upon first reading so that time and energy evidently would be required in order to determine whether the apparent obscurity is due to the author's confusion or to profound originality. Since an investigator has limited time and energy, he may in good conscience decide not to study the proposal carefully unless the credentials of the proposer indicate competence. Respect of this kind for the scientific establishment does not imply the abandonment of one's own judgment, and it is reasonable if learning about nature is assumed to be generally progressive.[24] However, one should not rigidify a reasonable guideline. Even an obscurely presented theory by an unknown author may draw the reader in by exhibiting some intellectual freshness, but at this point the subjective judgment of the individual is evidently irreplaceable. Even if the hypothesis of an outsider is dismissed by the establishment, that particular avenue of inquiry is not thereby definitively blocked if the society as a whole is sufficiently tolerant and flexible. When there is freedom of research and communication and adequate provision of leisure, the outsider may exhibit his seriousness by mastering the subject sufficiently to make his presentation lucid by professional standards or by designing and performing his own experiments;[25] and he may be able to seek the advice of acknowledged experts, who are sometimes more sympathetic in face-to-face encounters than in reading a written page. Citadels of entrenched scientific opinions have been conquered often enough to indicate that receptivity to novel insight has been present in at least some members of the scientific establishment (and perhaps enough even to vindicate the Platonic assumption that the truth exercises a gentle but persuasive force upon the minds of men). At any rate, if the society becomes so inflexible, intolerant, and incurious as to prevent a dedicated outsider from receiving a hearing for his ideas, it is hard to see how a sharpened set of methodological rules would make the social climate more favorable to inquiry.

A more difficult part of the problem is to determine when "seriously proposed" should be applied to a single hypothesis and when it should rather be reserved for the disjunction of a family of hypotheses of which it is a member. Consider, for example, the family of hypotheses $\{h(\alpha)\}$ asserting that the electrostatic force between two point charges is of the form $F \sim r^{-\alpha}$, where r is the distance between the charges. The hypothesis $\alpha = 2$ was proposed by Priestley and D. Bernoulli (Whittaker 1951: p. 53), though it is named for Coulomb because of his delicate experiments confirming it. Suppose that another eighteenth-century scientist had proposed $\alpha = 2.0001$. Were both of these hypotheses to be considered "seriously proposed," or were both to be considered as merely arbitrary specifications of the seriously proposed family $\{h(\alpha)\}$, or was the first to be given the status of "seriously proposed" and the second not? Scientific practice of the time surely favored the third course, unless some special and satisfactory motivation had been given for the proposal $\alpha = 2.0001$. Methodologists sometimes rationalize scientific practice in cases like this by appealing to simplicity, and there can be no doubt that the exponent -2 appears intuitively to be simpler than the exponent -2.0001. In this essay I shall not undertake to survey the various analyses which have been made of the concept of simplicity (cf. Kyburg 1964: pp. 19–20), although, in section IV, I shall criticize Kemeny's suggestions and, in section V, I shall argue that factual considerations are relevant to the concept. At this point I shall only say that there appears to be general agreement on the inadequacy (from the standpoint of inductive logic) of all explications which have so far been given of the concept of simplicity, and that it is, therefore, desirable to explain the favoritism shown to $\alpha = 2$ along other lines. Without appealing to simplicity, one can give several reasons for singling out the inverse square hypothesis from the continuum of possibilities. Historically the most important reason was undoubtedly the analogy to the law of gravitation, but this reason merely shifts the problem to that of justifying the *exact* value $\alpha = 2$ in the gravitational force law. Another reason, which was heuristically important in Newton's proposal and, therefore, indirectly in the proposal of Coulomb's law, is that $\alpha = 2$ is the only hypothesis of the family $\{h(\alpha)\}$ which implies that the flux across the surface of a sphere with the point source at its center is independent of the radius of the sphere. In fact, there is a more general consequence of this kind, which was not known to Newton or to the natural scientists of the eighteenth century: that the flux across any simply connected surface enclosing the point source is the same. These mathematical consequences made it plausible—even before the development of electromagnetic theory in the nineteenth century—that an inverse square law could be readily embedded in a comprehensive formulation of the principles of electricity, whereas the hypothesis $\alpha = 2 + \epsilon$, where ϵ is too small to be detected by known experimental methods, could not be readily embedded in this way. More speculatively, one could say that if a deep theory is to be found connecting the electrostatic force law with the dimensionality of space, as the results concerning the flux through surfaces suggest, then force laws with integral exponents appear much more promising than the other members of $\{h(\alpha)\}$. However, if the electrostatic force law was proposed without any envisagement of a more comprehensive theory, then there was no motivation for going beyond descriptive adequacy, and it would have been appropriate to interpret the proposal $\alpha = 2$ as shorthand for something like "α has a a value which is indistinguishable from 2 by direct measurements of the electrostatic force between charges."

A general methodological guideline can be extracted from this example: if $\{h(\alpha_1, \ldots, \alpha_k)\}$ is a family of hypotheses with k parameters such that the point $(\alpha'_1, \ldots, \alpha'_k)$ in parameter space is associated with a comprehensive theory, which may be explicitly stated or only sketched, into which $h(\alpha'_1, \ldots, \alpha'_k)$ but none of the other members of the family would fit, then this hypothesis individually may be considered to be seriously proposed; the disjunction of all the other members of the family should be considered as seriously proposed but the individual components should not be. This guideline permits the machinery of the tempered personalist formulation of scientific inference to lead to the tentative acceptance of exact values of parameters only when the exact values are of theoretical interest. The vagueness of this guideline is evident, for there is nothing to prevent the proposer of a bizarre hypothesis from supplementing his proposal with a sketch of a bizarre comprehensive theory in which it might be embedded. Furthermore, the notion of a "sketch" of a comprehensive theory is extremely vague. At this point, however, I am dubious that anything of value can be accomplished by sharpening the methodological prescription; there is no substitute for an intelligent examination of the reasons for the special proposal $h(\alpha'_1, \ldots, \alpha'_k)$ and for a personal judgment about their plausibility.

It should be noted that when the predicate "seriously proposed" is applied to the disjunction of a family of hypotheses, the tempering condition can be amplified somewhat, so as to prescribe open-mindedness not only to the family as a whole but to its subfamilies. Thus, if the members of the family are $h(\alpha_1, \ldots, \alpha_k)$, then a reasonable amplification of the tempering condition would prescribe that the prior probability density $P_{X,b}[h(\alpha_1, \ldots, \alpha_k)|a]$ must be such as to permit a posterior probability which is strongly peaked about any preassigned point in the space of the parameters for some possible result r of the envisaged observations.[26] The peaking of the posterior probability distribution function in an unexpected way may stimulate new speculation, leading to the serious proposal of a point in parameter space which previously had not been classified as seriously proposed but which can be so considered in a new investigation.

E. Since the tempered personalist concept of probability is derivative with rather small changes from Jeffreys's working sense of "probability," its utility in formulating scientific inference can be exhibited parasitically by referring to the wealth of applications in his *Theory of Probability*. I shall give one example, in order to show how and with what modifications Jeffreys's calculations can be borrowed by tempered personalism.

In the classical sampling problem of Laplace, the information a asserts that there is a population of N objects of which an unknown number r have a specified property α, and from this population objects are drawn without replacement by a method designed to favor neither α nor $\bar{\alpha}$. Let $h(x)$ be the hypothesis that $r = x$ and d_{nm} be the proposition that in a sample of $n + m$ members n have the property α and m do not. Laplace appealed to the principle of indifference (which was then called the "principle of insufficient reason") in assigning equal prior probabilities to all constitutions of the population, that is, $P^L[h(x)|a] = 1/N + 1$ for $x = 0, \ldots, N$ (where the superscript L refers to Laplace's evaluation). From this assignment, together with the axioms of probability and the assumption that at each draw the probability of obtaining any object not yet removed is equal to the probability of obtaining any other, he demonstrated his "rule of succession"

$$P^L(s|d_{n0}\&a) = \frac{n + 1}{n + 2},$$

where s is the proposition that the next object to be drawn will have the property α. Laplace considered this result to provide a mathematical justification for inductive inference, since a large value of n yields a posterior probability close to 1 that the uniformity exhibited in the sample will extend to the new instance. However, the same set of premises implies that

$$P^L[h(N)|d_{n0}\&a] = \frac{n + 1}{N + 1},$$

and, therefore, the posterior probability of the generalization that all members of the population have the property α is small unless the sample is a large part of the population. Jeffreys correctly argues that scientific progress requires the possibility of tentatively accepting generalizations, and, therefore, he suggests (1961: p. 129) that Laplace's prior probability assignments be replaced by the nonuniform distribution

$$P^J[h(0)|a] = P^J[h(N)|a] = k,$$

$$P^J[h(x)|a] = \frac{1-2k}{N-1} \qquad \text{for } x = 1, \ldots, N-1,$$

for some appropriate real number k (the superscript J referring to Jeffreys's evaluation). It follows that

$$\frac{P^J[h(N)|d_{n0}\&a]}{\sum_{x=0}^{N-1} P^J[h(x)|d_{n0}\&a]} = \frac{n+1}{N-n} \cdot \frac{k}{1-2k}(N-1).$$

"Hence if n is large, the ratio is greater than $(n+1)k/(1-2k)$ whatever N may be, and the posterior probability that $r = N$ will approach 1, almost irrespective of N, as soon as n has reached $1/k$" (ibid.: p. 130). This reasoning is instrumentalistic and makes no obvious appeal to objectively determined probabilities. Jeffreys recognizes a range of possible choices of k which are all methodologically reasonable, his preference being $k = \frac{1}{4} + \frac{1}{2(N+1)}$, which he motivates by the following classification of possibilities:

(1) Population homogeneous on account of some general rule.
(2) No general rule, but extreme values to be treated on a level with others. Alternative (1) would then be distributed equally between the two possible cases, and (2) between its $n+1$ possible cases. (Ibid.: pp. 130–31)

In evaluating the probabilities of $h(0)$ and $h(N)$, the procedure of tempered personalism is similar to that of Jeffreys. Since these two hypotheses can be expected to be seriously proposed if the sampling problem occurs within the context of a scientific investigation, $P_{x,b}[h(0)|a]$ and $P_{x,b}[h(N)|a]$ must be large enough to permit the possibility of high posterior probabilities of $h(0)$ and $h(N)$. The tempering condition also requires similar open-mindedness to other seriously proposed hypotheses of the family $\{h(x)\}$, but Jeffreys's procedure is obviously also adaptable to such complications. Tempered personalism differs from Jeffreys's theory primarily in the role which the former assigns to subjective judgment; and subjectively the prior probabilities of $h(o)$ and $h(N)$ may be unequal, and the distribution over $h(1), \ldots, h(N-1)$ need not be in uniform. How much latitude is permitted to subjective judgment is determined by the tempering condition and the circumstances of the investigation. Thus, if only $h(o)$ and $h(N)$ are singled out for special consideration from among the $h(x)$, then the only other seriously proposed hypothesis is the general hypothesis of heterogeneity $H = h(1) \vee \ldots \vee h(N-1)$, and the tempering condition requires the prior probabilities of $h(o)$, $h(N)$, and H to be large enough to permit each to have the largest posterior probability for some d_{nm}. If the experiment envisaged will yield a sample of size n_0, and if the prior probability distribution over the disjunctive components of H is roughly uniform (thus satisfying the amplified version of the tempering condition which was stated at the end of Section III.D), then Jeffreys's formula (1) above permits an approximate calculation of a lower bound k_0 on $P_{x,b}[h(o)|a]$ and $P_{x,b}[h(N)|a]$, namely,

$$k_0 = \left[2 + (n_0 + 1) \cdot \frac{N-1}{N-n_0} \right]^{-1}.$$

In the realistic case in which $N >> n_0 >> 1$, one obtains a simple approximate expression for k_0: $k_0 = 1/n_0$.

I must admit in conclusion my uncertainty about the relation between tempered personalism and one other important part of Jeffreys's treatment of prior probabilities: his use of invariance considerations to assign prior probabilities to hypotheses with adjustable parameters when no particular values are seriously proposed (ibid.: chap. III, especially secs. 3.1 and 3.10). There are technical difficulties in Jeffreys's analysis as it is now presented, such as the appearance of nonnormalizable distribution functions (cf. Hacking 1965: pp. 203–05), and it is possible that when these are removed, the residue will yield nothing that cannot be obtained by subjective probability evaluations modified by the tempering condition. If, on the other hand, these invariance considerations are fruitful, it may be desirable to incorporate them in some form into the prescription of tempered personalism, since the basic idea of a rule "that is applicable under any non-singular transformation of the parameters, and will lead to equivalent results" (Jeffreys 1961: p. 192) is reasonable.

F. Sections II and III have been largely concerned with the way in which probabilities should be evaluated if the concept of probability is to be a useful instrument in theoretical investigations. Little has been said, however, on the question of how correct probability evaluations can be used to serve the primary purpose of theoretical inquiry, which is (if one adheres to the etymon θεωρία) to obtain a *view* of the world.

For a Copernican epistemologist, who is skeptical of achieving certainty about the principles of nature, a "view" is properly to be construed not as a set of propositions about various aspects of the world, but rather as a set of modalities in which propositions about the world are entertained.[27] The view as a whole is tentative, since it is subject to modification in the light of further experience and of new proposals, and moreover tentativeness is distributed in varying degrees among the parts, for usually a number of propositions (in addition to the ubiquitous "catchall" hypothesis *that something else is true*) are entertained simultaneously with different degrees of commitment. Thus the view which is actually attained by theoretical inquiry is characterized by intellectual tensions, rather than by the calmness which is the traditional connotation of θεωρία. Even the greatest achievements of natural science, which permit a large number of previously uncoordinated or anomalous facts to be seen as parts of a pattern, are not free from tentativeness—partly because of the remote possibility that apparent success has been due to a long series of coincidences, but more seriously because of the possibility that a highly successful theory may be displaced by a more general theory (perhaps with a radically different conceptual structure) which agrees approximately with the old theory over a limited range of circumstances but not in all of them (cf. Section II.D).

A general answer to the question about the use of probability in theoretical inquiry can be stated in terms of the foregoing characterization of a "view": it makes possible the systematization of a person's tentative entertainment of propositions about the world. Probability is useful for this general purpose in spite of the fact that scientists whose thinking about nature is judicious and orderly do not usually try to weigh their tentative commitments quantitatively, and also in spite of the fact that the most successful theories of natural science are so obviously better supported by the evidence than their rivals that detailed probability calculations are dispensable. In the first place, the axioms of probability are needed in order to give a unitary formulation of scientific inference, and the formulation in probabilistic terms of special procedures of inference, such as the hypothetico-deductive method, permits important qualifications and refinements (cf. Section II.A). Secondly, the evaluation of probabilities (and not merely the mathematical structure of probability theory) is valuable for systematizing the tentative entertainment of propositions whenever there is some delicacy in the relationship of evidence to hypotheses, for example, when considerations of possible experimental errors must be taken together with considerations about prior preferences among hypotheses, or when the evidence consists of complicated correlations. In the ordinary variety of research, as contrasted with the dramatic achievements of science which often preoccupy methodologists, this kind of delicacy of relationship is very common, and statistical analysis of data has proved indispensable. It is not surprising that the scientific work of Harold Jeffreys, whose *Theory of Probability* is by far the best treatise on the use of probability in scientific inquiry, has primarily been in the complicated and untidy field of geophysics. Finally, both in introspecting upon one's own commitments and in making comparisons with the commitments of others, it is important to determine as precisely as possible the loci of uncertainty and disagreement. The explicit use of the apparatus of probability theory, as contrasted with its tacit use in informal scientific inference, is valuable in this kind of analysis. Notably, if posterior probabilities are calculated by means of Bayes's theorem, then prior probabilities and likelihoods must be specified, and furthermore the sharp part of the body of assumptions and information must be made explicit. As explained in Section III.B, explicitness on these matters can be conducive to consensus among investigators by revealing differences among assumptions which are in need of supplementary investigations. The revelation of striking differences among prior probability evaluations is often symptomatic of prejudices or limitations of imagination on the part of one or more of the investigators. Even if scientific inference is formulated in terms of the (untempered) personalist concept of probability, the analysis of prior probabilities can be conducive to consensus by stimulating men to reexamine their own beliefs and introspect about possible prejudice or obtuseness. The formulation of scientific inference in terms of the tempered personalist concept of probability has, in this regard, the additional virtue of requiring open-mindedness toward the insights of other men.

This probabilistic account of a "view" of the world is incomplete without some remarks about the role of extremely well-confirmed hypotheses. Although scientific inference is based upon critical habits of thought, it nevertheless concludes on some occasions that one hypothesis is strikingly preeminent over its rivals. It often happens, for example, that the initial experiments which result in preferring a certain hypothesis are difficult to achieve and are not entirely convincing, but independent reconfirmations are abundant and relatively easy—and one has an impression of easy progress after emergence from tangled underbrush. There are various reasons for this kind of phenomenon: the initial experiments may show what precautions are necessary, as in Lavoisier's careful accounting for all reaction products in confirming the conservation of mass in chemical reactions; or the initial experiments may indicate the irrelevance of factors which previously were distracting, as in the demonstration of the etiology of malaria; or there may be conceptual clarification, as in Galileo's combination of experiment and analysis. Whatever the reason may be for the avalanche of independent confirmations of a hypothesis, grouping them all together as if they were parts of a single investigation would result in overwhelmingly large ratios of the posterior probability of the successful hypothesis to the posterior probabilities of all its rivals. The dispensability of numerical evaluations of probabilities for recognizing the most striking achievements of science—which anti-Bayesians correctly point out—is an obvious consequence of this effect.

Suppose not only that the ratio of the posterior probability of h_1 to that of $h_i (i = 2, \ldots, n)$ is very large, but also that the number n of seriously proposed hypotheses is moderate (which is a reasonable assumption, in view of the discussion of seriously proposed hypotheses in section III.D). In that case $P_{X,b}(h_1|e\&a) = 1-\epsilon$, where

$$\epsilon = \sum_{i=2}^{n} \epsilon_i = \sum_{i=2}^{n} P_{X,b}(h_i|e\&a)$$

is a positive real number much less than 1. Then h_1 can be "accepted" as part of the body of assumptions and information which is relied upon in subsequent investigations.[28] In the notation of the "local" concept of probability introduced in section III.B, this acceptance consists of replacing the proposition a by a' (equivalent to $a\&h_1$) and working with a new local probability function $P_{X,b'}(q|r\&a')$, where b' differs from b by absorbing the data e of the previous investigation and possibly other new background information of negligible importance. (However, if e is directly relevant to the new investigation, it cannot be relegated in this way to background information; as pointed out in section III.B, the separation of the total body of information and assumptions into parts a and b is an idealization.) Suppose, for the purpose of closer analysis, that instead of accepting h_1 the residue of uncertainty of the preceding investigation were explicitly taken into account in the new investigation in the following way:

$$P_{X,b'}(q|r\&a) = P_{X,b'}[q\&(h_1\lor \ldots \lor h_n)]|r\&a)$$

$$= \sum_{i\ 1}^{n} P_{X,b'}(q\&h_i|r\&a)$$

$$= \sum_{i=1}^{n} P_{X,b'}(q|r\&a\ \&\ h_i) \cdot P_{X,b'}(h_i|r\&a)$$

$$\cong \sum_{i=1}^{n} P_{X,b'}(q|r\&a\&h_i) \cdot P_{X,b}(h_i|a\&e)$$

$$= (1 - \epsilon)P_{X,b'}(q|r\&a') + \sum_{i=2}^{n} \epsilon_i P_{X,b'}(q|r\&a\&h_i),$$

where the approximate equality in the next to the last step is contingent upon the irrelevance of the data r of the new investigation to the h_i. It is evident that the correction obtained in this way to the "local" probability function $P_{X,b'}(q|r\&a')$ is small, because of the magnitudes of ϵ and the ϵ_i. Furthermore, the correction term is hard to evaluate, since the overwhelming confirmation of h_1 implies that the posterior probabilities $\epsilon_2, \ldots, \epsilon_{n-1}$ are almost certain to be extremely small, and only the posterior probability ϵ_n of the catchall hypothesis h_n is likely to be nonnegligible; but since h_n is the proposition that *something other* than the definite proposals $h_1, \ldots, h_{n-1}$ is true, the subjective degree of commitment to a proposition q relative to $r\&a\&h_n$ (which is required in evaluating the tempered personalist probability $P_{X,b'}[q|r\&a\&h_n]$) is likely to be more than ordinarily indefinite. Thus, the comparison of a "local" probability evaluation with a treatment which takes into account the overlap between two investigations reveals a motivation for the local concept and also exhibits its approximate accuracy.

The acceptance of h_1 is not tantamount to an unqualified commitment to its truth. The neglected ϵ is preserved in a residual attitude of tentativeness toward h_1, which involves the willingness to reopen an investigation concerning h_1 if suitable motivation is provided, such as persistent failure to make good progress in subsequent investigations which take h_1 for granted. A reinvestigation of h_1, however, must be quite different from the initial investigations which confirmed and reconfirmed it, for h_1 cannot be simply dismissed without some explanation of its previous remarkable success; the explanation may take the form of a superseding theory of greater generality than h_1, or perhaps it will consist in the revelation of a deep-lying systematic error in the preceding investigations, but in any case the set of seriously proposed hypotheses in the reinvestigation must differ somewhat from the set $h_1, \ldots, h_n$ considered earlier. Even if ϵ were 0, a literal belief in the truth of h_1 would not be justified, for $P_{X,b}(h_1|e\&a) = 1$ means that a rational degree of commitment for X to h_1, *relative to a&b*, is 1. There will surely be some tentativeness in X's entertainment of a; and furthermore, as explained in section II.D, commitment to a specific hypothesis is somewhat weaker than belief in its literal truth (for indeed, if this were not the case, then it would be difficult to see how sufficient evidence could ever be accumulated to make the probability of the catchall hypothesis much smaller than 1).

A final remark about the role of highly confirmed hypotheses lies on the border line between the methodology and the psychology of science. There is no a priori reason why critical procedures of analysis, such as the deliberate design of experiments for testing the deductive consequences of theories, should not always lead to the rejection of all definite proposals in favor of the catchall hypothesis, or at best (by means of suitable appearance-saving assumptions) to suspense among many alternatives. However, it is extremely unlikely that if the outcomes of scientific investigations were uniformly negative or indecisive, critical habits of thought would ever have displaced the acceptance of loose explanations of natural phenomena provided by primitive religion and mythology.[29] The existence of highly confirmed hypotheses, which have been reconfirmed when reinvestigated and have been successfully used as assumptions underlying further investigations, demonstrates the fertility of procedures which are, as Popper especially has insisted, largely eliminative. The historical achievement of overwhelmingly successful hypotheses at various levels of generality has been essential for instilling confidence in critical thought concerning matters remote from mundane affairs (where a certain amount of critical thinking is a component of common sense), and also for counteracting the tendency of critical thought to lapse into sterile cynicism for lack of substantial results. A sophisticated scientific methodology can profit greatly from a detailed analysis by historians of science of this complex interplay of achievement and confidence.[30]

IV. The Sensitivity of Scientific Inference

A. In this section I shall consider to what extent an a priori justification can be given for the tempered personalist formulation of scientific inference. It is a delicate matter to make a priori claims in favor of a method of confirmation while disclaiming a priori knowledge about the world, and one invariably finds that excessive claims in any respect must be paid for by undesirable concessions and assumptions in other respects. An illuminating way to approach this problem is to examine the family of arguments called "pragmatic justifications of induction," which has often been regarded as the most promising a priori defense of inductive inference. All versions of this argument purport to establish, with various qualifications, that by persisting indefinitely in the application of a properly formulated method of inductive inference, we can discriminate true from false hypotheses, however the world is constituted. A brief examination of three of the most interesting pragmatic justifications—those of Peirce, Reichenbach, and Kemeny—will indicate the difficulties of achieving a satisfactory defense along this line. Because Peirce is the most obscure of the three, but also in my opinion the most suggestive of an alternative analysis, I shall disregard historical order and discuss his argument last.

The a priori claims that I shall make in favor of the tempered personalist formulation of scientific inference are relatively modest. It is a method which can navigate systematically, though not infallibly, between those errors due to overskepticism and those due to credulousness; and it can do so in finite intervals of time without requiring the long run. It is designed to complement and support any powers which human beings may possess for making intelligent guesses about nature—though I do not suppose that tempered personalism is in any sense a logic of discovery. Finally, it is capable of assimilating and employing, without abandonment of its basic structure, any methodological device which analysis or experience indicates to be valuable. I feel that one may fairly sum up these claims by ascribing "sensitivity to the truth" to the tempered personalist formulation of scientific inference and also to the informal processes of confirmation used by scientists from which tempered personalism is extracted. There is likely to be disagreement on whether a justification of scientific inference is achieved in this way, but I doubt whether appreciably stronger claims can be established without taking into account the actual constitution of the world.

B:1. Reichenbach conceives the central problem of induction very narrowly: as consisting in the determination of the limits of relative frequencies in infinite empirical sequences. Letting a_n be the number of elements having a specified property among the first n members of a specified sequence of entities and letting $f^n = a_n/n$, he correctly reasons as follows:

If the sequence has a limit of the frequency, there must exist an n such that from there on the frequency f^i $(i > n)$ will remain within the interval $f^n \pm e$, where e is a quantity that we choose as small as we like, but that, once chosen, is kept constant. Now if we posit that the frequency f^i will remain within the interval $f^n \pm e$, and if we correct this posit for greater n by the same rule, we must finally come to the correct result. (1949: pp. 445–46)

He then supplements this conditional claim for the eventual success of the procedure of "positing" by an argument that nothing is lost by supposing the antecedent to be true:

Inductive positing in the sense of a trial-and-error method is justified as long as it is not known that the attempt is hopeless, that there is no limit of the frequency. Should we have no success, the positing was useless; but why not take our chances? (Ibid.: p. 363)

The suggestion that the long-run success of induction can be proved only conditionally, but that it is rational to accept the unconfirmable antecedent of the conditional argument, is reminiscent of earlier epistemological proposals, notably Pascal's wager and Kant's treatment of "regulative principles." It will also be seen that Peirce anticipated the employment of such *as if* argumentation for justifying scientific inference. But Reichenbach deserves great credit for bringing this mode of argumentation into prominence and making the explicit point that it provides a new approach to the problem of Hume (cf. 1938: pp. 356–57).

There are, however, a number of crucial defects in Reichenbach's theory of induction. No satisfactory suggestion has been made for treating the probability or the acceptability of theories in terms of the frequency concept of probability, although Reichenbach's intent is to provide a rationale for scientific inference. The ontological status of an infinite empirical sequence of future events of a given kind is dubious. The application of the frequency concept of probability to an individual case in which the outcome is uncertain requires that the case be embedded in a reference sequence, and this can always be done in infinitely many different ways. The mutual relevance of the limit of a sequence of frequencies and the structure of any specified finite initial segment of the sequence is dubious, unless a reason can be given for considering the segment to be a "good sample" of the entire sequence; but Reichenbach's theory prevents him from making any judgment about the goodness of a sample except in a state of advanced knowledge, when the structure of an infinite sequence of infinite sequences is known (1949: p. 443). Finally, the "straight rule" of positing that the limiting frequency is close to f^n is only one of an infinite set of methods which asymptotically lead to the correct value of the limiting frequency if it exists; and each of these methods has equal claim to being a guide to rational decisions about the future.[31] I shall not discuss these defects in detail, however, since they have been thoroughly explored by Popper (1961), Russell (1948), Burks (1951), Lenz (1958), Katz (1962), and others.

2. Kemeny's work on simplicity (1953) has the great virtues of detaching the pragmatic argumentation from a preoccupation with limits of relative frequencies and of applying it to inductive inferences concerning quite general classes of scientific hypotheses. He assumes that a denumerable class of hypotheses h_i is considered concerning a certain question and that one and only one member h of the class is true. A sequence of experiments is envisaged such that the possible outcomes of the first n experiments are $e_j{}^n$, the actual outcome being e^n. Kemeny wishes to give a methodologically defensible rule for "selecting" a hypothesis on the basis of e^n. (The notion of "selecting" is not explained, but apparently he intends something like making a tentative commitment as to which hypothesis is true. He does not discuss other types of conclusions of inductive reasoning, such as ordering hypotheses or assigning to them degrees of credibility, though his remark in 1953, page 407, as well as his other works on induction, indicate his awareness of their importance.) The rule of selection which he proposes after rejecting several alternatives is:

Select the simplest hypothesis compatible with the observed values. (If there are several, select any one of them.) (Ibid.: p. 397)

Kemeny defines compatibility in terms of a measure $m(h_i, e_j{}^n)$ of the *deviation* between a hypothesis and the observed results, but he says little about the choice of m except to require that "the deviation between a given hypothesis and the observed results tends to 0 if and only if the hypothesis is the true one" (ibid.: p. 394). He adopts a convention, which is frequent in statistical practice, of taking h to be compatible with $e_j{}^n$ if upon assumption of h there is at least a 1 percent probability that the outcome of the first n experiments will deviate from h_i by as much as $e_j{}^n$ does. The crucial and ingenious innovation in Kemeny's work is the proposal of four conditions under which a set of hypotheses is said to be "ordered according to simplicity." None of the four depends upon factual assumptions about nature or upon subjective judgments. The essential condition is that for every hypothesis h_i there is an integer N_i such that if $n \geq N_i$, then the compatibility of h_i with e^n implies that any other hypothesis as simple as h_i, or simpler, is incompatible with e^n.[32] Kemeny's conditions for a simplicity ordering permit him to assert the theorem:

If the true hypothesis is one of the hypotheses under consideration, then—given enough experiments—we are 99 percent sure of selecting it. (Ibid.: p. 401)

This theorem provides the justification for inductive procedures which incorporate Kemeny's selection rule (together with his explications of "compatibility" and "simplicity"). He claims that his procedure is similar to that of working scientists, who do not try to fit their hypotheses exactly to the data, but rather gradually try out more complex hypotheses—whatever that means to them—when the data is too much out of line with the simpler ones.

Kemeny's exposition is quite elliptical, and as a result some points are obscure which probably could be cleared up without difficulty. Thus, he should have stated that an ordering according to simplicity is relative to a sequence of experiments. Also, he does not explicitly state what concept of probability he is employing; apparently he does not regard this as a crucial issue because the only probabilities involved are likelihoods (of e_j^n conditional upon h_i), and there is more general agreement about evaluating these—whatever their significance—than in evaluating prior probabilities. Nevertheless, explicit assumptions need to be made about the likelihoods and also about the measure of deviation in order to justify the following statement, which is crucial to his argument:

As we know from statistics, as n increases, the deviations allowed by the compatibility requirement decrease. Hence for high n we can find an interval around the observed values (an interval that can be made as small as required by increasing n) such that all compatible hypotheses lie within this interval. (Ibid.: p. 400)

Indeed, until these points are clarified the criteria for constituting a simplicity ordering remain obscure.

A more important matter, however, is the extent to which Kemeny's theorem does justify his kind of inductive procedure. First of all, the theorem is noneffective: one cannot tell, without already knowing which is the true hypothesis, how many eyperiments are "enough experiments," since the integer N_i in the explication of "simplicity ordering" depends on i—as it obviously must if the hypotheses are permitted to come "closer" to one another as one proceeds in the ordering. It should be emphasized that "we are 99 percent sure of selecting it" does not mean that $P(h^n|e^n) = .99$, and in fact Kemeny does not claim to provide machinery for evaluating $P(h^n|e^n)$. In short, Kemeny's procedure shares with that of Reichenbach the disadvantage of requiring that one live to "the ripe old age of denumerable infinity" [33] in order to draw an inductive conclusion with confidence. Furthermore, as Kemeny recognizes, his criteria for a simplicity ordering do not determine the order uniquely, and evidently the use of different orderings would in general entail different selections of hypotheses at each stage of experimentation; and, as Katz emphasizes (1962: pp. 86 ff.), there seems to be little hope of choosing an optimum ordering by means of a condition that N_i should be made as small as possible.[34] Finally, Kemeny's theorem is conditional in form ("If the true hypothesis is one of the hypotheses under consideration . . ."), but the kind of *as if* argumentation which Reichenbach applied to the antecedent "If the sequence has a limit . . ." is not legitimate here. It may very well happen that the truth is a member of a different class of hypotheses, and it is not the case that nothing would be lost by pretending that the antecedent is true. Many of the dramatic episodes in the history of science consisted first in the suggestion of the plausibility of a previously unconsidered class of hypotheses and then in the exhibition that the truth probably is to be found in the new class.[35]

3. It is difficult to summarize and evaluate Peirce's theory of scientific inference, partly because he changed his opinions in important respects during his career without writing a definitive statement of his latest doctrine and partly because of obscurities of exposition. I suspect that these textual difficulties reflect the intellectual difficulties which Peirce experienced in attempting to develop to his own satisfaction the ingenious and attractive idea that one can be certain of asymptotic approach to the truth by means of the scientific method. However, the tension between his commitment to this idea and his self-criticism may have been responsible for a number of insights which are valuable even if his central idea is unworkable.

Peirce's broadest justification of the scientific method is that it approaches the truth because of its submissiveness to reality; it is a method

by which our beliefs are determined by nothing human, but by some external permanency—by something upon which our thinking has no effect. But which, on the other hand, unceasingly tends to influence thought; or in other words, by something Real. (5.384. The last sentence is a later addition to the passage by Peirce and is placed in a footnote by the editors.)

Again broadly speaking, the scientific method achieves this submissiveness by systematically and self-critically correcting beliefs in the light of experience (for example, 7.78).

Peirce's characterization of the scientific method (7.80–88) and elsewhere is very rich, and it includes not only an analysis of modes of inference but also fine considerations of heuristics, of the relation of science to metaphysics, and of the ethics of inquiry. He is evidently strongly drawn by the idea of an asymptotic approach to the truth by means of the scientific method *as a whole* (for example, 7.77), even though he recognizes the sporadic character of some elements in the method, especially the proposal of hypotheses (which he variously refers to as "abduction," "presumption," "hypotheses," and "retroduction"). However, the only clear example of an infallible asymptotic approach which he offers is the simple one which is the heart of Reichenbach's treatment of scientific inference: the evaluation of the limit of relative frequencies in infinite sequences of events (for example, 2.650, 6.100, 7.77, 7.120). Since this kind of inference ("statistical" or "quantitative" induction) is only one of the three kinds of induction which he recognizes, and since induction taken generically is not the whole of the scientific method, even sympathetic commentators on Peirce have found that his demonstrations fall far short of realizing his general program (for example, Murphey 1961, Burks 1964, Lenz 1964, Madden 1964).

Peirce may have underestimated the gap in the realization of his program by overestimating the amount of information about the statistical structure of an infinite sequence that can be obtained from a finite segment of it. He does not maintain, like Reichenbach, that the success of induction in dealing with a particular sequence depends upon the existence of a limit of the sequence; nor does he resort to the argument that nothing is lost by acting as if the limit exists. Instead, he claims that

if experience in general is to fluctuate irregularly to and fro, in a manner to deprive the ratio sought of all definite value, we shall be able to find out approximately within what limits it fluctuates and if, after having one definite value, it changes and assumes another, we shall be able to find that out, and in short, whatever may be the variations of this ratio in experience, experience indefinitely extended will enable us to detect them, so as to predict rightly, at last, what its ultimate value may be, if it have any ultimate value, or what the ultimate law of succession of values may be, if there be any such ultimate law, or that it ultimately fluctuates irregularly within certain limits, if it do so ultimately fluctuate. (6.40)

Although he says disappointingly little about the means for extracting so much information in general circumstances, I find two hints about his thinking. One is that the structure of erratic sequences is to be investigated "with the aid of retroduction and of deductions from retroductive suggestions" (2.767), which indicates that he held no illusions about the existence of an algorithm for the purpose. The other is an appeal to a constructivist theory of the infinite, on the basis of which he seems to make excessive claims for the effectiveness of inductive procedures:

Whatever has no end can have no mode of being other than that of a law, and therefore whatever general character it may have must be describable, but the only way of describing an endless series is by stating explicitly or implicitly the law of the succession of one term upon another. But every such term has a finite ordinal place from the beginning and therefore, if it presents any regularity for all finite successions from the beginning, it presents the same regularity throughout. (5.170)

The first of these passages seems to me very sensible, but in view of the chance character of the proposal of hypotheses, it weakens rather than strengthens the program of establishing the long-range infallibility of the scientific method. The second passage does indeed seem to support his program, except that I see no way that it can be construed so as not to be fallacious: for whatever the true sequence $\{f^i\}$ of fractions may be, it is not the case that there exists an n (even an unknown n) such that $\{f^i\}$ is the only law-governed sequence with the initial segment $f^1, \ldots, f^n$.

I am inclined to believe that the statements of Peirce which throw the most light upon inductive inference are those which qualify his central idea or are tangential to it. For example, he presents an *as if* argument, though its locus is different from Reichenbach's, for it is a justification of the process of abduction rather than of the assumption that a sequence of relative frequencies converges:

I now proceed to consider what principles should guide us in abduction. . . . Underlying all such principles there is a fundamental and primary abduction, a hypothesis which we must embrace at the outset, however destitute of evidentiary support it may be. That hypothesis is that the facts in hand admit of rationalization, and of rationalization by us. That we must hope they do, for the same reason that a general who has to capture a position or see his country ruined, must go on the hypothesis that there is some way in which he can and shall capture it. (7.219. See also 5.145, 5.357, 6.529, 7.77)

It should also be noted that if statistical induction must be investigated "with the aid of retroduction," as he says in a passage cited earlier, then the justification of induction derivatively depends upon an *as if* argument.

One important qualification of Peirce's central argument for justifying induction is his requirement that the set of events upon which an estimate of probability (in the sense of relative frequency in the long run) is based should be randomly chosen from the population under investigation (for example, 2.726). The principle of random sampling permits Peirce to escape from one of the objections raised above against Reichenbach's theory—that the limit of a sequence of frequencies and the structure of a specified segment of the sequence are mutually irrelevant. Peirce is able to speak of the probable error at any finite stage of the process of investigation (cf. 2.770). However, the concept of a "random" or "fair" sample is implicitly probabilistic, and, therefore, its employment in the process of estimating a probability appears to be an inversion from the standpoint of the frequency theory of probability (as Reichenbach points out in 1949: p. 446). There is a further complication. According to Peirce, a sample is random if it is "taken according to a precept or method which, being applied over and over again indefinitely, would in the long run result in the drawing of any one set of instances as often as any other set of the same number" (2.726). But his doctrine of dispositions repudiates any identification of a "would-be" with what actually happens (for example, 2.664), and, therefore, it is difficult for him to provide a criterion for randomness of sampling without employing a nonfrequency concept of probability. The concept of "verisimilitude" (2.663) comes close to being such a concept, and he goes so far as to say (in a letter) that "all determinations of probability ultimately rest on such verisimilitudes" (8.224). This line of thought, which appears late in his career, is unfortunately not developed very fully.

The problem of establishing criteria for randomness led Peirce to insert ethical considerations into the inductive process itself. For example, "the drawing of objects at random is an act in which honesty is called for; and it is often hard enough to be sure that we have dealt honestly with ourselves in the matter, and still more hard to be satisfied of the honesty of another" (2.727). He also makes the following intriguing suggestion of a minimal presupposition regarding the reliability of data:

I am willing to concede, in order to concede as much as possible, that when a man draws instances at random, all that he knows is that he *tries* to follow a certain precept; so that the sampling process might be rendered generally fallacious by the existence of a mysterious and malign connection between the mind and the universe, such that the possession by an object of an *unperceived* character might influence the will toward choosing it or rejecting it. . . . I grant then, that even upon my theory some fact has to be supposed to make induction and hypothesis valid processes; namely, it is supposed that the supernal powers withhold their hands and let me alone, and that no mysterious uniformity or adaptation interferes with the action of chance. (2.749)

This passage suggests an *as if* argument to the effect that we have nothing to lose by assuming our experimental evidence not to be distorted by factors which are unperceivable by us.

To summarize, I find at least four methodological ideas of great value in Peirce's papers on scientific inference: that the scientific method achieves its successes by submission to reality, that a hopeful attitude toward hypotheses proposed by human beings is indispensable to rational investigation of the unknown, that a usable criterion of fair sampling involves subjective and ethical considerations, and that it is rational to make certain weak assumptions about the fairness of the data in order to permit inquiry to proceed. His suggestions on the instinctive basis of abduction go beyond methodology and will be discussed in section V.

C. Of the three treatments of inductive inference discussed above, Peirce's comes closest to exhibiting how the informal methods of confirmation actually used by scientists are sensitive to the truth, but even it requires corrections in various respects. Implicit in these informal processes is a wonderfully balanced and sinuous strategy with a strength that can be recognized a priori, despite the possibility that it will yield the truth only if the world is not too deceptively constituted. The primary virtue which I claim for tempered personalism is that it succeeds in catching much of the essence of this strategy: maintaining a balance between open-mindedness toward proposals and a critical attitude toward them; mediating between tenacity concerning plausible hypotheses, even in the face of a moderate amount of adverse evidence, and skepticism concerning artificial explanations to shore them up; utilizing the formally structured reasoning of probability theory in conjunction with informal and intuitive thinking; paying respect to the community of investigators and also to the intellectual conscience of the individual; disentangling questions for stepwise investigation while appreciating their interconnectedness.

Some of the strength which I attribute to tempered personalism can be conveniently exhibited by a comparison with Kemeny's proposals. His procedure partially catches the balanced strategy of informal scientific inference, for his rule of selection combines a tenaciousness regarding preferred hypotheses (by using a tolerant standard of compatibility between hypothesis and evidence) with a critical attitude toward them (by rejecting a more preferred hypothesis in favor of a less preferred one, if the former turns out to be incompatible with the data).[36] Although his procedure and tempered personalism are similar in this very important respect, there are obvious differences between them.

Tempered personalism has no rule for "selecting" a hypothesis at every stage of experimentation, nor even at the end of an investigation, which may very well conclude with the approximate equality of several seriously proposed hypotheses. The nearest approach to a rule of selection in tempered personalism is the process of "accepting" a hypothesis which has posterior probability close to 1 (discussed in section III.F), and the tentative nature of this acceptance has been pointed out. In Kemeny's procedure tentativeness is manifested only in the possibility that new evidence will lead to the selection of a different hypothesis, whereas in tempered personalism tentativeness is suffused throughout a person's theoretical view of the world.[37]

The most profound difference between the two procedures lies in their principles of preference among hypotheses antecedent to observation. Kemeny gives preference to the earlier members of a predesignated infinite sequence of hypotheses—the order of the sequence being arbitrary, subject to the limitations discussed in section IV.B. The principle of preference in tempered personalism, which is essentially contained in the tempering condition, is always relativized to a particular person and a particular time: he must assign nonnegligible prior probability to each seriously proposed hypothesis $h_1, \ldots, h_n$ of which he is aware concerning the matter under investigation, whereas all other hypotheses must be treated as disjunctive components of the catchall hypothesis and, hence, must be assigned prior probabilities which are generally many orders of magnitude smaller than those assigned to $h_1, \ldots, h_n$ (and indeed are infinitesimal in the case of a continuum of disjunctive components); the ordering and the relative weighting of $h_1, \ldots, h_n$ are subjective, within the limits set by the tempering condition. Although the principle of preference in tempered personalism seems unsystematic from the standpoint of any predesignated ordering of hypotheses, it has several great advantages: the most obvious is that tempered personalism dispenses with an assumption that the truth lies in a predesignated class of hypotheses. It was pointed out in section IV.B that historically assumptions of this kind have been false and methodologically one has no justification for acting as if such an assumption were true. Tempered personalism permits an investigator to transcend any initially limited class of hypotheses simply by admitting as seriously proposed some hypothesis not belonging to that class. Even if a hypothesis seriously proposed by a very perceptive scientist does fall within Kemeny's predesignated class, a very long interval—perhaps longer than the duration of the human race—might have to elapse (at current rates of performing experiments) before the predesignated ordering of hypotheses would permit it a chance to be selected. The mechanical consideration of hypotheses in an order established a priori is, therefore, an abnegation of the clues which nature may provide unexpectedly as an investigation develops and of the powers which men immersed in a subject may possess of utilizing these clues. At the present stage in my treatment of scientific inference, I wish to avoid making factual assumptions about the world and, therefore, cannot say anything about the actual occurrence of reliable clues or about the capacity of investigators to utilize them fruitfully. I can say at this stage, however, that unless we act *as if* good approximations to the truth will occur among the hypotheses which will be seriously proposed within a reasonable interval, we are in effect despairing of attaining the objective of inquiry.

Kemeny's procedure disregards the proposals put forth by investigators on intuitive grounds because it is not only a method of confirmation but something of a "logic of discovery"; for once the class of admissible hypotheses has been chosen and ordered and a measure of compatibility has been defined, it provides an algorithm for selecting a hypothesis h as a function of the data e_j". But mechanizing the process of selection leaves no role for intelligent guessing about nature except in the initial choice of the sequence of hypotheses—a sweeping operation which requires much greater intuitive powers than do the individual conjectures of men immersed in specific problems. Tempered personalism, on the other hand, is in no way a "logic of discovery" but rather supports whatever powers human beings may have for making intelligent guesses. It is well adapted to the possibility that these powers are exhibited sporadically and in very different degrees by different people, and that they are refined and stimulated by the progress of knowledge. In contrast to a method which imposes an a priori ordering upon hypotheses, tempered personalism can take advantage of the possibility that a profound scientific discovery can have the effect of making hypotheses which previously would have been extremely remote in any plausible ordering seem, at least from a subjective standpoint, "natural" and simple. (The outstanding example is the discovery by Galileo and Newton of the relation between force and acceleration, whereby second-order differential equations became a familiar conceptual tool; the solutions to some of the commonest of these equations appear extremely complex from the standpoint of a naïve direct ordering of functional relationships.) Tempered personalism thus avoids the skepticism toward human abductive powers implicit in any formal scheme which treats on the same footing seriously proposed hypotheses, frivolously proposed hypotheses, and unsuggested hypotheses. By giving preferential treatment to seriously proposed hypotheses but insisting upon open-mindedness within this preferred class, the tempering condition provides a safeguard against one of the major types of error that could be committed by a method of confirmation: the error of rejecting, because of a priori commitments, a true hypothesis which some one has been fortunate enough to put forth. In this way tempered personalism incorporates Saki's great methodological maxim: "In baiting a mouse-trap with cheese, always leave room for the mouse."

Because of the informality of its principle of preference among hypotheses, tempered personalism is in a sense on a metalevel relative to any formal procedure for prior ordering and weighting. If a formal ordering of hypotheses, such as one of Kemeny's or that of Jeffreys and Wrinch, is seriously proposed, and if the first n hypotheses in the ordering have been assigned extremely low posterior probabilities in previous investigations, the $n + 1^{th}$ could be considered a seriously proposed hypothesis and, hence, would be assigned a nonnegligible prior probability in a new investigation. In this way the formal ordering would be interleafed with the order of intuitive proposals. The interleafing would permit the systematic exploration of a predesignated class of hypotheses, concurrently with excursions into possibilities which are very remote in the ordering or which lie outside it. The informal excursions would be fruitful if scientists are sufficiently imaginative and if the experimental data are sufficiently suggestive, whereas the plodding exploration of the ordered class of hypotheses would occupy idle equipment and would provide exercises in technique, with the possibility of unexpected striking confirmations, during intervals in which imagination is barren—though one may properly doubt whether the institution of scientific research would survive if these intervals became excessively long.

D. In order to complete the discussion of the sensitivity of the tempered personalist formulation of scientific inference to the truth, it is necessary to show that its receptivity toward seriously proposed hypotheses is adequately balanced by a capacity to evaluate them critically. Two elements of its apparatus are intended to serve this purpose: the catchall hypothesis, which says that none of the specific alternatives under consideration is true, is taken as seriously proposed and is, therefore, assigned a nonnegligible prior probability; and the posterior probability of each hypothesis is dependent, because of Bayes's theorem, upon the likelihood relative to it of the evidence actually gathered. The first of these has been discussed at several places in section III, but little has yet been said about likelihoods.

Prima facie the evaluation of the likelihoods $P_{x,b}(e|h_i \& a)$ is a subjective process, limited only by the axioms of probability, for the tempering condition refers explicitly only to assignments of prior probabilities. However, it will be seen that the tempering condition is relevant to the evaluation of likelihoods and restricts the subjectivity of these evaluations. Furthermore, the methodological arguments in favor of the tempering condition provide a partial justification for relying upon likelihoods in critically judging hypotheses.

Typically, the information and assumptions contained in a permit the evaluation of the likelihoods $P_{X,b}(e_m|h_i\&a)$ to be derivable from judgments of indifference concerning a set of possible experimental outcomes, once certain known or presumed differences are dismissed as irrelevant to their occurrence.[38] For example, in the Laplacean sampling problem (considered in section III.E, but without attention to likelihood evaluations), a includes the information that the population of interest consists of N objects, of which r are selected successively by some process; and h_i asserts that i of the members of the population have the property α. Ordinarily the known and presumed differences among the $N - x$ objects remaining in the population after the selection of the first x members of the sample are dismissed as irrelevant to the selection of the $x + 1^{th}$. In particular, the individuating differences whereby the members are identifiable as "object 1," "object 2," . . . , "object N" are dismissed as irrelevant, as is the difference between possession and nonpossession of the property of interest α. By dismissing these differences, a judgment of indifference is possible, that is, if s_x asserts that the first x objects selected were objects j_1, . . . , j_x, and if e_m and $e_{m'}$ respectively assert that the mth and m'^{th} objects will be selected at the $x + 1^{th}$ draw (where neither m nor m' equals any of the j_1, . . . ,j_x), then

$$P_{X,b}(e_m|h_i\&a\&s_x) = P_{X,b}(e_{m'}|h_i\&a\&s_x).$$

The evaluation of $P_{X,b}$ $(d_{mn}|h_i\&a)$—which are the likelihoods involved in section III.E—follows straightforwardly by combinatorial analysis and the axioms of probability. A similar but more complex analysis can be carried through in more realistic problems, where the h_i are competing scientific hypotheses, a includes auxiliary assumptions about natural laws and boundary conditions and also about the statistical characteristics of the measuring instruments employed, and e_m and $e_{m'}$ are replaced by propositions concerning individual instrument readings. The evaluation of likelihoods in problems involving measurement depends upon a tacit judgment of indifference that the experimental error of each reading is drawn without bias from a population of errors associated with the instrument; and this judgment is possible only if the special circumstances of the investigation, which conceivably could cause a systematic error, are dismissed as irrelevant.

The tempering condition is relevant to the dismissal of differences among cases. Indeed, it permits a kind of social check upon personal judgments about the data. Someone may seriously make the proposal g that the method of selecting a sample is favorable or unfavorable in a certain manner to objects having the property α. For example, ecological experimentation requires sampling from all the members of a species in a given area by such means as trapping, and the susceptibility to being caught may be supposed to be correlated with the property α. The tempering condition applies to g as to all serious proposals, for as Jeffreys remarks, "There is no epistemological difference between the Smith effect and Smith's systematic error; the difference is that Smith is pleased to find the former, while he may be annoyed at the latter" (1961: p. 300). The prescription that g be given nonnegligible prior probability could be followed by doubling the set of seriously proposed hypotheses in the investigation at hand—that is, by taking as seriously proposed not only $h_1, \ldots, h_n$ but also $h_1\&g, \ldots, h_n\&g$. However, in view of the methodological advantages of disentangling problems (discussed in section III.A), it is generally preferable to conduct an auxiliary investigation concerning the bias of the sampling procedures.

Even after all serious proposals concerning possible bias have been checked and appropriate corrections in the sampling procedure (or in the analysis of sampling data) have been made, there always remain special circumstances concerning a sample which are unnoticed by investigators or which are habitually dismissed as trivial, and one might skeptically suspect that a systematic error is associated with them. The proper answer to this kind of skepticism is that a possible source of systematic error which no one has proposed for serious consideration is merely one of the infinite set of possible but unsuggested alternatives to the seriously proposed hypotheses in the investigation, and, hence, it should be treated as a component of the catchall hypothesis. It was argued previously that seriously proposed hypotheses should be treated differently from all others in order to give intelligent proposals a chance of acceptance into the corpus of scientific knowledge. In the case of hypotheses regarding systematic errors, an additional reason can be given for this differential treatment: that unless the unsuggested hypotheses are given very small prior probabilities, no evidence could be utilized as reliable grounds for critically judging the seriously proposed hypotheses. A certain amount of tentative trust is a prerequisite for

critically probing and testing, whereas sweeping skepticism is methodologically sterile. Jeffreys gives a similar warning:

A separate statement of the possible range of the systematic error may be useful
if there is any way of arriving at one, but it must be a separate statement and
not used to increase the uncertainty provided by the consistency of the observations themselves, which has a value for the future in any case. In induction
there is no harm in being occasionally wrong; it is inevitable that we shall be.
But there is harm in stating results in such a form that they do not represent
the evidence available at the time when they are stated, or make it impossible
for future workers to make the best use of that evidence. (Ibid.: p. 302)

One qualification should be added at this point, even though it is quite
obvious: that the recommendation to treat as statistically irrelevant those
experimental circumstances which are not seriously proposed as relevant
must not be construed as an excuse for slovenliness in experimentation or
for relaxation of the vigilance of the investigator regarding his own
prejudices. That samples may be skewed, perhaps unconsciously and in
subtle ways, so as to favor the preferred hypothesis of the experimenter,
should be a standing serious proposal of a possible source of systematic
error. (See Peirce 2.727, quoted in section IV.B.)

   E. There was no reference to the "long run" in the claims made above
that the tempered personalist formulation of scientific inference has a
kind of sensitivity to the truth. An examination of a situation which has
been fully analyzed in the literature—the Laplacean sampling problem
with replacement—will suffice to indicate that consideration of the long
run does not lead to an essential strengthening of the foregoing claims.
In the sampling problem with a finite population but with replacement,
no apodictic statement can be made on the basis of a finite or infinite
number of draws other than that h_0 is false if at least one object with
property α is drawn, and h_N is false if at least one object lacking this
property is drawn.

   It is well known that if probability theory is formulated so as to permit
the treatment of infinite sets of alternatives (by using the axiom of complete additivity—note 8), then there exist disjoint sets S_i of infinite binary
sequences of α's and $\bar{\alpha}$'s such that $P_{X,b}(s\epsilon S_j|h_i\&a)$ is 1 if $i = j$ and 0 if
$i \neq j$, where s is the sequence actually obtained in an infinite sequence of
draws (for example, Kac 1959: pp. 18–22). Bayes's theorem then yields
$P_{X,b}(h_i|s\epsilon S_j\&a) = 1$ if $i = j$ and 0 if $i \neq j$. But it is also well known that
these results do not permit nontrivial inferences to be made with cer-
tainty about the composition of the population, since the probability 0
of a proposition does not entail its falsehood. The kind of conditional
apodictic statement which Reichenbach's analysis permits is tangential

to the point of interest here. He would say that if *s* has a definite statistical structure (which for him means that the sequence of relative frequencies of α's in finite segments converges), then for any $\epsilon > 0$ there exists an *n* such that the relative frequency of α's in every segment of length greater than *n* differs from the limit by less than ϵ. But the question under investigation concerns the composition of the *population,* and the sequence of draws from it is only a means for learning about that. To deny that this is so, and to say that the only question of human interest or of scientific interest is the structure of *s,* implies a commitment to a kind of phenomenalism, which is incompatible with the Copernican point of view. One more remark is too obvious to require elaboration: that nontrivial inductive problems in the natural sciences, in which the seriously proposed hypotheses imply propositions about the statistical distribution over possible observations in a potential infinity of situations, are closer in character to the sampling problem with replacement than without replacement.

The one point in the ascription of sensitivity to tempered personalism where reference to the long run may be required is its capability of assimilating other methodological devices. It was seen, for example, that any one of Kemeny's orderings of hypotheses according to simplicity can be interleafed with an ordering of seriously proposed hypotheses; but because of the intrinsically long-run objective of Kemeny's procedure, no advantage of this interleafing can be exhibited in any finite segment of the sequence.

The foregoing considerations indicate the futility of trying to establish without qualification that *the tempered personalist formulation of scientific inference is a method which, if persisted in indefinitely, can discriminate true from false hypotheses* (call the italicized statement "Ψ"). Nevertheless, it is tempting to take Reichenbach's pragmatic justification as a paradigm and to make a conditional apodicitic claim—that is, *if conditions* (1), (2), . . . *are satisfied, then* Ψ—and to do so in such a way that a methodological argument can be given for acting as if conditions (1), (2), etc., are true, whether or not their truth can ever be checked. Indeed, a step in that direction was already taken in section IV.C, in reasoning that "unless we act *as if* good approximations to the truth will occur among the hypotheses which will be seriously proposed within a reasonable interval, we are in effect despairing of attaining the object of inquiry." This reasoning supplies a natural condition (1) for the desired conditional claim. Unfortunately, in spite of long reflection on the matter, I see no way of completing the list of conditions in a way that is neither hopelessly obscure nor hopelessly hobbled.

V. Circumstances Favorable to Induction

A. The foregoing a priori analysis of the tempered personalist formulation of scientific inference showed several ways in which errors could
be committed in spite of systematic apparatus to prevent them, and gave
reasons for believing that any formulation of scientific inference would
be fallible in the same way. Moreover, since scientific inference is not a
method of discovery, it cannot exclude the possibility of a nonerroneous,
but thoroughly demoralizing, rejection of every specific scientific hypothesis that will be proposed during an indefinitely long period of time concerning some aspect of nature. Consequently, if we wish to understand
the fact that we are now in possession of a vast and detailed and apparently reliable system of scientific knowledge, the explanation cannot be
entirely methodological but must refer to facts about the world.

What kind of explanation can reasonably be expected for the success
of scientific inquiry is far from clear, for consider how many general and
particular facts are relevant to any scientific discovery: the laws of the
domain in which the discovery is made, the laws governing the instruments used in observations, the psychological principles of perception
and concept formation, the facts of geology and meteorology that permit
observations to be made, the sociological facts about the milieu in which
the research was undertaken, and the crucial biographical facts about
the discoverers. In a sense the only adequate explanation of the existence
of the system of scientific knowledge would consist of an encyclopedia
of the natural and social sciences, together with accounts of the contingencies in the history of science. In spite of this holistic consideration,
it surely should be possible to give good partial explanations by intelli-
gently following important threads in the total fabric; and grounds for
optimism regarding such enterprises are provided by the existence of
illuminating work in the history, sociology, and psychology of science.

The first purpose of this section is to propose a special line of partial
explanation, which consists in *exhibiting circumstances complementary
to and supporting the logical structure of scientific inference.* There are
several crucial junctures in the formulation of scientific inference at which
the methodological prescriptions are counsels of desperation unless circumstances are favorable to inquiry. At these junctures scientific inference
proceeds as if human faculties and the natural environment are favorable
in various ways to determining the truth: as if the background of information and assumptions (of which only a small part can be critically
tested) is on the whole reliable, as if problems can be disentangled for

stepwise treatment, and as if some good approximations to the truth are to be found among the hypotheses seriously proposed within some reasonable interval of time. The partial explanation of the success of scientific inquiry which I have in mind consists in showing that nature is indeed favorable to inquiry at these junctures. The explanation should be to some extent independent of a detailed characterization of the environment and of human faculties, because the strategy of scientific inference is designed to be sinuous and adaptable to a wide range of possible worlds. Within reasonable limits it should be possible to make a plausible case that if a certain fact (or even a law) had been otherwise, the truth about the situation could nevertheless have been discovered. An explanation along these lines has a unitary character even though it evidently involves a rather special selection of topics from psychology and other sciences and, therefore (like all partial explanations of the success of the natural sciences), must tacitly refer to a comprehensive body of knowledge in these fields.

The second purpose of the section is to examine the methodological consequences of the envisaged partial explanation of the success of scientific inquiry. The primary consequence is to provide an a posteriori justification for various characteristics of the tempered personalist formulation of scientific inference. Thus, the "localization" of problems is seen to be well suited to investigations in the actual world. Also the role of subjective judgment in evaluating probabilities and in other aspects of scientific inference is to some extent sanctioned, for the adaptation of human beings to the natural environment ensures that a crude rationality is operative in our mental activity even when we are unable to make it articulate. An entirely different consequence is that antecedent knowledge about the world can provide guidelines which supplement both subjective judgment and the a priori structure of scientific inference. Some examples will be given of guidelines for deciding whether or not to classify a hypothesis as seriously proposed, and for evaluating probabilities on the grounds of indifference and simplicity.

The problem of circularity is evidently raised by the fact that a body of natural knowledge, which was itself tentatively established by means of scientific inference, is methodologically significant. An analysis of the sense in which scientific inference is circular and an argument that the circularity is nonvicious will conclude the essay.

B. I shall begin the discussion of the circumstances which support the structure of scientific inference by considering the existence of reliable prescientific knowledge and then shall outline some of the reasons for believing that nature is also favorable to scientific inference at other junctures. It will be seen throughout that this favorableness is easily indicated in a sweeping and impressionistic manner but that a deep and precise analysis is difficult.

1. Although scientific and prescientific elements are inextricably mixed in the belief system of a scientist, it is possible to discern in this system a common-sense picture of the everyday world which on the whole is independent of his professional training and research.[39] This picture is indispensable as a background to controlled scientific inquiry, since many of the terms occurring in the evidence, and some of those occurring in the assumptions a and in $h_1, \ldots, h_n$ as well, presuppose knowledge about the spatio-temporal structure of the everyday world, correlations among classes of appearances, object constancy under change of viewing conditions, etc.[40] The fact that we are at home in the everyday world, to the extent of coping with most of the practical problems of staying alive, suffices to show in a coarse way that this picture is reliable. But in order to achieve a fine understanding of its reliability, we need to know exactly what the content of this picture is, how much of it is culturally determined, by what processes it is acquired, and whether any part of it can be characterized as innate. These are deep questions in psychology (though overlapping somewhat with biology and with the social sciences), and all of them are at present largely unsettled.[41]

Nevertheless, I have the impression (based on a very limited knowledge of the literature) that an intricate but coherent set of answers to these questions is beginning to emerge. The evidence is accumulating that there is a common cross-cultural core to pictures of the everyday world. This thesis was challenged by Whorf (1941 and elsewhere), who emphasized the great variation among languages in such fundamental respects as the syntactical treatment of time, action, and objects. However, the connection between language and world view is probably much looser than Whorf maintained (cf. Hockett 1954). Furthermore, the view that linguistic differences are revelatory of profound divergences in metaphysics is undermined by the mass of evidence in favor of the existence of formal linguistic universals (Chomsky 1965: pp. 29 ff). A plausible proposal about the ontology of the common core was made by Strawson (1959): that objects, which are many-faceted, quasi-permanent things, have a more fundamental status than events and appearances, and that there exist persons, who are located in space and time like objects and who can interact with them, but are unlike objects in being the subjects of feelings, thoughts, and intentions. Yet this way of putting the matter may be excessively abstract and intellectual, and it should probably be supplemented by considering the sensorimotor and perceptual components of common sense (cf. Flavell 1963: pp. 129–50). Both the existence of a common core of representations of the world and the depth of its biological grounding are indicated by clinical findings that circumstances which seriously impede the internalization of basic commonsense principles—for example, artificial handling which insulates an infant from persons or interference with the normal sequence of events in a way that affects the child's concept of objects—tend to induce psychoses (for example, Vernon 1962: pp. 182–83). We also have some insight into the way in which the common core of representations of the everyday world is ripe, so to speak, for incorporation into very diverse comprehensive conceptual systems: for example, objects are many-faceted and are universally recognized as being capable of deceptive appearances, and, furthermore, they probably are universally understood to be involved in causal relations, but these characteristics of objects are invitations to intellectual elaboration in the form of explanations, theories, myths, and cosmologies.[42]

The bare thesis of the existence of cultural universals is compatible with alternative theories of the genesis of concepts, specifically, both with suitable versions of the doctrine of innate ideas and with suitable theories of the operation of very general mechanisms of learning (for example, association) upon sensory input which is ordered only by the regularities in the environment. In language learning the additional consideration that the grammatical rules mastered by children are complicated whereas the sensory input is relatively meager strongly favors the thesis of innate ideas (Chomsky 1965 and elsewhere). It seems probable, however, either that special cognitive mechanisms (for example, for acquisition of grammar and for learning the geometry of the visual field) coexist and interact with more generalized learning apparatus, or else that the organism effectively exercises a general learning strategy by being able to switch from one specific mechanism to another and to collate the outputs of various mechanisms. Either hypothesis accounts in a general way for the conjunction of efficiency in achieving intricate skills with flexibility in solving problems under variable conditions. (A particularly good discussion of the second of these two hypotheses is in Sutherland 1959.)

Whatever the correct account of special and general cognitive mechanisms will be, it is to be expected that the maturation of reasoning powers and the growth of a picture of the everyday world will turn out to be interdependent processes. As a result, a thorough investigation of the derivation of the reliable body of information required for scientific inference will throw light upon the process of intellectual development which culminates in the capacity to perform controlled scientific inferences (or more accurately, which culminates in capabilities which can be shaped to this end if cultural and individual circumstances are favorable). The most elaborate experimental and theoretical work on this interdependence has been done by Piaget, who divides the intellectual development of the child into three major periods (and various subperiods) and finds that the techniques and the tentative views of the world which are crystallized in one period are prerequisites for the development of intelligence in later periods (1952; also Inhelder and Piaget 1958, and Flavell 1963). The interdependence of the development of reasoning powers and the growth of a world picture is expressed in a sophisticated way by Bruner, who acknowledges (much more than Piaget) both the role of special cognitive mechanisms and the influence of culture (1966: chap. II and p. 321). It is evident that theories of cognition along the lines indicated by Piaget and Bruner are very remote from Hume's doctrine of "experimental reasoning," according to which a single principle of "custom" guides the inferences of animals, children, and philosophers,[43] and remote also from the varieties of learning theory which were dominant until quite recently.

2. The circumstances which are favorable to the disentanglement of problems are diverse—and certainly cannot be studied as part of a single discipline, in the way that the existence of a reliable body of prescientific knowledge can to a large extent be considered part of the subject matter of developmental psychology. Perhaps the most important of these circumstances is the deep-lying principle that a sharp distinction can be made between physical laws and boundary conditions (cf. Wigner 1964). Without the distinction between laws and boundary conditions, it is hard to see how physics could be studied systematically, since the art of experimentation consists largely of choosing boundary conditions in favorable ways, by achieving relative isolation from disturbing factors and by arranging spatial symmetries, so as to check hypotheses about regularities which are supposed to hold generally. Furthermore, the laws themselves are, at least to very good approximation, invariant under time translation—an invariance which provides the deepest factual basis for the conformity of the future to the past. The distinction between laws and boundary conditions is also essential in disentangling problems concerning the operation of the physical instruments used in

biological research from problems concerning biological functions. I do not mean to prejudge the speculations of philosophers like Whitehead (1929: p. 162) that the intrinsic behavior of physical particles is modified by their incorporation into organisms. However, the evidence is overwhelming that if there is such a modification, it is extremely small and, therefore, of negligible importance, at least in the biological problems currently under investigation, in interpreting the biological data obtained by means of physical apparatus. Further preconditions for controlled experimentation in all the natural sciences are that the forces between physical systems fall off rapidly with their distance and that the laws of nature permit physical and biological configurations which are highly stable against small perturbations; otherwise, every experiment would have to take account of detailed astronomical, meteorological, and geological information.

An entirely different factor in the disentanglement of problems is psychological and is related to the general reliability of our common-sense picture of the world, namely, that to first approximation we have a good sense of the relevance and irrelevance of various factors to phenomena of interest. This crude sense of relevance is often wrong, and some of its errors, such as overestimating the influence of "wonders" in the heavens upon terrestrial events, have hampered the development of knowledge. Nevertheless, when this sense of relevance is controlled by critical intelligence, it makes crucial observations possible by enabling men to disregard the innumerable details that are potentially distracting in experimentation.

3. The last of the circumstances noted in section V.A which support the logical structure of scientific inference is the occurrence of good approximations to the truth among the hypotheses which men have seriously proposed.[44] The evidence of this occurrence is mostly contained in the history of science. It consists, first, of the record of striking confirmations and reconfirmations of many individual hypotheses which have been subjected to severe tests and, secondly, of the fact that on the whole this record has been cumulative and progressive. To be sure, there are historical examples of hypotheses which for long times were considered to be highly confirmed but which afterward were judged to be erroneous in the light of new evidence and further analysis. However, the usual pattern since the achievements of physical explanation in the seventeenth century has been that a well-confirmed hypothesis is not only reconfirmed by further tests but also fits fairly well into the general scientific world view—Kuhn's "normal science" (1962: chaps. II–IV).

Even when discrepancies have developed between individual highly confirmed hypotheses and the prevailing world view, and have been resolved only by profound intellectual changes which deserve the name of "scientific revolution," the continuity of scientific knowledge is to some extent maintained by the existence of "correspondence" relations between old and new theories (cf. section II.D and note 17).

In order to obtain a deeper explanation of the existence of a vast system of scientific knowledge, it is necessary to understand the natural basis of the process which Peirce calls "abduction." Such an understanding would add little to the weight of evidence from the history of science in support of the proposition that good approximations to the truth have occurred among the hypotheses which have been seriously proposed. However, the question is evidently of great intrinsic interest, and it is also relevant to one of the methodological consequences to be discussed in section V.C (the sanctioning of subjective judgment).

Peirce is deeply impressed by the smallness of the number of fallacious guesses which men of genius—Kepler being his favorite example—have had to make concerning many phenomena before coming upon approximately correct ones. As a result, he finds the history of science to be incomprehensible unless the human mind possesses a *lume naturale* which results from the influence of the pervasive laws of nature (for example, 5.604). He appeals to the evolutionary history of the race in order to explain how this influence fostered human abductive powers; for example, he writes:

You cannot seriously think that every little chicken, that is hatched, has to rummage through all possible theories until it lights upon the good idea of picking up something and eating it. On the contrary, you think the chicken has an innate idea of doing this; that is to say, that it can think of this, but has no faculty of thinking of anything else. The chicken you say pecks by instinct. But if you are going to think every poor chicken endowed with an innate instinct toward a positive truth, why should you think that to man alone this gift is denied? If you carefully consider with an unbiassed mind all the circumstances of the early history of science and all the other facts specifically bearing on the question, . . . I am quite sure that you must be brought to acknowledge that man's mind has a natural adaptation to imagining correct theories of some kinds, and in particular to correct theories about forces, without some glimmer of which he could not form social ties and consequently could not reproduce his kind. In short, the instincts conducive to assimilation of food, and the instincts conducive to reproduction, must have involved from the beginning certain tendencies to think truly about physics, on the one hand, and about psychics, on the other. It is somehow more than a figure of speech to say that nature fecundates the mind of man with ideas which, when these ideas grow up, will resemble their father, Nature. (5.591)

I find Peirce's argument broadly convincing and inspiring but disappointing in its lack of details. To fill in the details, it would be necessary, for example, to investigate the question of the existence of innate ideas and, if they exist, to reconstruct their evolutionary development, presumably by studies of the comparative psychology of animals since paleontology is not informative about such matters. Some conjectures along these lines concerning the idea of causality have been made by Simpson (1963) and Barr (1964). Furthermore, since the intelligent proposal of hypotheses is one of the least mechanical of mental operations, it is important to understand not only the relatively fixed parts of our intellectual equipment but also the imagination and its play with possibilities. There is impressive evidence that the minds of gifted men who are immersed in their subjects survey and in some sense evaluate a wide range of possibilities on a subconscious level before consciously making serious proposals (Hadamard 1945), but little seems to be known about the machinery of this process and even less about its evolutionary background.

One respect in which the evolutionary account of human abductive powers requires supplementation is in understanding the formation of good hypotheses about matters which are remote from direct practical concerns. An evolutionary explanation of features of our cognitive faculties postulates a sequence of mutations and selections in the history of the species which had the effect of solving with partial success certain problems of surviving and propagating. One characteristic of the line of evolution leading to homo sapiens seems to be the development of a high degree of realism in ordinary judgments with respect to space, time, forces, objects, and persons (cf. Freud's "reality principle"). However, realism of this practical variety does not ensure the capability of making good guesses on theoretical questions remote from ordinary concerns, such as the microscopic constitution of matter; and it does not even ensure the ability to correct certain anthropocentric tendencies of common sense, such as the attribution of sensed qualities to physical things and of purpose to physical forces—attributions which may be biologically valuable in spite of being "unrealistic." In part, the transition from realism in practical affairs to realism in theorizing about nature can be accounted for by the impulse of curiosity and by the critical functions of intelligence, both of which fit into an evolutionary account of abductive powers since both are biologically useful in obvious ways. But this transition cannot be fully understood without reference to favorable characteristics of the laws of physics and of certain important contingent facts about the physical environment, for example, how the behavior of planets and of the approximately rigid stones available in our geological epoch are suggestive of the mechanics of point particles, how the range of temperatures we can control is sufficient to reveal many clues about the chemical composition of substances, and how the properties of visible light are a guide to the quantum properties of elementary particles.[45] Incidentally, it is an uncertain extrapolation to suppose that nature will continue to favor human abductive powers in domains increasingly remote from our immediate concerns. The formulation of good hypotheses in atomic physics required great flights of imagination, and the truth about the subatomic domain may be orders of magnitude stranger and less accessible to human conjecture. (Nevertheless, the methodological arguments presented in sections III and IV for assigning nonnegligible prior probabilities to seriously proposed hypotheses, including those concerning remote domains, is unaffected by such doubts.)

Some remarks are necessary at this point to prevent a possible misapprehension. In discussing the factual circumstances which support the logical structure of scientific inference, the emphasis has been on physical, biological, and psychological factors favorable to scientific inquiry. However, I do not mean to disguise the cultural character of our system of scientific knowledge. Many cultures have had sophisticated technologies, governments, arts, etc., without coming close to an institutionalization of scientific research or the proliferation of scientific knowledge. One, therefore, cannot expect considerations of developmental psychology, evolutionary biology, and other natural sciences to explain more than the *potentiality* of sustained scientific thought. In particular, one should be skeptical of speculations that intellectual history parallels or is foreshadowed by the intellectual development of the individual (for example, Flavell 1963: pp. 252–56). A striking counterexample is the wonderfully abstruse and nonanthropocentric theorizing by Pythagoras at a very early stage in the history of science. Another example, which is damaging to many theories of scientific development, is the highly critical and sophisticated character of the cosmological argumentation of the Epicureans, as contrasted with the carelessness of their treatment of terrestrial phenomena (Cornford 1965: chap. 2). If we wish to understand how the potentiality for scientific knowledge has been realized, we cannot dispense with detailed sociological and historical studies.

The program which has been outlined is part of the "the way down," that is, part of the enterprise of understanding how creatures like our-selves, capable of the kind of experience and endowed with the kind of faculties that make natural knowledge possible, can exist and function. To use a rough analogy, this understanding is comparable to understand-ing the functioning of optical instruments, such as lenses, in terms of the physical theories of light and of dielectric media. The roughness of the analogy is due in part to the fact that these physical theories are well established and precise, whereas genetic psychology and other sciences concerned with human behavior are still in their infancy. A more impor-tant reason for the roughness of the analogy, however, is that the acquisition of natural knowledge by human beings is an incomparably more complex phenomenon than the interaction of light with lenses. Indeed, because of the innumerable cultural and biographical factors which are relevant, the development of natural knowledge must be regarded as a matter of chance (entirely apart from the problem of determinism). Somewhat less roughly, our understanding of the acquisi-tion of natural knowledge can be analogized to our understanding of the evolution of a specific anatomical feature of an animal. No evolutionary biologist claims to be able to predict from first principles that a species with a particular feature will evolve and survive for a long time. Nev-ertheless, when a biologist is presented with a particular feature of an animal, he can say illuminating things about its integration with other anatomical and physiological characteristics, about its role in enabling the species to fill a certain ecological niche, about its superiority in definite respects to the features of variant species which are rare or extinct, and also about the possibility that it is not biologically useful to the species but is present because of genetic linkage to useful features (cf. Huxley 1942 and Rensch 1960 for examples of the varied and intricate types of explanation which are fruitful in evolutionary theory). Similarly, the discussion above shows in outline that it is possible to understand the existence of a remarkably coherent body of natural knowledge by referring to general features of the environment, to the importance of realism in the life strategy of homo sapiens, to the ability of native human intelligence to achieve realism in important respects, to the psychological linkage between the biologically useful functions of intelligence and its capacity to acquire theoretical knowledge of no direct biological value, and finally to the remarkable potentiality of human intelligence for undergoing systematic refinement and increasing its sensitivity to the truth by the development of the scientific method.

Mutatis mutandis, the patterns of explanation of specific anatomical features of a species and of the existence of natural knowledge are very similar.

C. Scientific methodology is part of "the way up," that is, part of the enterprise of showing that and how human beings can obtain an understanding of the universe beyond themselves. It has two distinct tasks: (i) to formulate the content of scientific method—to state as precisely as possible whatever can be formulated as rules and also to determine what part, if any, of scientific method cannot be formulated in this way, but must be embedded in intuition or habit; and (ii) to investigate the rationale or justification for scientific method and, in particular, for the part of scientific method which can be characterized as scientific inference. Thus, the Bayesian analysis of the hypothetico-deductive method in section II and the presentation of the tempered personalist formulation of scientific inference in section III belong to the first of these tasks, whereas the argument in section IV that the tempered personalist formulation of scientific inference has a kind of sensitivity to the truth belongs to the second.

Studying the acquisition of knowledge from the standpoint of "the way down," as in section V.B, has important consequences for "the way up"—primarily for the second task of methodology, though indirectly for the first task as well. Broadly speaking, one finds that there are other factors besides the bare logical structure of scientific inference which contribute to its sensitivity to the truth. There are, for example, certain characteristics of the tempered personalist formulation of scientific inference which can be justified on a priori grounds only as desperate expedients, but which a posteriori can be seen to be well suited to investigation of the actual world. Thus, the "local" use of the machinery of probability theory was prescribed in section III because subjective judgment becomes confused and indefinite without localization. But the second of the three circumstances considered in section V.B. indicates that localization is appropriate to investigation of nature and will rarely cause irreversible errors.[46] Similarly, a body of common-sense knowledge must be utilized in order to permit consequences to be drawn from scientific hypotheses which can be confronted sharply with experience; and the rough reliability of our prescientific pictures of the world, which was the first of the three circumstances, ensures that this process will not usually be a source of errors. In short, these two circumstances sharpen the accuracy of the critical function performed by scientific inference. The prescription of assigning nonnegligible prior probabilities to all seriously proposed hypotheses is justifiable on a priori grounds as a necessary condition for accepting the truth about some phenomenon if it is ever proposed. But in view of the third circumstance discussed in section V.B, that human beings possess remarkable abductive powers, the hypotheses seriously proposed by men immersed in their subject are statistically much more likely to contain a good approximation to the truth than an equal number of hypotheses selected from among all possibilities by a random process.

The considerations of section V.B also show that the role of subjective judgment in the tempered personalist formulation of scientific inference is more than merely a desperate expedient. According to sections II and III, the logical structure of scientific inference is quite meager, and, therefore, a major role must be assigned at various points in the inferential process to subjective judgment, in order to ensure that definite conclusions can result from inquiry. It was also argued that the a priori principles of scientific inference exercise control over subjective judgment, so that large classes of errors can reliably be avoided and consensus can be eventually reached by investigators who differ greatly in their evaluation of prior probabilities. However, subjective judgment is not merely something to be kept in check, and its role in scientific inference is sanctioned (with reservations) by characteristics of human nature and the environment. How well our intelligence is adapted to the natural environment is exhibited by the rough reliability of our prescientific picture of the world, by our fairly good native sense of the relevance or irrelevance of various factors to phenomena of interest, and by the abductive powers of men who are immersed in a subject matter. Since intelligence is at work in our subjective judgments, even when its *modus operandi* cannot be adequately described, these judgments can enhance the sensitivity of scientific inference.[47] Incidentally, this helps explain why the achievement of scientific knowledge does not depend upon sophistication regarding scientific method. Locke's famous jibe, that "God has not been so sparing to men, to make them barely two-legged creatures, and left it to Aristotle to make them rational" (*Essay*, bk. IV, chap. xvii, sec. 4) applies to inductive inference as well as to deductive.

It must be acknowledged that the a posteriori justification of scientific inference is limited. There are no grounds for a conditional apodictic claim to the effect that if the propositions about human beings and the environment which were asserted in section V.B are correct, then eventually nonerroneous conclusions would be reached (or asymptotically approached) by means of scientific inference. The occurrence of large random sampling errors, regardless of the size of the sample, is compatible not only with logic but also with the known laws of nature. Furthermore, our knowledge of circumstances generally favorable to scientific inference is compatible with the occurrence of systematic errors which will remain undetected for an indefinitely long time in the investigation of special phenomena. However, the program of section V.B gives strong grounds for optimism that the persistence of such systematic errors, in spite of careful experimentation and analysis, is improbable. In particular, the examination of the acquisition of scientific knowledge from the standpoint of "the way down" provides evidence that there is no undetectable systematic error introduced by the uneliminable presence of the human subject in scientific inference.

D. The foregoing considerations in defense of subjective judgment should not disguise the fact that subjective judgment is often distorted by proprietary interest in a theory, by prejudice, and by obtuseness; and, therefore, guidelines which provide an alternative to or a check upon subjective judgment are desirable. If the vein of a priori principles is exhausted, then these guidelines can only be obtained from knowledge of the actual world. The investigation of the acquisition of scientific knowledge from the standpoint of "the way down" provides a justification for a posteriori supplementation of scientific method. The manifest cumulative character of scientific knowledge has been seen to be almost certainly not the result of systematic error, and, therefore, the use of a tentative body of scientific knowledge to suggest guidelines in further investigations is unlikely to have the effect of propagating or compounding a systematic error. In this way the program of section V.B indirectly contributes to the first task of scientific methodology, which is to formulate the content of scientific method.

I shall give several examples of guidelines in deciding whether hypotheses are seriously proposed and in evaluating prior probabilities. A related problem, which I shall not discuss but which deserves attention, is the derivation of guidelines concerning the acceptance of hypotheses into the tentative body of scientific knowledge.

Two examples will show how guidelines for classifying hypotheses as seriously proposed or not are obtained from antecedent scientific knowledge. The first is the reevaluation of entire classes of hypotheses as a result of the triumph of classical mechanism. The conceptual clarification of dynamics and the accurate explanation of many celestial and terrestrial motions led physical scientists generally to discount hypotheses formulated qualitatively and in terms of sensuous properties of bodies and to pay serious attention to hypotheses formulated mathematically and in terms of nonsensuous properties. Less dramatic, but of comparable importance, was the discrediting of magic by the gradual extension of critical thinking into emotionally sensitive areas. A consequence was the devaluation of certain classes of hypotheses which are characteristic of magic, especially those motivated by surface analogies (for instance, that rain can be induced by bloodletting and that the waxing moon is favorable to conception).[48] In addition to such large-scale reevaluations of classes of hypotheses, there is usually an understanding among investigators, based upon their knowledge of the recent history of the subject, that certain types of hypotheses are likely to be fertile and others to be sterile. Consequently, specialized knowledge provides specialized guidelines for deciding whether a hypothesis is seriously proposed.

It may be objected that guidelines based upon a tentative body of scientific knowledge are likely to be excessively conservative in just the way that the tempering condition is intended to prevent. This possibility need not be a danger, however, if good sense is used. It is reasonable for an expert to refuse to take seriously a proposal in conflict with a theory which is well confirmed, especially when the proposer evidently is ignorant of it or does not fully understand it. Even if the proposal itself has never been directly tested, the expert may see that because of the established theory the testing would be routine and hence not worth his attention unless it would unexpectedly lead to anomalous results. On the other hand, it is important to take seriously a challenge to an established theory which is not careless about the evidence whereby the theory was established. Thus, conceptual analysis may show that a hypothesis which challenges one established theory is, nevertheless, compatible, contrary to appearances, with the deeper parts of the accepted body of knowledge; or the proposal of the hypothesis may be accompanied by an examination of weak points in the evidence upon which the established theories are based; or the approximate agreement within a limited domain between the predictions of established theories and those of the new hypothesis may be exhibited, thus accounting from a new point of view for the success of the former. The essential point is that without setting limits to the content of subsequent conjectures, the tentative body of scientific knowledge generally serves as an indispensable base of reference for them.

Within the limitations of the axioms of probability and the tempering condition, the evaluation of tempered personalist probabilities is left to

subjective judgment. However, at least two guidelines for these evaluations are commonly recognized: (1) the principle of indifference and
(2) the principle that the simpler of two hypotheses has higher prior
probability than the less simple. Much of the discussion of these two
principles seems to be based upon the assumption that if they are valid,
then the grounds for their validity must be a priori; but my contention is
that their valid content is largely derived from antecedent knowledge
about the world.

1. Some comments on the principle of indifference are found in note
38. It is pointed out that a formulation of the principle which is weak
enough to be a priori—namely, the equiprobability of propositions asserting various outcomes, once all individuating differences among the
outcomes are suppressed—is superfluous in scientific inference, whereas
more stringent forms of the principle depend upon the dismissal as
irrelevant of classes of differences. Antecedent knowledge is used in
several different ways in making judgments of irrelevance.

In evaluating "geometrical" probabilities, the differences among equal
volumes in a parameter space with a properly defined measure are
judged to be irrelevant, but antecedent knowledge is required to determine whether a measure is proper. The well-known paradox of Bertrand
(1889: p. 5; Keynes 1921: pp. 41–42), which results from applying the
principle of indifference to alternative measures on a certain continuum
(namely, the class of chords of a given circle), was intended to show
that there are no unambiguous objective prior probabilities in such cases.
However, it may also happen that in these cases personal probability
evaluations are confused or vacillating. If so, then the body of assumptions *a* must be augmented (thus essentially changing the character of
the problem) in order to arrive at a situation in which personal probability evaluations become unambiguous, which in turn makes possible the
evaluation of tempered personalist probabilities. The augmentation of *a*
does not necessarily require that statistical data be gathered. In fact, the
standard resolution of the Bertrand paradox (for example, Borel 1965:
pp. 87–88) consists in specifying the physical means by which an element of the continuum of interest is selected. The probability distribution over this continuum is then determined—via the axiom of complete
additivity and other probability axioms, together with physical principles
which are contained in the total body of information *a&b*—by a distribution over a different continuum, namely, the points within the given
circle. From the standpoint of personalism the problem is thereby solved
if the personal probability distribution over the latter continuum is

unambiguous. However, the considerations at the beginning of section V.C remain relevant. Our adaptation to the environment is reflected in our subjective probability evaluations and partly explains why practical decisions based upon these evaluations are fairly successful. In particular, a uniform personal probability distribution over the points within a circle results from the assumption of a Euclidean metric and the associated measures of area and volume, which are deeply embedded in the background information of every normal human being.

In more complicated problems it is often impossible to obtain a reduction to an unambiguous personal probability distribution without resorting to sophisticated antecedent knowledge. Notably, in classical statistical mechanics one wishes to find probability distributions of dynamical variables characteristic of systems of particles moving under given constraints and interactions. These distributions can always be determined in principle (that is, neglecting formidable mathematical difficulties) by assuming that equal volumes in the phase space of a closed system have equal prior probabilities (Tolman 1938: pp. 59–61). One of the deep problems of statistical physics is to derive this assumption from weaker ones (cf. Farquhar 1964), but at present its primary justification must still be said to be the great explanatory power of the statistical physical theories based upon it. The justification of the assumption is, therefore, typical of scientific inference concerning general theories. The important point for the considerations of the present section, however, is that the attempt to evaluate the personal probability of a physical system to be found in a specified region of phase space can be expected to result in *subjective confusion* unless something of the order of generality of this basic assumption of statistical mechanics has been confirmed, or unless at least clear alternative assumptions of this order of generality are under consideration and are all assigned personal probabilities. In short, it is naïve to expect a man to be able to form unambiguous subjective judgments (and a fortiori sensible ones) about such matters unless he entertains sophisticated statistical conjectures, which in turn presuppose an extensive knowledge of physical theory. It was argued in section II that probability evaluations are "local," and in the present context locality requires that the total body of information and assumptions $a\&b$ contain a certain amount of physical theory. That explicit antecedent knowledge is dispensable in making personal probability evaluations regarding more mundane questions indicates the richness of the background information of ordinary men concerning such questions.

In problems concerning gambling devices, which constitute the locus classicus for applications of the principle of indifference, the role of antecedent knowledge is entirely different. Most adults in our culture know, for example, that there is no correlation between the color of a playing card and its susceptibility to being drawn from a deck, unless the circumstances are abnormal (for example, the drawer is cheating). Hence, if information about the usual behavior of playing cards is contained in $a\&b$, if e is the proposition that a single drawing will occur from a standard deck of cards all turned face down, and if $r(j)$, $b(j)$, $d(j)$ are respectively the propositions that the jth card is red, is black, and is the one to be drawn, then

$$
\begin{aligned}
(1) \qquad & P_{X,b}[d(j)|a\&e\&r(j)] \\
& = P_{X,b}[d(j)|a\&e\&b(j)] \\
& = P_{X,b}[d(j)|a\&e].
\end{aligned}
$$

But how important is the antecedent information about playing cards for this judgment? Suppose, for example, that $a\&b$ is replaced by $a'\&b'$, in which no information about playing cards is contained. Unless X has some unusual beliefs about the efficacy of specific colors, his probability evaluations can be expected to satisfy equation ($1'$) which results from (1) by replacing a and b by a' and b'. Or suppose that c is the proposition that the drawer has the facility and the intention to cheat, but it contains no information as to which color is advantageous to the drawer. Then past experience about the behavior of playing cards is of no obvious use to X, and, nevertheless, one can expect his judgment to be

$$
\begin{aligned}
(2) \qquad & P_{X,b}[d(j)|a\&e\&c\&r(j)] \\
& = P_{X,b}[d(j)|a\&e\&c\&b(j)] \\
& = P_{X,b}[d(j)|a\&e\&c].
\end{aligned}
$$

The parallels among (1), ($1'$), and (2) indicate that antecedent knowledge is not needed for supplementing a subjective judgment concerning a *single* drawing. Where antecedent information about playing cards is important is in leading X to assign a low prior probability (that is, prior relative to a, in the local sense of section II) to the proposition that a correlating mechanism exists; and this assignment has the consequence that the judgment of irrelevance of black or red when no previous drawings have been made dominates the evaluation of the posterior probability even when the results of a small number of drawings are known. In other words, the effect of antecedent information in this case is to increase the applicability of a subjective judgment of irrelevance. This conclusion can be greatly generalized, since we have a large but diffuse body of knowledge that, in stochastic situations in which sub-classes of events are differentiated from one another in physically minor ways, the frequencies of the various subclasses are approximately equal. Additional knowledge is required to judge reliably whether a character-istic is indeed minor, and, furthermore, the possibility always remains that a special mechanism favors the occurrence of a subclass which is distinguished by a prima facie negligible characteristic. In spite of these qualifications, however, our diffuse knowledge of such stochastic situa-tions strongly reenforces the methodological prescription of section IV against extreme suspiciousness of the presence of unknown correlations.

2. After surveying the recent literature on simplicity, Kyburg re-marked that, in spite of much effort, "the whole discussion of simplicity has been curiously inconclusive" (1964: pp. 19–20). I suspect that one reason for this situation is that methodologists have tended to ascribe too large a role to simplicity, namely, the role of being the primary criterion for comparing all hypotheses compatible with the experimental data. This role is too great a burden for the concept of simplicity to bear, and, indeed, one of the advantages of the tempered personalist formulation of scientific inference is that it uses a different primary criterion for compar-ing hypotheses, namely, that of being or not being seriously proposed. Simplicity considerations are undoubtedly important in scientific investi-gations, but usually, if not always, within contexts that are circumscribed by assumptions or by antecedent knowledge. The circumscription often has the effect of suggesting, either unequivocally or with the possibility of small variations, the ordering of a favored family of hypotheses. Thus, if the motion of a planet is supposed to be a superposition of circular motions, then it is reasonable to identify an ordering according to decreasing simplicity with an ordering according to the number of epicycles. And if a theory of interacting fields is assumed to be Lorentz invariant, then the orders of the tensors entering into the interaction term are plausible indices of the complexity of the theory. Without the guide-lines suggested by antecedent knowledge or by stringent assumptions, so many alternatives suggest themselves as candidates for "ordering accord-ing to simplicity" (as pointed out in the discussion of Kemeny in section IV.B) that the concept becomes methodologically valueless. Moreover, proposed orderings which purport to be a priori usually derive whatever intuitive appeal they possess from a tentative picture of the world. A case in point is the simplicity postulate proposed by Wrinch and Jeffreys (1921; also Jeffreys 1931: p. 45), but later disavowed by Jeffreys, accord-

ing to which the highest prior probabilities are assigned to those laws which can be cast in the form of differential equations of low order and degree. This assignment is reasonable only if the basic natural regularities govern rates of change, as in Newtonian physics, rather than configurations, as in Kepler's astronomy, and only if continuity is a fundamental feature of nature. Far too much knowledge of physics is implicit in Wrinch and Jeffreys's proposal to admit the claim that "it would represent the initial knowledge of a perfect reasoner arriving in the world with no theoretical knowledge whatever" (Jeffreys 1961: p. 49).

Weyl's statement, "The required simplicity is not necessarily the obvious one, but we must let nature train us to recognize the true inner simplicity" (1949: p. 155) seems to me to approach the heart of the matter. It does justice to the history of science in a way that no formal theory of simplicity comes close to doing, and it expresses compactly the Platonic conceptions which I think are essential to an adequate naturalistic epistemology. However, one may be dubious about the methodological value of Weyl's statement; for if the secrets of nature must be known before simplicity can be understood, then it is hard to see how considerations of simplicity can sharpen a person's probability evaluations in the course of investigating nature. A partial answer is that evidence and analysis may indicate a general respect in which nature is simple without yielding the detailed laws. Nature has, in fact, "trained" us to recognize several kinds of inner simplicity: geometrical invariance principles such as Lorentz invariance, simplicity of the nature and behavior of elementary particles, and simplicity arising from the statistics of large numbers of particles. Nevertheless, this extensive body of knowledge leaves many physical questions open. For example, exactly which nongeometrical invariance principles govern the elementary particles? At the same time, it often suggests an obvious ordering of a family of hypotheses, for example, an ordering according to the magnitude of n of the hypotheses that the appropriate group for describing the nongeometrical invariances of strongly interacting particles is $SU(n)$ (see Carruthers 1966). There is another methodological consequence of Weyl's statement, which concerns heuristics as much as confirmation. Suppose that a hypothesis of great generality is highly confirmed, even though it is not manifestly simpler than its disconfirmed rivals. Then a point of view should be sought from which the simplicity of the successful hypothesis is evident. Such points of view have indeed been found upon analysis and have provided some of the most striking advances in scientific understanding. An outstanding example was the exhibition of the simplicity of Maxwell's electrodynamics within the theory of special relativity. The fertility of deliberate quests for points of view from which established principles appear simple is evidence that Weyl is pointing to an objective feature of the world and is not merely using poetic diction to describe our habituation to novel theories.

On a more mundane level Goodman makes an important and sensible contribution to the analysis of simplicity:

> Formulation of general standards for comparing the simplicity of hypotheses is a difficult and neglected task. Here brevity is no reliable test; for since we can always, by a calculated selection of vocabulary, translate any hypothesis into one of minimal length, the simplicity of the vocabulary must also be appraised. I am inclined to think that the standards of simplicity for hypotheses derive from our classificatory habits as disclosed in our language, and that the relative entrenchment of predicates underlies our judgment of relative simplicity. (1961: p. 151; see also 1955: p. 119)

The argument which he gives for rejecting brevity as a reliable test for simplicity can evidently also be directed against other formal criteria, such as number of parameters or number of quantifiers, unless these criteria are supplemented by appraisal of the basic vocabulary. However, I think that Goodman's analysis can be deepened by considering the concept of entrenchment from a naturalistic standpoint. Ordinary language habits reflect an extensive common-sense knowledge of the world, which is roughly reliable in that it permits normal human beings to survive under normal circumstances (cf. Jeffrey 1965: pp. 176–77). Furthermore, as pointed out in section V.B, studies in biology and genetic psychology have provided at least some insight into the basis of this reliability, for example, the role of innate ideas which presumably are the product of natural selection and the fact that the learning processes of children are critical in crude but important ways. When a man judges the hypothesis that all emeralds are green to be simpler than the hypothesis that all emeralds are grue (where "grue" is a predicate which "applies to all things examined before t just in case they are green but to other things just in case they are blue" [Goodman 1955: p. 74]) and assigns far higher prior probability to the former in spite of the fact that from a logical point of view the "green"-"blue" and the "grue"-"bleen" vocabularies are coordinate (ibid.: p. 79), he is relying upon his common-sense picture of the world. He is not hopelessly conservative in so doing, for there are anomalies within common-sense pictures of the world which have had the social consequence of stimulating the scientific theorizing whereby common sense is transcended. However, most philosophers will agree that the route whereby common sense is transcended—and Weyl's view of nature is achieved—does not pass via gruelike concepts, even though further analysis would be required to demonstrate this decisively.

E. I shall now briefly summarize the theory of scientific inference which has been proposed and make some comments about its rather complicated structure, in particular about the circularity which arises because of a posteriori contributions to methodology.

The first part of the theory is a formulation of scientific inference in terms of the tempered personalist concept of probability. The formulation has a meager logical structure, consisting only of the axioms of probability and the tempering condition (with the possibility of some supplementation). An a priori methodological justification can be given for this logical structure. The axioms of probability are necessary conditions for orderly thinking about propositions with unknown truth values, whereas the tempering condition is only a way of prescribing open-mindedness within a probabilistic scheme of inference; and open-mindedness is a necessary condition for the acceptance of true hypotheses in case they should ever be proposed. Despite these a priori considerations, the logical structure of this formulation is inseparable from the context of actual inferential processes—in contrast to the situation in deductive logic. This is shown by the preferred status of hypotheses which are actually proposed and by the crucial role which had to be assigned to subjective judgment in order to ensure that a person can use the apparatus of probability theory in arriving at a view of the world. Although the logical structure is meager, it suffices to provide systematic (but not infallible) protection against major classes of possible errors. In this way a sensitivity to the truth can be claimed for scientific inference on a priori grounds, though the exact content of this sensitivity was not determined. For example, no attempt was made to survey models of the universe with the intention of making a dichotomy between those in which errors of acceptance and rejection of hypotheses could be eliminated in the long run and those in which they could not; and in view of the role of subjective judgment in scientific inference, it is hard to see how this would be a well-posed problem for statisticians, even if the set of models were accurately described.

The second part of the theory complements the first part by investigating the context of actual inferential processes which the tempered personalist formulation refers to but does not characterize. The relevant scientific knowledge that we now possess indicates that this context is favorable to scientific inference: that even without special cultivation human intelligence is sufficiently well adapted to the environment to be quite apt at discriminating, on the basis of experience, between those hypotheses which are good approximations to the truth and those which are not, and that the informal methods of confirmation used by scientists (which I claim are made explicit in the tempered personalist formulation of scientific inference) improve this native adaptation and extend it beyond the area of direct biological utility. In this way an examination of the acquisition of scientific knowledge from the standpoint of "the way down" greatly strengthens the claim that scientific inference has a sensitivity to the truth and thereby contributes to "the way up." Furthermore, this general knowledge of the favorableness of nature to scientific inquiry provides a rationale for relying upon special bodies of scientific knowledge to augment the scientific method with a posteriori principles, for example, with guidelines for the evaluation of probabilities.

The epistemological status of the propositions discussed in section V.B, to the effect that nature is favorable to scientific inference, is complex. In a sense they are undoubtedly presupposed in the operations of native intelligence. Any nontrivial process of induction makes use of a background of common-sense knowledge and, therefore, implicitly assumes the rough reliability of that knowledge; any actual inquiry localizes a problem and, therefore, assumes the propriety of localization; and any actual inquiry (as contrasted with bloodlessly going through the motions of inquiry) is optimistic about the ability of human beings to propose approximately true hypotheses. However, the explicit formulation of these propositions requires reflection upon inquiry; and their confirmation, as is evident from the sketches of section V.B, requires extensive physical, biological, and psychological investigations. Furthermore, such investigations are essential for eliminating the vagueness of the naïve formulations of these propositions: to determine *how* reliable common-sense knowledge is, *how* amenable nature is to the disentanglement of problems, and *how* good the abductive powers of human beings are. Scientific inference is, therefore, required to clarify and confirm these propositions; on the other hand, the truth of these propositions is presupposed by a strong justification of scientific inference, that is, a justification which goes beyond the counsel of desperation which a priori considerations provide. This is the circularity which arises in the theory of scientific inference that has been proposed.[49]

I claim, however, that the circularity is nonvicious in the following sense: *the theory as a whole is open to critical evaluation in the light of experience, for the reciprocal support of a methodology and a scientific world picture does not render it impregnable to criticism.*[50] This open-ness is partly due to the open-mindedness which is incorporated into the logical structure of the formulation of scientific inference. The negation of any of the propositions of section V.B could be seriously proposed and then, in accordance with the tempering condition, would have to be assigned a sufficiently high prior probability to permit its survival under experimental scrutiny. It is logically possible then that experimental evidence would lead to a high posterior probability of the negation of one of the propositions upon which the a posteriori contributions to methodology were based. Whether it is psychologically possible is a different matter. Experience which would lead to the rejection of the proposition that common sense is roughly reliable would be intolerable to creatures like us and would almost certainly induce disillusionment with critical thinking at the least, and more likely insanity or death. Similar consequences could be expected from experience which leads to the thorough rejection, even in matters which are vital to our biological functioning, of the propositions that problems can be disentangled and that human beings have good abductive powers. However, the judg-ments that these would be the probable outcomes are a posteriori, derived from our knowledge of human nature, and, therefore, do not contradict the logical openness of the theory of scientific inference. Furthermore, it does seem to be psychologically possible for creatures like ourselves to undergo experience which disconfirms the propositions that in matters remote from direct biological concern, problems can be disentangled and human beings have good abductive powers.

There is also an openness implicit in the derivation of guidelines from specific bodies of scientific knowledge. The reciprocal support of meth-odology and the scientific picture of the world is compatible with a gradual correction of both. This dialectic would undoubtedly terminate in confusion (or worse, as indicated in the preceding paragraph) if certain basic propositions assumed in the early stages of the dialectic were radically false. But correction is possible and fruitful precisely because these basic propositions are true only with qualifications: for example, common-sense knowledge is not completely reliable, and the loci of its unreliability become evident with the growth of scientific knowledge. It is even possible that the dialectic would lead to the adoption of a scientific method radically different from that of contempo-rary scientists. Suppose, for example, that psychological evidence would come to support Aristotle's contention that the properly prepared intel-lect is able to grasp the universal form of any species (*On the Soul*, bk.

III, chapts. 4–8); then Aristotle's method of obtaining major premises by intuitive induction (*Posterior Analytics*, bk. II, chap. 19) would be justified, and the current scientific method would become a stepping-stone to a more direct one. One of the virtues claimed in section IV for the tempered personalist formulation of scientific inference is an ability to subsume methodological devices which experience or analysis indicate to be powerful. A corollary is the possibility of its being dominated by that which it subsumes, as the Mongolians by the conquered Chinese.

The possibility remains that the dialectical interplay of science and methodology will come to an end. In this unlikely event it is to be hoped that the reason will be the attainment of essentially complete knowledge of the place of human beings in nature. There would then be an exact meshing of methodology and of the scientific picture of the world, not because embarrassing questions will have been evaded and critical experiments left unperformed, but because the perspective will be complete and all pieces will fit into place. Conceivably, however, such a meshing could occur in spite of deep-lying errors in the scientific picture, which failed to be exposed in spite of strenuous experimental probing. There would, so to speak, be a probabilistic version of the Cartesian deceiving demon. Despite this conceivable eventuality, a full commitment to a methodology and a scientific picture which mesh in this way would be rational, for no stronger justification is compatible with the status of human beings in the universe. After all, in spite of the advances which have been made upon Hume's treatment of inductive inference, his basic lesson remains valid: that the conditions of human existence which force us to resort to inductive procedures also impose limitations upon our ability to justify them.

NOTES

1. To my knowledge the first use of this term in an epistemological sense was Kant's "Copernican Revolution," but in view of his doctrine that "the order and regularity in the appearances, which we entitle *nature,* we ourselves introduce" (Critique A 125), his epistemology is anti-Copernican in my sense of the term. Two contemporary philosophers who use "Copernican" as I do are Feigl (1950: pp. 40–41) and Smart (1963: p. 151), but I surmise that nineteenth-century precedents can be found. It is reasonable to understand "naturalistic" as having a wider extension than "Copernican" and applying to any epistemological theory which is based upon a study of man's place in nature. Thus Aristotle's epistemology is naturalistic, though I would not classify it as Copernican. In characterizing my own point of view, I shall use both descriptions: "Naturalistic" to emphasize

the continuity of philosophical and scientific investigations and "Copernican" to emphasize the insignificance of human beings in the universe.

2. The possibility of nonvicious circularity in the structure of knowledge was first suggested to me by Weiss (1938: chap. 1).

3. But not *infinitely* long-range. See sec. IV.

4. To paraphrase Goethe:
>*Grau ist alle Theorie,*
>*Doch grauer Methodologie.*

5. Structure is emphasized here partly because in sec. II I reject the Platonic conception of probability proposed by Keynes, according to which probabilities depend only upon certain internal relations among propositions and concepts.

6. My recommendation of the use of the hypothetico-deductive method in philosophical investigations (1965: p. 261) now seems to me to be an oversimplification. An excellent example of the kind of detailed study which is needed is Stein's examination of classical physical geometry (1967).

7. A few with whose work I am acquainted are Fodor (1966), Harris (1965), Hirst (1959), Mandelbaum (1965), Piaget (1950), Popper (1962), Quine (1957), Sellars (1963), and Smart (1963). A near contemporary whose work is too little noticed by later naturalistic epistemologists is Dewey (1929 and elsewhere).

8. For a powerful mathematical development of probability theory it is necessary to replace axiom (iii) by the strong axiom of complete addivity: $\Sigma P(h_i|e) = P(H|e)$, where e entails $\sim (h_i \& h_j)$, for $i \neq j$, and H is the infinite disjunction of the h_i.

9. A judicious evaluation of Popper's position from a Bayesian point of view is made by Salmon (1966: pp. 252–55).

10. See, for example, Hacking's remarks on personalism, which is a version of Bayesian probability theory (1965: p. 208).

11. Passmore (1957: p. 348) perceptively notes Keynes's debt to Moore, who interprets ethical concepts in this manner.

12. Even the very limited Platonism in probability theory which I once accepted (1955: pp. 11, 27–28) no longer seems tenable to me.

13. This is DeFinetti's definition of "coherence." I have proposed a slightly stronger definition (1955: p. 9).

14. This argument is elaborated by Polanyi (1958: chap. 2), whose personalist epistemology is in the spirit of the personalist probability theorists but is of much wider scope.

15. In Carnap's notation "*P*" represents the explicandum of which a *c*-function is an explicatum. It should be noted that Carnap regards the explanation of $P(h,e)$ in terms of a betting quotient as accurate only when the betting stakes are small compared with *X*'s total fortune, and he gives another explanation, in terms similar to but not exactly the same as betting (namely, estimating the subjective utility of a benefit bestowed in case *h* is true), which he regards as more accurate (1963: p. 967).

16. I am indebted to Prof. Joseph Agassi for a discussion on this point.

17. When conditions (i) and (ii) hold, the true theory is related to *h* by a kind of "correspondence principle."

18. It is often convenient to exclude from *i* all sharp data bearing on the hypotheses of interest and to include these data as part of the evidence used to compute the posterior probabilities. Idealizing the structure of the investigation in this way permits a more uniform application of Jeffreys's prescription of openmindedness: the prior probabilities of all the seriously proposed hypotheses h_i can then be evaluated without considering sharp data, whereas the latter are taken into ac-

count in likelihood calculations, on which there is generally much more inter-subjective agreement than on prior probabilities.

19. This expression was used by L. Savage in a discussion.

20. Among the recognized senses of the verb "to temper" which suggest my technical usage are the following: "to restrain within due limits, or within the bounds of moderation," "to bring (steel) to a suitable degree of hardness and elasticity or resiliency . . . ," "to bring into harmony, attune" (*The Oxford Universal Dictionary*, 3rd edition).

21. This suggests that the theory of probability which I am proposing as a modification of personalism might properly be called "socialism," had this term not been preempted.

22. In effect this question is raised by Putnam in the following passage: "Consider a total betting system which includes the rule: if it is ever shown that a hypothesis S is not included in the simplicity ordering corresponding to the betting system at time *t*, where *t* is the time in question, then modify the betting system so as to 'insert' the hypothesis S at a place *n* corresponding to one's intuitive judgment of the 'complexity' of the hypothesis S. This rule violates two principles imposed by Carnap. First of all, it violates the rule that if one changes one's degree of confirmation in one's life, then this should be wholly accounted for by the change in E, that is the underlying *c*-function itself must not be changed. Secondly, it can easily be shown that even if one's bets at any one time are coherent, one's total betting strategy through time will not be coherent. But there is no doubt that this is a good rule nonetheless . . ." (1963a: p. 10).

23. The reasoning of this paragraph can be paralleled to show that the coherence of sets of betting quotients must also be considered as a "local" concept, applicable only when a definite set of outcomes is envisaged and bet upon. In order to use the condition of coherence for the purpose of relating a set of subjectively acceptable betting quotients when $h_1, \ldots, h_n$ are the only outcomes envisaged to a set when $h_1, \ldots, h_{n+1}$ are envisaged, it would be necessary to evaluate a betting quotient for bets upon the proposition *that the outcome* h_{n+1} *will be added to those already envisaged*—and this appears to be a psychological impossibility.

24. Although a posteriori considerations are out of place in the first stage of my treatment of scientific inference, I shall anticipate the second stage and assert a proposition which seems to be overwhelmingly supported by the history of science: that the sensitivity of scientific method to the truth is not diminished by a reluctance to classify a hypothesis as seriously proposed unless the proposer has taken pains to relate it clearly to the currently accepted body of knowledge. On the whole Burke's maxim is apropos: "People will not look forward to posterity, who never look back to their ancestors" (*Reflections on the Revolution in France*).

25. Feyerabend (1963) recommends a "principle of proliferation" as an antidote to intellectual sterility. But in scientific method, as in the civil law, there should be a "principle of paternity," according to which the man who engenders a hypothesis has some responsibility for supporting it; and, as in civil affairs, this principle would to some extent curb the principle of proliferation.

26. Because $\int P_{x,b}[h(a_1, \ldots, a_k)|a]da_1 \ldots da_k$ must be $\leqq 1$, while some of the a_i may have infinite range, a closer parallel to the statement of the tempering condition in the discrete case—such as setting a lower bound to the integral over a fixed volume in parameter space—is in general not possible.

27. The meaning of "view" is, of course, a matter of convention. An alternative to the meaning adopted in the text is "a comprehensive set of propositions about various aspects of the world." If this sense is preferred, then a person who is in doubt

about many things does not have a single view of the world, but rather he entertains a large number of views with varying weights. Were the two different senses of "view" spelled out with ideal care and detail (perhaps in terms of Leibnizian possible worlds), then one might be able to express the state of a man's theoretical knowledge equivalently in two different ways, and either would serve equally well in a rational reconstruction of scientific inference. However, the equivalence of the two modes of expression would be difficult to exhibit not only for practical reasons but in principle because of the obscurities in our concepts. Both senses of "view" are idealizations in that they neglect the presence of conceptual obscurities which are unavoidable as long as our knowledge is incomplete. When the analysis of the content of a person's theoretical knowledge is less than ideal, there is a strong reason for preferring the sense of "view" adopted in the text: subjective uncertainties are for the most part felt "locally" concerning relatively small sets of propositions. Unless the contents of views in the second sense were specified with great precision, the assignation of weights to them would not permit one to infer where the local uncertainties lay.

28. A good discussion of the acceptance of hypotheses, which stresses different points from the ones made here, is found in Jeffreys's section entitled "Deduction as an Approximation" (1961: pp. 365–68). Kyburg (1964: pp. 29–30) surveys some of the extensive literature on the problem of acceptance. It should be noted, however, that most of this literature is concerned with induction in a practical context, and as Carnap points out (1963: p. 973), the problem of acceptance in the context of theory is quite different.

29. See Scheibe and Sarbin (1965) for a discussion of the momentum of a superstition in the absence of a countervailing explanation.

30. There is much to be salvaged from the Hegelian thesis that a dialectical process is at work in the development of intellectual disciplines, but painstaking research in the history of science and philosophy is required to determine what is salvageable. (See Feyerabend's essay in this volume.)

31. Salmon (1961, 1963) attempts to show that the straight rule is the only one of this infinite set of methods which also satisfies the reasonable "principle of linguistic invariance"; but Hacking (1965) has proved that other members of this set also satisfy the principle.

32. This is a slight modification of condition (3) in Kemeny (1953: p. 403), made for the purpose of avoiding the unessential explanation of a technical term.

33. Attributed to Carnap by Salmon (1965).

34. A little progress along these lines is possible, however, by noting that good statistical discrimination between two hypotheses requires more data the "closer" the hypotheses are to each other. Consequently, it is desirable to choose orderings which keep very "close" hypotheses as far apart as possible. This desideratum ought to be formulable precisely so as to yield a *partial ordering* of the class of orderings.

35. One might attempt to answer this objection by taking the class of considered hypotheses to consist of all noncontradictory hypotheses formulable in a given language. But, as Putnam points out (1963: p. 775), in a nontrivial language this class cannot be effectively enumerated.

36. See also Putnam (1963: pp. 772, 775), who suggests that *corrigibility* and *tenacity* are essential characteristics of a good inductive method. He differs from Kemeny, however, by proposing inductive methods in which "the acceptance of a hypothesis depends on which hypotheses are actually proposed, and also on the *order* in which they are proposed" (ibid.: pp. 771, 775). The preferred status of actually proposed hypotheses is a crucial feature which both my treatment of

scientific inference (derivative from Jeffreys) and Putnam's have in common. However, my formulation of scientific inference is Bayesian, whereas his is not. One consequence of this difference is that acceptance plays a less central role in my treatment than in his, for a Bayesian recognizes the possibility that a delimited investigation may terminate with approximately equal posterior probabilities assigned to rival hypotheses.

37. Although the problems of inductive inference in the context of practice are not within the scope of this paper, I think it is worth remarking that a theoretical view of the world which consists of varying degrees of commitment to rival hypotheses lends itself better to practical application than a view which always tentatively selects one hypothesis as true.

38. The judgments of indifference that are made once all known and presumed differences among the possible outcomes are dismissed as irrelevant do not, as I see the matter, need to be based upon a "principle of indifference." The judgments $P_{x,b}(e_m|h_i \& a) = P_{x,b}(e_{m'}|h_i \& a)$ are immediate results of dismissing the differences between e_m and $e_{m'}$ as irrelevant, without the mediation of such a principle. Nor do I see what further justification can be given for this judgment other than the two grounds for dismissal of differences which were discussed above—namely, an investigation, with a negative conclusion, of the serious proposal that a certain difference is relevant, and the methodological argument against an insatiable suspiciousness of the presence of systematic errors. In other words, there seems to be no need for a principle of indifference which states the equiprobability of the propositions asserting various outcomes, once *all individuating differences* among these outcomes are suppressed; and, indeed, there are semantic difficulties even in *clearly formulating* a stringent principle of this kind. To be sure, there are some clearly formulated rules of equiprobability, notably Carnap's axioms of invariance. According to him, these axioms "represent the valid part of the principle of indifference, whose classical form, e.g., in the system of Laplace, was too general and too strong and was therefore correctly rejected by later authors" (1963: p. 975). These axioms are, in effect, systematic attempts to dismiss as irrelevant certain types of differences among propositions (though Carnap prefers to speak in terms of the sentences of a definite language rather than in terms of propositions). For example,

> A7. The values of $c(h,e)$ remain unchanged under any finite permutation of individuals.
>
> . . . .
>
> A8. The value of $c(h,e)$ remains unchanged under any permutation of the predicates of any family. (Ibid: p. 975)

It is, I believe, legitimate and important to systematize the dismissal of differences, but I can see no other kind of justification of these axioms of invariance than the two grounds mentioned above for dismissal of differences among cases. However, the process of dismissal of differences may not be amenable to the kind of generalization which Carnap is attempting. In particular, some of the criticisms which have been made against (A8) (e.g., Salmon 1961: p. 249) indicate that the irrelevance of differences among predicates of families must be investigated piecemeal, a family at a time.

39. The common sense of our own culture is, of course, very much affected by the results of science, as Dewey points out (1938: p. 75).

40. Even those philosophers who maintain that the observation reports of scientists could in principle be formulated in a phenomenalistic language usually recognize that a "thing" language is somehow much more convenient (for example, Carnap

1950: pp. 23–24). However, I think that a thorough psychological study of the ontogenesis of common-sense knowledge will (and to a large extent already does) show the impossibility in principle of a purely phenomenalistic observation language, thus supporting the contentions of such antiphenomenalists as Sellars (1963: pp. 83–84 and elsewhere) and Strawson (1959).

41. Also, although there is no doubt that in some sense the pictures of the everyday world are reliable, the exact sense and the exact extent are far from clear. One finds great variation among philosophers of science in characterizations of the relationship between common-sense and scientific pictures of the world—with Eddington, for example, emphasizing their discrepancies (1928: pp. xi–xiii and elsewhere), and Bohr, at the opposite extreme, insisting upon the continuity of the concepts of theoretical physics with common-sense concepts and indeed with the "forms of perception" (1934: p. 1). Clarification of these matters is difficult, since it requires an explicit characterization of the content of the common core of representations of the everyday world, a comparison with relevant scientific knowledge, and some sensible criterion of what constitutes reliability. I shall not attempt this clarification here but shall only mention a few considerations. (1) For the purpose of biological adaptation a "reality principle" is needed only in coping with events on a human scale, not with microscopic or with cosmic events. (2) Some of the features of everyday pictures of the world which are most profoundly erroneous from a scientific point of view, such as the attribution of sensed colors to physical objects, are practically advantageous (cf. Whitehead 1929, pt. II: chap. 8). (3) The common core of representations of the world is surely vague in view of its incorporability into very different conceptual systems. (4) The effect of this vagueness is mitigated in practical activity by the "inarticulate intelligence" of our sensorimotor behavior (Polanyi 1958: pp. 71 ff.; Piaget 1952: passim).

42. Although this remark is vague in content and conjectural concerning cognitive operations, I think that it is epistemologically important. It implies, for example, a very different relationship between observational and theoretical terms from that set forth by Carnap (1956), Hempel (1965), and Nagel (1961: chap. 6). However, systematic applications of current psychological knowledge about concept formation (and probably also some investigations beyond the present frontiers of the subject) are necessary for a thorough understanding of the status of theoretical terms in science.

43. Hume anticipates an objection by considering the question: "Since all reasonings concerning facts or causes is derived from custom, it may be asked how it happens that men so much surpass animals in reasoning, and one man so much surpasses another" (*Inquiry:* sec. IX, n. 1). The answer which he gives is very fine within the limits of his method of investigation, but is symptomatic of the thoroughly nonexperimental character of his study of psychology.

44. A proposition with similar content is Wigner's "empirical law of epistemology" which asserts "the appropriateness and accuracy of the mathematical formulation of the laws of nature in terms of concepts chosen for their manipulability, the 'laws of nature' being of almost fantastic accuracy but of strictly limited scope" (1967: p. 233).

45. An intriguing way to see what has to be done to account for human abductive powers is to compare human beings with Rothstein's "wiggleworms." He attempts "to sketch how science might develop for a race of blind, deaf, highly intelligent worms living in black, cold, sea-bottom muck, and possessing only senses of touch, temperature, and a kind of taste (i.e., a chemical sense)" (1962: p. 28). Rothstein emphathizes with his worms impressively, and he concludes that if an

objectively real world exists, then they would sooner or later discover the same laws governing its behavior that we do. However, his account of the discoveries of handless creatures, with extremely limited abilities to perform controlled experiments, strains our credulity. It forces us to reflect on what privileged children of nature we are, how lavish our environment is in clues and suggestions, and how often we are presented with situations of almost laboratory purity for exhibiting fundamental processes. The human condition may be miserable in many respects; but for the purpose of extracting natural principles from the tangle of appearances, it seems to be very good indeed.

46. The methodological prescription of localizing investigations provides a partial answer to Duhem's holistic conception of natural science (1954: pp. 187–90), but this answer is supplemented by the evidence that nature is favorable to the disentanglement of problems.

47. I agree with much of Polanyi's position concerning the personal factor in scientific knowledge (Polanyi 1958: especially pt. 1), but I think he does not say enough about the circumstances which make the personal factor effective. Rescher (1961) makes some good comments on the reliability of subjective probability evaluations, though his treatment of the concept of probability (a kind of hybrid of the personalist and logical theories) seems to me obscure.

48. Analogy remains an essential part of scientific heuristics, but (to paraphrase a statement of Weyl's concerning simplicity) we must let nature train us to recognize good analogies. As Maxwell pointed out (1952: pp. 156–57), analogies of structure are especially fruitful, and his own proposal of the existence of the displacement current is a classical case in point.

49. This circular argument is quite different in detail from that proposed by Black (1966). However, I strongly concur with his idea that the effectiveness of a circular justification of induction depends upon a global point of view: "Perhaps the place to find a connection between induction and intelligible human interests in arriving at the truth is in the entire *practice* and not in some artificially dissected component of it" (ibid.: p. 199).

50. I have also discussed nonvicious circularity in a review (1954: pp. 657–58) and in greater detail in an unpublished thesis (1953: pp. 119–27).

REFERENCES

Aczél, J. (1966) *Lectures on Functional Equations and Their Applications.* New York, Academic Press.

Aristotle (1941) "On the Soul," in *The Basic Works of Aristotle,* ed. R. McKeon. New York, Random House.

———— (1941a) "Posterior Analytics," in *The Basic Works of Aristotle,* ed. R. McKeon. New York, Random House.

Barr, H. J. (1964) "The Epistemology of Causality from the Point of View of Evolutionary Biology," *Philosophy of Science,* XXXI:286–88.

Bertrand, J. (1889) *Calcul des probabilités.* Paris, Gauthier-Villars.

Black, Max (1966) "The Raison D'Être of Inductive Argument," *The British Journal for the Philosophy of Science,* XVII:177–204.

Bohr, Niels (1934) *Atomic Theory and the Description of Nature.* Cambridge, University Press.

Borel, Emile (1965) *Elements of the Theory of Probability.* Englewood Cliffs, N.J., Prentice-Hall.

Bruner, Jerome (1966) *Cognitive Growth.* New York, Wiley.

Burks, Arthur (1951) "Reichenbach's Theory of Probability and Induction," *Review of Metaphysics,* IV:377–93.

———— (1964) "Peirce's Two Theories of Probability," in *Studies in the Philosophy of Charles Sanders Peirce* (2d ser.), eds. Edward Moore and Richard Robin. Amherst, University of Massachusetts Press. Pp. 141–50.

Carnap, Rudolf (1950) "Empiricism, Semantics, and Ontology," *Revue Internationale de Philosophie,* XI:20–40.

———— (1952) *The Continuum of Inductive Methods.* Chicago, University of Chicago Press.

———— (1956) "The Methodological Character of Theoretical Concepts," in *Minnesota Studies in the Philosophy of Science,* I, eds. H. Feigl and M. Scriven. Minneapolis, University of Minnesota Press. Pp. 38–76.

———— (1963) "Replies and Systematic Expositions," *The Philosophy of Rudolf Carnap,* ed. P. A. Schilpp. La Salle, Open Court.

———— (1968) "A Basic System of Inductive Logic." Dittographed. Los Angeles, University of California.

Carruthers, P. (1966) *Introduction to Unitary Symmetry.* New York, Wiley.

Chomsky, Noam (1965) *Aspects of the Theory of Syntax.* Cambridge, M.I.T. Press.

Cornford, Frances (1965) *Principium Sapientiae.* New York, Harper and Row.

Cox, Richard (1946) "Probability, Frequency, and Reasonable Expectation," *American Journal of Physics,* XIV:1–13.

———— (1961) *The Algebra of Probable Inference.* Baltimore, Johns Hopkins Press.

DeFinetti, Bruno (1937) "La prévision: ses lois logiques, ses sources subjectives," *Annales de l'Institut Henri Poincaré,* VII:1–68. English translation (1964) in *Studies in Subjective Probability,* eds. Henry Kyburg and Howard Smokler. New York, Wiley.

Dewey, John (1929) *Experience and Nature* (2d ed.). La Salle, Open Court.

———— (1938) *Logic: The Theory of Inquiry.* New York, Holt, Rinehart and Winston.

Duhem, Pierre (1954) *The Aim and Structure of Physical Theory.* Princeton, Princeton University Press.

Eddington, Arthur (1928) *The Nature of the Physical World.* New York, Macmillan.

Edwards, W., Lindman, H., and Savage, L. J. (1963) "Bayesian Statistical Inference for Psychological Research," *Psychological Review,* LXX:193–242.

Farquhar, I. E. (1964) *Ergodic Theory in Statistical Mechanics.* New York, Interscience.

Feigl, Herbert (1950) "Existential Hypotheses: Realistic vs. Phenomenalistic Interpretations," *Philosophy of Science,* XVII:35–62.

Feyerabend, Paul (1963) "How to Be a Good Empiricist," in *Philosophy of Science: The Delaware Seminar,* II, ed. B. Baumrin. New York, Interscience. Pp. 3–39.

Flavell, J. H. (1963) *The Developmental Psychology of Jean Piaget.* Princeton, N.J., Van Nostrand.

Fodor, Jerry (1966) "Could There Be a Theory of Perception?" *Journal of Philosophy,* XLIII:369–80.

Gödel, Kurt (1946) "Russell's Mathematical Logic," in *The Philosophy of Bertrand Russell,* ed. P. A. Schilpp. Menasha, Wis., George Banta Press. Pp. 123–53.

Good, I. J. (1950) *Probability and the Weighing of Evidence.* London, C. Griffin.

Goodman, Nelson (1955) *Fact, Fiction, and Forecast.* Cambridge, Harvard University Press.

—— (1961) "Safety, Strength, Simplicity," *Philosophy of Science,* XXVIII:150–51.

Hacking, Ian (1965) *Logic of Statistical Inference.* Cambridge, University Press.

—— (1965a) "Salmon's Vindication of Induction," *Journal of Philosophy,* LXII:269–71.

Hadamard, Jacques (1945) *Psychology of Invention in the Mathematical Field.* New York, Dover.

Harris, E. E. (1965) *Foundations of Metaphysics in Science.* New York, Humanities Press.

Hempel, C. G. (1965) *Aspects of Scientific Explanation and Other Essays in the Philosophy of Science.* New York, Free Press.

Hirst, R. J. (1959) *The Problems of Perception.* New York, Macmillan.

Hockett, Charles (1954) "Chinese Versus English: An Exploration of the Whorfian Theses," in *Language in Culture,* ed. Harry Hoijer. American Anthropological Association, Memoir no. 79 106–23. Reprinted (1959) in *Readings in Anthropology,* I, ed. Morton Fried. New York, Crowell.

Hume, David (1955) *An Inquiry Concerning Human Understanding,* ed. Charles Hendel. New York, Liberal Arts Press.

Huxley, Julian (1942) *Evolution: The Modern Synthesis.* New York, Harper.

Inhelder, B., and Piaget, J. (1958) *The Growth of Logical Thinking from Childhood to Adolescence.* New York, Basic Books.

Jeffrey, Richard (1965) *The Logic of Decision.* New York, McGraw-Hill.

Jeffreys, Harold (1931, 2d ed., 1937) *Scientific Inference.* Cambridge, University Press.

—— (1961) *Theory of Probability* (3rd ed.). Oxford, Clarendon Press.

Kac, Mark (1959) *Probability and Related Topics in Physical Sciences.* New York, Interscience.

Kant, Immanuel (1929) *Critique of Pure Reason,* trans. Norman Kemp Smith. New York, Macmillan.

Katz, Jerrold (1962) *The Problem of Induction and Its Solution.* Chicago, University of Chicago Press.

Kemeny, John (1953) "The Use of Simplicity in Induction," *Philosophical Review,* LXII:391–408.

Keynes, John M. (1921) *A Treatise on Probability.* London, Macmillan.

Koopman, B. O. (1940) "The Axioms and Algebra of Intuitive Probability," *Annals of Mathematics,* ser. 2, XLI:269–92.

Kuhn, Thomas (1962) *The Structure of Scientific Revolutions.* Chicago, University of Chicago Press.

Kyburg, Henry (1964) "Recent Work in Inductive Logic," *American Philosophical Quarterly,* I, no. 4:1–39.

Lenz, John W. (1956) "Carnap on Defining 'Degree of Confirmation,' " *Philosophy of Science,* XXIII:230–36.

—— (1958) "Problems for the Practicalists' Justification of Induction," *Philosophical Studies,* IX:4–8.

—— (1964) "Induction as Self-Corrective," in *Studies in the Philosophy of Charles Sanders Peirce* (2d ser.), eds. Edward Moore and Richard Robin. Amherst, University of Massachusetts Press. Pp. 151–62.

Locke, John (1894) *Essay Concerning Human Understanding,* ed. A. C. Fraser. 2 vols. Oxford, Clarendon Press.

McKeon, Richard (1951) "Philosophy and Method," *Journal of Philosophy,* XLVIII:653–82.

Madden, E. H. (1964) "Peirce on Probability," in *Studies in the Philosophy of*

Charles Sanders Peirce (2d ser.), eds. Edward Moore and Richard Robin. Amherst, University of Massachusetts Press. Pp. 122–40.

Mandelbaum, Maurice (1965) *Philosophy, Science, and Sense Perception.* Baltimore, Johns Hopkins Press.

Maxwell, James C. (1952) *Scientific Papers.* New York, Dover.

Murphey, Murray (1961) *The Development of Peirce's Philosophy.* Cambridge, Harvard University Press.

Nagel, Ernest (1961) *The Structure of Science.* New York, Harcourt, Brace and World.

—— (1963) "Carnap's Theory of Induction," in *The Philosophy of Rudolf Carnap,* ed. P. A. Schilpp. La Salle, Open Court. Pp. 785–826.

Passmore, John (1957) *A Hundred Years of Philosophy.* London, Duckworth.

Peirce, Charles Sanders. *Collected Papers.* Vol. I–VI (1931–35), eds. Charles Hartshorne and Paul Weiss. Cambridge, Harvard University Press. Vol. VII–VIII (1958), ed. Arthur Burks. Cambridge, Harvard University Press. (Note: In the Peirce references I have followed the standard practice of indicating the volume and paragraph number of *Collected Papers.*)

Piaget, Jean (1950) *Introduction à l'épistémologie génétique.* 3 vols. Paris, Presses Universitaires de France.

—— (1952) *The Origins of Intelligence in Children.* New York, Norton.

Polanyi, Michael (1958) *Personal Knowledge.* Chicago, University of Chicago Press.

Popper, Karl (1961) *The Logic of Scientific Discovery.* New York, Science Editions.

—— (1962) *Conjectures and Refutations.* New York, Basic Books.

Putnam, Hilary (1963) "'Degree of Confirmation' and Inductive Logic," *The Philosophy of Rudolf Carnap,* ed. P. A. Schilpp. La Salle, Open Court. Pp. 761–84.

—— (1963a) "Probability and Confirmation," The Voice of America Forum Lectures. Washington D.C., U.S. Information Agency. Reprinted (1967) in *Philosophy of Science Today,* ed. S. Morgenbesser. New York, Basic Books.

Quine, W. V. (1957) "The Scope and Language of Science," *The British Journal for the Philosophy of Science,* VIII:1–17. Reprinted (1966) in *The Ways of Paradox.* New York, Random House.

Raiffa, H., and Schlaifer, R. (1961) *Applied Statistical Decision Theory.* Boston, Harvard University Graduate School of Business Administration.

Ramsey, Frank (1931) "Truth and Probability," in *The Foundations of Mathematics and Other Logical Essays,* ed. Frank Ramsey. London, Rutledge & Kegan Paul. Reprinted (1964) in *Studies in Subjective Probability,* eds. Henry Kyburg and Howard Smokler. New York, Wiley.

Reichenbach, Hans (1938) *Experience and Prediction.* Chicago, University of Chicago Press.

—— (1949) *The Theory of Probability.* Berkeley and Los Angeles, University of California Press.

Rensch, Bernhard (1960) *Evolution Above the Species Level.* New York, Columbia University Press.

Rescher, Nicholas (1961) "On the Probability of Nonrecurring Events," in *Current Issues in the Philosophy of Science,* eds. Herbert Feigl and Grover Maxwell. New York, Holt, Rinehart, and Winston. Pp. 228–37.

Rothstein, Jerome (1962) "Wiggleworm Physics," *Physics Today,* September: 28–38.

Russell, Bertrand (1948) *Human Knowledge: Its Scope and Limits.* New York, Simon & Schuster.

Salmon, Wesley (1961) "Vindication of Induction," in *Current Issues in the Philoso-*

phy of Science, eds. Herbert Feigl and Grover Maxwell. New York, Holt, Rinehart, and Winston. Pp. 245–56.

—— (1963) "Inductive Inference," in *Philosophy of Science: The Delaware Seminar,* II, ed. B. Baumrin. New York, Interscience. Pp. 341–70.

—— (1965) "What Happens in the Long Run?" *Philosophical Review,* LXXXIV:373–78.

—— (1966) "The Foundations of Scientific Inference," in *Mind and Cosmos: Essays in Contemporary Science and Philosophy,* ed. Robert Colodny. Pittsburgh, University of Pittsburgh Press. Pp. 135–275.

Savage, Leonard J. (1954) *Foundations of Statistics.* New York, Wiley.

—— (1961) "The Foundations of Statistics Reconsidered," in *Proceedings of the Fourth Berkeley Symposium on Mathematical Statistics and Probability.* Berkeley, University of California Press. Pp. 575–85. Reprinted (1964) in *Studies in Subjective Probability,* eds. Henry Kyburg and Howard Smokler. New York, Wiley.

Scheibe, K., and Sarbin, T. (1965) "Towards a Theoretical Conceptualization of Superstition," *The British Journal for the Philosophy of Science,* XVI:143–58.

Sellars, Wilfrid (1963) *Science, Perception and Reality.* New York, Humanities Press.

Shimony, Abner (1953) *A Theory of Confirmation.* Ph.D. dissertation, Yale University.

—— (1954) "Braithwaite on Scientific Method," *Review of Metaphysics,* VII:644–50.

—— (1955) "Coherence and the Axioms of Confirmation," Journal of Symbolic Logic, XX:1–28.

—— (1965) "Quantum Physics and the Philosophy of Whitehead," in *Philosophy in America,* ed. Max Black. London, Allen and Unwin. Pp. 240–61. Reprinted (1965) in *Boston Studies in the Philosophy of Science,* vol. 2, eds. Robert Cohen and Marx Wartofsky. New York, Humanities Press.

Simpson, George G. (1963) "Biology and the Nature of Science," *Science,* CXXXIX:81–88.

Smart, J. J. C. (1963) *Philosophy and Scientific Realism.* New York, Humanities Press.

Stein, Howard (1967) "Newtonian Space-Time," *The Texas Quarterly,* Autumn: 174–200.

Strawson, Peter (1959) *Individuals.* London, Methuen.

Sutherland, N. S. (1959) "Stimulus Analyzing Mechanisms," *Mechanization of Thought Processes,* vol. 2. London, National Physical Laboratory Symposium No. 10.

Tolman, Richard (1938) *The Principles of Statistical Mechanics.* London, Oxford University Press.

Vernon, M. D. (1962) *The Psychology of Perception.* Baltimore, Penguin Books.

Weiss, Paul (1938) *Reality.* Princeton, Princeton University Press.

Weyl, Hermann (1949) *Philosophy of Mathematics and Natural Science.* Princeton, Princeton University Press.

Whitehead, A. N. (1929) *Process and Reality.* New York, Macmillan.

Whittaker, Edmund (1951) *History of the Theories of Aether and Electricity: The Classical Theories.* London, Thomas Nelson.

Whorf, Benjamin (1941) "The Relation of Habitual Thought and Behavior to Language," in *Language, Personality, and Culture: Essays in Memory of Edward Sapir,* ed. Leslie Spier. Menasha, Wis., Sapir Memorial Publication Fund. Pp.

75–93. Reprinted (1956) in *Language, Thought, and Reality: Selected Writings of Benjamin Lee Whorf*, ed. John Carroll. Cambridge, M.I.T. Press.

Wigner, Eugene (1964) "Events, Laws of Nature, and Invariance Principles," in *The Nobel Prize Lectures*. Amsterdam, N.Y., Elsevier.

——— (1967) *Symmetries and Reflections*. Bloomington, Indiana University Press. (A reprint of the preceding reference is included.)

Wrinch, D., and Jeffreys, H. (1921) "On Certain Fundamental Principles of Scientific Inquiry," *Philosophical Magazine* (6th ser.), XLII:369–90.

WESLEY C. SALMON
Indiana University

Statistical Explanation

> It is needless to remark that those teachers of logic who have
> not yet waked up to the doctrine of probability, which for
> more than a century has been the logic of the exact sciences,
> will pass into another stage of fossilization without knowing
> what modern logic is.
> —C. S. Peirce
> *Values in a Universe of Chance*

EVER SINCE HIS CLASSIC PAPER with Paul Oppenheim, "Studies in the
Logic of Explanation," first published in 1948,[1] Carl G. Hempel has
maintained that an "explanatory account [of a particular event] may be
regarded as an argument to the effect that the event to be explained . . .
was to be expected by reason of certain explanatory facts" (my italics).[2]
It seems fair to say that this basic principle has guided Hempel's work on
inductive as well as *deductive* explanation ever since.[3] In spite of its
enormous intuitive appeal, I believe that this precept is incorrect and
that it has led to an unsound account of scientific explanation. In this
paper I shall attempt to develop a different account of explanation and
argue for its superiority over the Hempelian one. In the case of inductive
explanation, the difference between the two treatments hinges funda-

This paper grew out of a discussion of statistical explanation presented at the
meeting of the American Association for the Advancement of Science, held in
Cleveland in 1963, as a part of the program of Section L organized by Adolf
Grünbaum, then vice-president for Section L. My paper, "The Status of Prior
Probabilities in Statistical Explanation," along with Henry E. Kyburg's comments and
my rejoinder, were published in *Philosophy of Science*, XXXII, no. 2 (April 1965).
The original version of this paper was written in 1964 in an attempt to work out
fuller solutions to some problems Kyburg raised, and it was presented at the
Pittsburgh Workshop Conference in May 1965, prior to the publication of Carl G.
Hempel, *Aspects of Scientific Explanation* (New York: Free Press, 1965).
I should like to express my gratitude to the National Science Foundation for
support of the research contained in this paper.

mentally upon the question of whether the relation between the explanans and the explanadum is to be understood as a relation of *high
probability* or as one of *statistical relevance.* Hempel obviously chooses
the former alternative; I shall elaborate an account based upon the latter
one. These two alternatives correspond closely to the "concepts of firmness" and the "concepts of increase of firmness," respectively, distinguished by Rudolf Carnap in the context of confirmation theory.[4] Carnap
has argued, convincingly in my opinion, that confusion of these two
types of concepts has led to serious trouble in inductive logic; I shall
maintain that the same thing has happened in the theory of explanation.
Attention will be focused chiefly upon inductive explanation, but I shall
try to show that a similar difficulty infects deductive explanation and
that, in fact, deductive explanation can advantageously be considered as
a special limiting case of inductive explanation. It is my hope that, in the
end, the present *relevance* account of scientific explanation will be
justified, partly by means of abstract "logical" considerations and partly
in terms of its ability to deal with problems that have proved quite
intractable within the Hempelian schema.

1. The Hempelian Account

Any serious contemporary treatment of scientific explanation must, it
seems to me, take Hempel's highly developed view as a point of departure. In the famous 1948 paper, Hempel and Oppenheim offered a
systematic account of deductive explanation, but they explicitly denied
that all scientific explanations fit that pattern; in particular, they called
attention to the fact that some explanations are of the inductive variety.
In spite of fairly general recognition of the need for inductive explanations, even on the part of proponents of Hempel's deductive model,
surprisingly little attention has been given to the problem of providing a
systematic treatment of explanations of this type. Before 1965, when he
published "Aspects of Scientific Explanation," [5] Hempel's "Deductive-
Nomological vs. Statistical Explanation" [6] was the only well-known extensive discussion. One could easily form the impression that most
theorists regarded deductive and inductive explanation as quite similar
in principle, so that an adequate account of inductive explanation would
emerge almost routinely by replacing the universal laws of deductive
explanation with statistical generalizations, and by replacing the deductive relationship between explanans and explanandum with some sort of
inductive relation. Such an attitude was, of course, dangerous in the
extreme, for even our present limited knowledge of inductive logic points

to deep and fundamental differences between deductive and inductive logical relations. This fact should have made us quite wary of drawing casual analogies between deductive and inductive patterns of explanation.[7] Yet even Hempel's detailed examination of statistical explanation [8] may have contributed to the false feeling of security, for one of the most significant results of that study was that both deductive and inductive explanations must fulfill a *requirement of total evidence.* In the case of deductive explanations the requirement is automatically satisfied; in the case of inductive explanations that requirement is nontrivial.

Accordingly, the situation in May 1965, at the time of the Pittsburgh Workshop Conference, permitted a rather simple and straightforward characterization which would cover both deductive and inductive explanations of particular events.[9] Either type of explanation, according to Hempel, is an argument; as such, it is a linguistic entity consisting of premises and conclusion.[10] The premises constitute the explanans, and the conclusion is the explanandum. The term "explanadum event" may be used to refer to the fact to be explained; the explanandum is the statement asserting that this fact obtains. The term "explanatory facts" may be used to refer to the facts adduced to explain the explanandum event; the explanans is the set of statements asserting that these explanatory facts obtain.[11] In order to explain a particular explanandum event, the explanatory facts must include both particular facts and general uniformities. As Hempel has often said, general uniformities as well as particular facts can be explained, but for now I shall confine attention to the explanation of particular events.

The parallel between the two types of explanation can easily be seen by comparing examples; here are two especially simple ones Hempel has offered: [12]

(1) *Deductive*
This crystal of rock salt, when put into a Bunsen flame, turns the flame yellow, for it is a sodium salt, and all sodium salts impart a yellow color to a Bunsen flame.

(2) *Inductive*
John Jones was almost certain to recover quickly from his streptococcus infection, for he was given penicillin, and almost all cases of streptococcus infection clear up quickly upon administration of penicillin.

These examples exhibit the following basic forms:

(3) *Deductive*
All F are G.

$$\frac{x \text{ is } F.}{x \text{ is } G.}$$

(4) *Inductive*
Almost all F are G.

$$\frac{x \text{ is } F.}{x \text{ is } G.}$$

There are two obvious differences between the deductive and inductive examples. First, the major premise in the deductive case is a universal generalization, whereas the major premise in the inductive case is a statistical generalization. The latter generalization asserts that a high, though unspecified, proportion of F are G. Other statistical generalizations may specify the exact numerical value. Second, the deductive schema represents a valid deductive argument, whereas the inductive schema represents a correct inductive argument. The double line in (4) indicates that the conclusion "follows inductively," that is, with high inductive probability. Hempel has shown forcefully that (4) is *not* to be construed as a deduction with the conclusion that "x is almost certain to be G." [13]

By the time Hempel had provided his detailed comparison of the two types of explanation, certain well-known conditions of adequacy had been spelled out; they would presumably apply both to deductive and to inductive explanations: [14]

(i) The explanatory argument must have correct (deductive or inductive) logical form. In a correct deductive argument the premises entail the conclusion; in a correct inductive argument the premises render the conclusion highly probable.

(ii) The premises of the argument must be true. [15]

(iii) Among the premises there must occur essentially at least one lawlike (universal or statistical) generalization. [16]

(iv) The requirement of total evidence (which is automatically satisfied by deductive explanations that satisfy the condition of validity) must be fulfilled. [17]

Explanations that conform to the foregoing conditions certainly satisfy Hempel's general principle. If the explanation is deductive, the explanandum event was to be expected because the explanandum is deducible from the explanans; the explanans necessitates the explanandum. If the explanation is inductive, it "*explains* a given phenomenon by showing that, in view of certain particular facts and certain statistical laws, its occurrence was to be expected with high logical, or inductive, probability." [18] In this case the explanandum event was to be expected because the explanans confers high probability upon the explanandum; the explanatory facts make the explanandum event highly probable.

2. Some Counterexamples

It is not at all difficult to find cases that satisfy all of the foregoing requirements, but that certainly cannot be regarded as genuine explanations. In a previously mentioned paper [19] I offered the following inductive examples:

(5) John Jones was almost certain to recover from his cold within a week, because he took vitamin C, and almost all colds clear up within a week after administration of vitamin C.

(6) John Jones experienced significant remission of his neurotic symptoms, for he underwent extensive psychoanalytic treatment, and a substantial percentage of those who undergo psychoanalytic treatment experience significant remission of neurotic symptoms.

Both of these examples correspond exactly with Hempel's inductive example (2) above, and both conform to his schema (4). The difficulty with (5) is that colds tend to clear up within a week regardless of the medication administered, and, I understand, controlled tests indicate that the percentage of recoveries is unaffected by the use of vitamin C. [20] The problem with (6) is the substantial spontaneous remission rate for neurotic symptoms of individuals who undergo no psychotherapy of any kind. Before we accept (6) as having any explanatory value whatever, we must know whether the remission rate for psychoanalytic patients is any greater than the spontaneous remission rate. I do not have the answer to this factual question.

I once thought that cases of the foregoing sort were peculiar to inductive explanation, but Henry Kyburg has shown me to be mistaken by providing the following example:

(7) This sample of table salt dissolves in water, for it has had a dissolving spell cast upon it, and all samples of table salt that have had dissolving spells cast upon them dissolve in water. [21]

It is easy to construct additional instances:

(8) John Jones avoided becoming pregnant during the past year, for he has taken his wife's birth control pills regularly, and every man who regularly takes birth control pills avoids pregnancy.

Both of these examples correspond exactly with Hempel's deductive example (1), and both conform to his schema (3) above. The difficulty with (7) and (8) is just like that of the inductive examples (5) and (6). Salt dissolves, spell or no spell, so we do not need to explain the dissolving of this sample in terms of a hex. Men do not become pregnant, pills or no pills, so the consumption of oral contraceptives is not required to explain the phenomenon in John Jones's case (though it may have considerable explanatory force with regard to his wife's pregnancy).

Each of the examples (5) through (8) constitutes an argument to show that the explanandum event was to be expected. Each one has correct (deductive or inductive) logical form—at least it has a form provided by one of Hempel's schemata.[22] Each one has true premises (or so we may assume for the present discussion). Each one has a (universal or statistical) generalization among its premises, and there is no more reason to doubt the lawlikeness of any of these generalizations than there is to doubt the lawlikeness of the generalizations in Hempel's examples. In each case the general premise is essential to the argument, for it would cease to have correct logical form if the general premise were simply deleted. We may assume that the requirement of total evidence is fulfilled in all cases. In the deductive examples it is automatically satisfied, of course, and in the inductive examples we may safely suppose that there is no further available evidence which would alter the probability of John Jones's recovery from either his cold or his neurotic symptoms. There is nothing about any of these examples, so far as I can see, which would disqualify them in terms of the foregoing criteria without also disqualifying Hempel's examples as well.

Folklore, ancient and modern, supplies further instances that would qualify as explanatory under these conditions:

(9) The moon reappeared after a lunar eclipse, for the people made a great deal of noise, banging on pots and pans and setting off fireworks, and the moon always reappears after an eclipse when much noise occurs.[23] (Ancient Chinese folklore)

(10) An acid-freak was standing at the corner of Broadway and Forty-second Street, uttering agonizing moans, when a passerby asked him what was the matter. "Why, man, I'm keeping the wild tigers away," he answered. "But there are no wild tigers

around here," replied the inquirer. "Yeah, see what a good job I'm doing!" (Modern American folklore)

A question might be raised about the significance of examples of the foregoing kind on the ground that conditions of adequacy are normally understood to be necessary conditions, not sufficient conditions. Thus, the fact that there are cases which satisfy all of the conditions of adequacy but which are not explanations is no objection whatever to the necessity of such conditions; this fact could be construed as an objection only if the conditions of adequacy are held to be jointly sufficient. In answer to this point, it must be noted that, although the Hempel-Oppenheim paper of 1948 did begin by setting out conditions of adequacy, it also offered a definition of deductive-nomological explanation, that is, explicit necessary and sufficient conditions.[24] The deductive examples offered above satisfy that definition. And if that definition is marred by purely technical difficulties, it seems clear that the examples conform to the spirit of the original definition.

One might, however, simply acknowledge the inadequacy of the early definition and maintain only that the conditions of adequacy should stand as necessary conditions for satisfactory explanations. In that case the existence of a large class of examples of the foregoing sort would seem to point only to the need for additional conditions of adequacy to rule them out. Even if one never hopes to have a set of necessary conditions that are jointly sufficient, a large and important set of examples which manifest a distinctive characteristic disqualifying them as genuine explanations demands an additional condition of adequacy. Thus, it might be said, the way to deal with the counterexamples is by adding a further condition of adequacy, much as Hempel has done in "Aspects of Scientific Explanation," where he enunciates the *requirement of maximal specificity*.[25] I shall argue, however, that this requirement does not exclude the examples I have offered, although a minor emendation will enable it to do so. Much more significantly, I shall maintain that the requirement of maximal specificity is the wrong type of requirement to take care of the difficulties that arise and that a much more fundamental revision of Hempel's conception is demanded. However, before beginning the detailed discussion of the requirement of maximal specificity, I shall make some general remarks about the counterexamples and their import—especially the inductive ones—in order to set the stage for a more enlightening discussion of Hempel's recent addition to the set of conditions of adequacy.

3. Preliminary Analysis

The obvious trouble with our horrible examples is that the "explanatory" argument is not needed to make us see that the explanandum event was to be expected. There are other, more satisfactory, grounds for this expectation. The "explanatory facts" adduced are irrelevant to the explanandum event despite the fact that the explanandum follows (deductively or inductively) from the explanans. Table salt dissolves in water regardless of hexing, almost all colds clear up within a week regardless of treatment, males do not get pregnant regardless of pills, the moon reappears regardless of the amount of Chinese din, and there are no wild tigers in Times Square regardless of our friend's moans. Each of these explanandum events has a high prior probability independent of the explanatory facts, and the probability of the explanandum event relative to the explanatory facts is the same as this prior probability. In this sense the explanatory facts are irrelevant to the explanandum event. The explanatory facts do nothing to enhance the probability of the explanandum event or to make us more certain of its occurrence than we would otherwise have been. This is not because we know that the fact to be explained has occurred; it is because we had other grounds for expecting it to occur, *even if we had not already witnessed it.*

Our examples thus show that it is not correct, even in a preliminary and inexact way, to characterize explanatory accounts as arguments showing that the explanandum event was to be expected. It is more accurate to say that an explanatory argument shows that the probability of the explanandum event relative to the explanatory facts is substantially greater than its prior probability.[26] An explanatory account, on this view, increases the degree to which the explanandum event was to be expected. As will emerge later in this paper, I do not regard such a statement as fully accurate; in fact, the increase in probability is merely a pleasant by-product which often accompanies a much more fundamental characteristic. Nevertheless, it makes a useful starting point for further analysis.

We cannot, of course, hope to understand even the foregoing rough characterization without becoming much clearer on the central concepts it employs. In particular, we must consider explicitly what is meant by "probability" and "prior probability." There are several standard views on the nature of probability, and I propose to consider briefly how the foregoing characterization of explanation, especially inductive explanation, would look in terms of each.

a. According to the *logical interpretation*, probability or degree of confirmation is a logical relation between evidence and hypothesis.[27] Degree of confirmation statements are analytic. There is no analogue of the rule of detachment (*modus ponens*), so inductive logic does not provide any basis for asserting inductive conclusions. There are certain methodological rules for the application of inductive logic, but these are not part of inductive logic itself. One such rule is the requirement of total evidence. For any given explanandum, taken as hypothesis, there are many true degree of confirmation statements corresponding to different possible evidence statements. The probability value to be chosen for practical applications is that given by the degree of confirmation statement embodying the total available (relevant) evidence. This is the value to be taken as a fair betting quotient, as an estimate of the relative frequency, or as the value to be employed in calculating the estimated utility.

I have long been seriously puzzled as to how Carnap's conception of logical probability could be consistently combined with Hempel's views on the nature of explanation.[28] In the first place, for purposes of either deductive or inductive explanation, it is not clear how the universal or statistical generalizations demanded by Hempel's schemata are to become available, for inductive logic has no means of providing such generalizations as detachable conclusions of any argument. In the second place, degree of confirmation statements embodying particular hypothesis statements and particular evidence statements are generally available, so inductive "inferences" from particulars to particulars are quite possible. In view of this fact, it is hard to see why inductive explanation should require any sort of general premise. In the third place, the relation between explanans and explanandum would seem to be a degree of confirmation statement, and not an argument with premises and conclusions in the usual sense. This last point seems to me to be profoundly important. Carnap's recognition that much, if not all, of inductive logic can proceed without inductive arguments—that is, without establishing conclusions on the basis of premises—is a deep insight that must be reckoned with. As a consequence, we must seriously question whether an explanation is an argument at all. This possibility arises, as I shall try to show later, even if one does not adhere to a logical interpretation of probability. Indeed, within an account of explanation in terms of the frequency interpretation, I shall argue contra Hempel that *explanations are not arguments*.[29]

For the logical interpretation of probability, it would seem, an explanation should involve an addition of new evidence to the body of total evidence resulting in a new body of total evidence relative to which the probability of the explanandum is higher than it was relative to the old body of total evidence. In the usual situation, of course, an explanation is sought only after it is known that the explanandum event has occurred. In this case, the body of total evidence already contains the explanandum, so no addition to the body of total evidence can change its probability. We must, therefore, somehow circumscribe a body of total evidence available prior to the occurrence of the explanandum event, relative to which our prior probability is to be taken. This body of evidence must contain neither the explanans nor the explanandum.[30]

The logical interpretation of probability admits degrees of confirmation on tautological evidence, and these probabilities are not only prior but also a priori. But a priori prior probabilities are not especially germane to the present discussion. We are not concerned with the a priori probability of a Bunsen flame turning yellow, nor with the a priori probability of a cold clearing up in a week, nor with the a priori probability of a remission of ones neurotic symptoms, nor with the a priori probability of finding a wild tiger in Times Square, etc. We are concerned with the probabilities of these events relative to more or less specific factual evidence. The prior probabilities are not logically a priori; they are prior with respect to some particular information or investigation.

b. According to the new *subjective* or *personalistic interpretation,* probability is simply orderly opinion.[31] The required orderliness is provided by the mathematical calculus of probability. One of the main concerns of the personalists is with the revision of opinion or degree of belief in the light of new evidence. The account of explanation suggested above fits very easily with the personalistic interpretation. At some time before the explanandum event occurs, the personalist would say, an individual has a certain degree of belief that it will occur—reflected, perhaps, in the kind of odds he is willing to give or take in betting on its occurrence. According to the present view of explanation, a personalist should require that the explanatory facts be such as to increase the prior confidence in the occurrence of the explanandum event. Of course, this view deliberately introduces into its treatment of scientific explanation a large degree of subjectivity, but this is not necessarily a defect in either the account of probability or the account of explanation.

c. According to the *frequency interpretation,* a probability is the limit of the relative frequency of an attribute in an infinite sequence of events.[32] A probability cannot, therefore, literally be assigned to a single event. Since there are many occasions when we must associate probabilities with single events, and not only with large (literally, infinite) aggregates of events, a way must be found for making the probability concept applicable to single events. We wager on single events—a single toss of the dice, a single horse race, a single deal of the cards. The results of our practical planning and effort depend upon the outcomes of single occurrences, not infinite sequences of them. If probability is to be a guide of life, it must be meaningful to apply probability values to single events. This problem of the single case has traditionally been one of the difficulties confronting the frequency theory, and we shall have to examine in some detail the way in which it can be handled. With regard to the single case, the frequency interpretation is on an entirely different footing from the logical and subjective interpretations. Neither of these latter interpretations faces any special problem of the single case, for statements about single events are perfectly admissible hypotheses for the logical theory, and the subjective theory deals directly with degrees of belief regarding single events.

The central topic of concern in this paper is the explanation of single events. If the frequency theory is to approach this problem at all, it must deal directly with the problem of the single case. To some, the fact that the frequency interpretation is faced with this special problem of the single case may constitute a compelling reason for concluding that the frequency interpretation is totally unsuitable for handling the explanation of single events. Such a conclusion would, I think, be premature. On the contrary, I would argue that a careful examination of the way in which the frequency interpretation handles the single case should prove extremely illuminating with respect to the general problem of inductive explanation. Although I have strong prejudices in favor of the frequency interpretation, this paper is, nevertheless, not the place to argue for them.[33] In this context I would claim instead that frequencies play such an important role in any theory of probability that an examination of the problem of the single case cannot fail to cast light on the problem of explanation regardless of one's persuasion concerning interpretations of the probability concept.

d. In recent years there has been a good deal of discussion of the *propensity interpretation* of probability.[34] This interpretation is so similar to the frequency interpretation in fundamental respects that everything I shall say about the problem of the single case for the frequency interpretation can be made directly applicable to the propensity interpretation by a simple translation: whereever I speak of "the problem of selecting the appropriate reference class" in connection with the frequency interpretation, read "the problem of specifying the nature of the chance setup" in reference to the propensity interpretation. It seems to me that precisely parallel considerations apply for these two interpretations of probability.

4. The Single Case

Let *A* be an unending sequence of draws of balls from an urn, and let *B* be the class of red things. *A* is known as the *reference class*, and *B* the *attribute class*. The probability of red draws from this urn, *P(A,B)*, is the limit of the relative frequency with which members of the reference class belong to the attribute class, that is, the limit of the relative frequency with which draws from the urn result in a red ball as the number of draws increases without any bound.[35]

Frequentists like John Venn and Hans Reichenbach have dealt with the problem of the single case by assigning each single event to a reference class and by transferring the probability value from that reference class to the single event in question.[36] Thus, if the limit of the relative frequency of red among draws from our urn is one-third, then we say that the probability of getting red on *the next draw* is one-third. In this way the meaning of the probability concept has been extended so that it applies to single events as well as to large aggregates.

The fundamental difficulty arises because a given event can be referred to any of a large number of reference classes, and the probability of the attribute in question may vary considerably from one of these to another. For instance, we could place two urns on a table, the one on the left containing only red balls, the one on the right containing equal numbers of red, white, and blue balls. The reference class *A* might consist of blind drawings from the right-hand urn, the ball being replaced and the urn thoroughly shaken after each draw. Another reference class *A'* might consist of draws made alternately from the left- and the right-hand urns. Infinitely many other reference classes are easily devised to which the next draw—the draw with which we are concerned —belongs. From which reference class shall we transfer our probability value to this single case? A method must be established for choosing the appropriate reference class. Notice, however, that there is no difficulty in selecting an attribute class. The question we ask determines the attribute class. We want to know the probability of getting red, so there is no further problem about the attribute class.

Reichenbach recommends adopting as a reference class "the narrowest class for which reliable statistics can be compiled." [37] This principle is, as Reichenbach himself has observed, rather ambiguous. Since increasing the reliability of statistics generally tends to broaden the class and since narrowing the class often tends to reduce the reliability of the statistics, the principle involves two desiderata which pull in opposite directions. It seems that we are being directed to maximize two variables that cannot simultaneously be maximized. This attempt to extend the meaning of the probability concept to single cases fails to provide a method for associating a unique probability value with a given single event. Fully aware of this fact, Reichenbach insisted that the probability concept applies *literally* only to sequences; talk about the probability of a single event is "elliptical" and the extended meaning is "fictitious." The choice of a reference class, he maintained, is often dictated by practical rather than theoretical considerations.

Although Reichenbach has not said so, it seems reasonable to suppose that he was making a distinction similar to that made by Carnap between the principles belonging to inductive logic and methodological rules for the application of inductive logic.[38] The requirement of total evidence, it will be recalled, is a methodological rule for the application of inductive logic. Reichenbach could be interpreted as suggesting analogously that probability theory itself is concerned only with limit statements about relative frequencies in infinite sequences of events, whereas the principle for selection of a reference class stands as a methodological rule for the practical application of probability statements. (A much stronger analogy between the requirement of total evidence and the principle for choosing a reference class for a single case will be shown below.) In fact, for Reichenbach, it might have been wise to withhold the term "probability" from single events, reserving his term "weight" for this purpose. We could then say that practical considerations determine what probability should be chosen to serve as a weight for a particular single event. The relative practical importance of reliability and precision would then determine the extent to which narrowness gives way to reliability of statistics (or conversely) in determining the appropriate reference class.

Although Reichenbach's formulation of the principle for the selection of reference classes is not entirely satisfactory, his intention seems fairly clear. In order to transfer a probability value from a sequence to a single case, it is necessary to have some basis for ascertaining the probability in that sequence. The reference class must, therefore, be broad enough to provide the required number of instances for examination to constitute evidence for an inductive inference. At the same time, we want to avoid choosing a reference class so broad that it includes cases irrelevant to the one with which we are concerned.

Statistical relevance is the essential notion here. It is desirable to narrow the reference class in statistically relevant ways, but not in statistically irrelevant ways. When we choose a reference class to which to refer a given single case, we must ask whether there is any statistically relevant way to subdivide that class. If so, we may choose the narrower subclass that results from the subdivision; if no statistically relevant way is known, we must avoid making the reference class any narrower. Consider, for example, the probability that a particular individual, John Smith, will still be alive ten years hence. To determine this probability, we take account of his age, sex, occupation, and health; we ignore his eye color, his automobile license number, and his last initial. We expect the relative frequency of survival for ten more years to vary among the following reference classes: humans, Americans, American males, forty-two-year-old American males, forty-two-year-old American male steeplejacks, and forty-two-year-old American male steeplejacks suffering from advanced cases of lung cancer. We believe that the relative frequency of survival for another ten years is the same in the following classes: forty-two-year-old American male steeplejacks with advanced cases of lung cancer, forty-two-year-old blue-eyed American male steeplejacks with advanced cases of lung cancer, and forty-two-year-old blue-eyed American male steeplejacks with even automobile license plate numbers who suffer from advanced cases of lung cancer.

Suppose we are dealing with some particular object or event x, and we seek to determine the probability (weight) that it has attribute B. Let x be assigned to a reference class A, of which it is a member. $P(A,B)$ is the probability of this attribute within this reference class. A set of mutually exclusive and exhaustive subclasses of a class is a *partition* of that class. We shall often be concerned with partitions of reference classes into two subclasses; such partitions can be effected by a property C which divides the class A into two subclasses, $A.C$ and $A.\bar{C}$. A property C is said to be *statistically relevant* to B within A if and only if $P(A.C,B) \neq P(A,B)$. This notion of statistical relevance is the fundamental concept upon which I hope to build an explication of inductive explanation.

In his development of a frequency theory based essentially upon the concept of randomness, Richard von Mises introduced the notion of a *place selection:* "By a place selection we mean the selection of a partial sequence in such a way that we decide whether an element should or should not be included without making use of the attribute of the element." [39] A place selection effects a partition of a reference class into two subclasses, elements of the place selection and elements not included in the place selection. In the reference class of draws from our urn, every third draw starting with the second, every kth draw where k is prime, every draw following a red result, every draw made with the left hand, and every draw made while the sky is cloudy all would be place selections. "Every draw of a red ball" and "every draw of a ball whose color is at the opposite end of the spectrum from violet" do not define place selections, for membership in these classes cannot be determined without reference to the attribute in question.

A place selection may or may not be statistically relevant to a given attribute in a given reference class. If the place selection is statistically irrelevant to an attribute within a reference class, the probability of that attribute within the subclass determined by the place selection is equal to the probability of that attribute within the entire original reference class. If every place selection is irrelevant to a given attribute in a given sequence, von Mises called the sequence *random*. If every property that determines a place selection is statistically irrelevant to B in A, I shall say that A is a *homogeneous reference class* for B. A reference class is homogeneous if there is no way, even in principle, to effect a statistically relevant partition without already knowing which elements have the attribute in question and which do not. Roughly speaking, each member of a homogeneous reference class is a random member.

The aim in selecting a reference class to which to assign a single case is not to select the narrowest, but the widest, available class. However, the reference class should be homogeneous, and achieving homogeneity requires making the reference class narrower if it was not already homogeneous. I would reformulate Reichenbach's method of selection of a reference class as follows: choose the broadest homogeneous reference class to which the single event belongs. I shall call this the *reference class rule*.

Let me make it clear immediately that, although I regard the above formulation as an improvement over Reichenbach's, I do not suppose that it removes all ambiguities about the selection of reference classes either in principle or in practice. In principle it is possible for an event to belong to two equally wide homogeneous reference classes, and the probabilities of the attribute in these two classes need not be the same. For instance, suppose that the drawing from the urn is not random and that the limit of the relative frequency of red for every *k*th draw (*k* prime) is 1/4, whereas the limit of the relative frequency of red for every even draw is 3/4. Each of these subsequences may be perfectly random; each of the foregoing place selections may, therefore, determine a homogeneous reference class. Since the intersection of these two place selections is finite, it does not determine a reference class for a probability. The second draw, however, belongs to both place selections; in this fictitious case there is a genuine ambiguity concerning the probability to be taken as the weight of red on the second draw.

In practice we often lack full knowledge of the properties relevant to a given attribute, so we do not know whether our reference class is homogeneous or not. Sometimes we have strong reason to believe that our reference class is not homogeneous, but we do not know what property will effect a statistically relevant partition. For instance, we may believe that there are causal factors that determine which streptococcus infections will respond to penicillin and which ones will not, but we may not yet know what these causal factors are. When we know or suspect that a reference class is not homogeneous, but we do not know how to make any statistically relevant partition, we may say that the reference class is *epistemically homogeneous*. In other cases, we know that a reference class is inhomogeneous and we know what attributes would effect a statistically relevant partition, but it is too much trouble to find out which elements belong to each subclass of the partition. For instance, we believe that a sufficiently detailed knowledge of the initial conditions under which a coin is tossed would enable us to predict (perfectly or very reliably) whether the outcome will be heads or tails, but practically speaking we are in no position to determine these initial conditions or make the elaborate calculations required to predict the outcome. In such cases we may say that the reference class is *practically homogeneous*.[40]

The reference class rule remains, then, a methodological rule for the application of probability knowledge to single events. In practice we attempt to refer our single cases to classes that are practically or epistemically homogeneous. When something important is at stake, we may try to extend our knowledge in order to improve the degree of homogeneity we can achieve. Strictly speaking, we cannot meaningfully refer to degrees of homogeneity until a quantitative concept of homogeneity has been provided. That will be done in section 6 below.

It would, of course, be a serious methodological error to assign a single case to an inhomogeneous reference class if neither epistemic nor practical considerations prevent partitioning to achieve homogeneity. This fact constitutes another basis for regarding the reference class rule as the counterpart of the requirement of total evidence. The requirement of total evidence demands that we use all available relevant evidence; the reference class rule demands that we partition whenever we have available a statistically relevant place selection by means of which to effect the partition.

Although we require homogeneity, we must also prohibit partitioning of the reference class by means of statistically irrelevant place selections. The reason is obvious. Irrelevant partitioning reduces, for no good reason, the inductive evidence available for ascertaining the limiting frequency of our attribute in a reference class that is as homogeneous as we can make it. Another important reason for prohibiting irrelevant partitions will emerge below when we discuss the importance of multiple homogeneous reference classes.

A couple of fairly obvious facts about homogeneous reference classes should be noted at this point. If all A's are B, A is a homogeneous reference class for B. (Somewhat counterintuitively, perhaps, B occurs perfectly randomly in A.) In this case, $P(A,B) = 1$ and $P(A.C,B) = 1$ for any C whatever; consequently, no place selection can yield a probability for B different from that in the reference class A. Analogously, A is homogeneous for B if no A's are B. In the frequency interpretation, of course, $P(A,B)$ can equal one even though not all A's are B. It follows that a probability of one does not entail that the reference class is homogeneous.

Some people maintain, often on a priori grounds, that A is homogeneous (not merely practically or epistemically homogeneous) for B only if all A's are B or no A's are B; such people are determinists. They hold that causal factors always determine which A's are B and which A's are not B; these causal factors can, in principle, be discovered and used to construct a place selection for making a statistically relevant partition of A. I do not believe in this particular form of determinism. It seems to me that there are cases in which A is a homogeneous reference class for B even though not all A's are B. In a sample of radioactive material a certain percentage of atoms disintegrate in a given length of time; no place selection can give us a partition of the atoms for which the frequency of disintegration differs from that in the whole sample. A beam of electrons is shot at a potential barrier and some pass through while others are reflected; no place selection will enable us to make a statistically relevant partition in the class of electrons in the beam. A beam of silver atoms is sent through a strongly inhomogeneous magnetic field (Stern-Gerlach experiment); some atoms are deflected upward and some are deflected downward, but there is no way of partitioning the beam in a statistically relevant manner. Some theorists maintain, of course, that further investigation will yield information that will enable us to make statistically relevant partitions in these cases, but this is, at present, no more than a declaration of faith in determinism. Whatever the final resolution of this controversy, the homogeneity of A for B does not logically entail that all A's are B. The truth or falsity of determinism cannot be settled a priori.

The purpose of the foregoing excursus on the frequency treatment of the problem of the single case has been to set the stage for a discussion of the explanation of particular events. Let us reconsider some of our examples in the light of this theory. The relative frequency with which we encounter instances of water-soluble substances in the normal course of things is noticeably less than one; therefore, the probability of water solubility in the reference class of samples of unspecified substances is significantly less than one. If we ask why a particular sample of unspecified material has dissolved in water, the prior weight of this explanandum event is less than one as referred to the class of samples of unspecified substances. This broad reference class is obviously inhomogeneous with respect to water solubility. If we partition it into the subclass of samples of table salt and samples of substances other than table salt, it turns out that every member of the former subclass is water-soluble. The reference class of samples of table salt is homogeneous with respect to water solubility. The weight for the single case, referred to this homogeneous reference class, is much greater than its prior weight. By referring the explanandum event to a homogeneous reference class and substantially increasing its weight, we have provided an inductive explanation of its occurrence. As the discussion develops, we shall see that the homogeneity of the reference class is the key to the explanation. The increase in weight is a fortunate dividend in many cases.

If we begin with the reference class of samples of table salt, asking why this sample of table salt dissolves in water, we already have a homogeneous reference class. If, however, we subdivide that reference class into hexed and unhexed samples, we have added nothing to the explanation of dissolving, for no new probability value results and we have not made the already homogeneous reference class any more homogeneous. Indeed, we have made matters worse by introducing a statistically irrelevant partition.

The original reference class of samples of unspecified substances can be partitioned into hexed and unhexed samples. If this partition is accomplished by means of a place selection—that is, if the hexing is done without reference to previous knowledge about solubility—the probabilities of water solubility in the subclasses will be no different from the probability in the original reference class. The reference class of hexed samples of unspecified substances is no more homogeneous than the reference class of samples of unspecified substances; moreover, it is narrower. The casting of a dissolving spell is statistically irrelevant to water solubility, so it cannot contribute to the homogeneity of the reference class, and it must not be used in assigning a weight to the single case. For this reason it contributes nothing to the explanation of the fact that this substance dissolves in water.

The vitamin C example involves the same sort of consideration. In the class of colds in general, there is a rather high frequency of recovery within a week. In the narrower reference class of colds for which the victim has taken vitamin C, the frequency of recovery within a week is no different. Vitamin C is not efficacious, and that fact is reflected in the statistical irrelevance of administration of vitamin C to recovery from a cold within a week. Subdivision of the reference class in terms of administration of vitamin C does not yield a more homogeneous reference class and, consequently, does not yield a higher weight for the explanandum event. In similar fashion, we know that noisemaking and shooting off fireworks are statistically irrelevant to the reappearance of the moon after an eclipse, and we know that our friend's loud moaning is statistically irrelevant to finding a wild tiger in Times Square. In none of these horrible examples do the "explanatory facts" contribute anything toward achieving a homogeneous reference class or to an increase of posterior weight over prior weight.

5. The Requirement of Maximal Specificity

In his 1962 essay. "Deductive-Nomological vs. Statistical Explanation," Hempel takes note of what he calls "the ambiguity of statistical systematization," [41] and he invokes a requirement of total evidence to overcome it.[42] This kind of ambiguity is seen to arise out of the possibility of referring a given event to several different reference classes, and Hempel explicitly compares his requirement of total evidence to Reichenbach's rule of selecting the narrowest reference class for which reliable statistics are available.[43] These requirements do not exclude any of the counterexamples introduced above, in particular, the "explanations" of John Jones's recovery from his cold (5) and of his remission of neurotic symptoms (6).

In the 1965 essay, "Aspects of Scientific Explanation," Hempel distinguishes the ambiguity of statistical explanation from the epistemic ambiguity of statistical explanation. The former ambiguity arises from the existence of different reference classes, with different relative frequencies of the attribute in question, to which a given single case may be referred. The latter ambiguity arises from the occurrence of items in our body of accepted scientific knowledge which lead us to refer the given single case to different reference classes. In order to cope with the problem of epistemic ambiguity, Hempel introduces the *requirement of maximal specificity for inductive-statistical explanations* as follows:

Consider a proposed explanation of the basic statistical form

(30)
$$p(G,F) = r$$
$$\frac{Fb}{Gb} \quad [r]$$

Let s be the conjunction of the premises, and, if K is the set of all statements accepted at the given time, let k be a sentence that is logically equivalent to K (in the sense that k is implied by K and in turn implies every sentence in K). Then, to be rationally acceptable in the knowledge situation represented by K, the proposed explanation (30) must meet the following condition (the requirement of maximal specificity): If $s.k$ implies that b belongs to a class F_1, and that F_1 is a subclass of F, then $s.k$ must also imply a statement specifying the statistical probability of G in F_1, say

$$p(G,F_1) = r_1$$

Here, r_1 must equal r unless the probability statement just cited is simply a theorem of mathematical probability theory.[44]

Hempel goes on to remark that

the requirement of maximal specificity, then, is here tentatively put forward as characterizing the extent to which the requirement of total evidence properly applies to inductive-statistical explanations. The general idea thus suggested comes to this: In formulating or appraising I-S explanation, we should take into account all that information provided by K which is of potential *explanatory* relevance to the explanandum event; i.e., all pertinent statistical laws, and such particular facts as might be connected, by the statistical laws, with the explanandum event.[45]

There are two immediate objections to the requirement of maximal specificity as thus formulated. First, like the requirement of total evidence, it fails to rule out such counterexamples as (5), the explanation of John Jones's recovery from his cold on account of taking vitamin C. This "explanation" is not blocked by the requirement, for, as we have noted, there is no narrower class for which the probability of recovery from a cold would be different. The trouble is that the class invoked in the "explanation" is too narrow; the class of people with colds—regardless of medication—already satisfies the requirement. The difficulty seems to arise from the emphasis upon *narrowness* in Reichenbach's formulation of his reference class rule. As I reformulate the rule, we seek the broadest homogeneous reference class. However, this problem is not serious, for it can be circumvented by a minor reformulation of Hempel's requirement. We need simply to add to Hempel's formulation the condition that F is the largest class that satisfies it. This addition would result in what might be called the *requirement of the maximal class of maximal specificity*.

Second, the final sentence in the requirement seems designed to achieve epistemic homogeneity, but it is not strong enough to do so. In explaining the reason for the qualification "r_1 must equal r unless the probability statement just cited is simply a theorem of mathematical probability theory," Hempel points out that $F.G$ is always a subset of F, but obviously $p(G,F.G) = 1$ by the mathematical calculus alone. Hence, without the qualification of the "unless" clause, the requirement of maximal specificity could never be nontrivially fulfilled.

When von Mises attempted to characterize randomness in precisely the sense that is at issue here, he pointed out that the relative frequency in a subsequence must be invariant if the subsequence is determined by a place selection, but that it need not be invariant for subsequences determined by other types of selections. For instance, let F be the class of draws from a particular urn and G the drawing of a red ball. Suppose, as a matter of fact, that the person drawing from the urn sometimes draws with his left hand and sometimes with his right, and that every draw with the left hand produces a red ball, whereas some draws with the right produce balls of other colors. Let H be the class of draws with the left hand. Under these conditions F does not fulfill the requirement of maximal specificity, for $F.H$ is a subclass of F in which G has a different probability than it does in F. Obviously, the fact that all H's are G is not a theorem of the probability calculus; it is a physical fact.

Now, to take a different example for purposes of comparison, suppose that everything is the same as in the foregoing case, except that $p(G,F) = p(G,H.F)$—that is, the frequency with which red appears is the same regardless of which hand is used for the draw, and, so far as H

is concerned, F satisfies the requirement of maximal specificity. Let J be the class of draws of a ball whose color is at the opposite end of the visible spectrum from violet. Clearly, all J's are G, but I consider this to be an extremely highly confirmed matter of fact, not a theorem of the probability calculus or of pure logic. It follows immediately that $p(G,F) \neq p(G,J.F) = 1$. Hence, by Hempel's formulation F still violates the requirement of maximal specificity, and so obviously will any class F except when either all F's are G or no F's are G. In the first case we wanted to say that the class F violates maximal specificity because of the empirical generalization that all F.H's are G; in the second case we were forced unwillingly to admit that F violates maximal specificity because of the empirical generalization that all F.J's are G. The difference is, roughly speaking, that we can know whether a draw is being made with the left hand before we see the result, but we cannot know that a draw results in a color at the opposite end of the visible spectrum from violet until we know what the result is. To characterize this difference precisely might be difficult, but it is required for von Mises's concept of randomness, for my concept of homogeneity, and, I believe, for a satisfactory reformulation of Hempel's requirement of maximal specificity.

When the foregoing two difficulties are overcome, I believe that Hempel's requirement of maximal specificity will become equivalent to my concept of epistemic homogeneity, and, I believe, it will rule out counterexamples of the sort that I have presented above. It will rule out the deductive examples as well as the inductive examples if we construe the deductive examples as limiting cases of inductive explanation and demand that they also satisfy the requirement of the maximal class of maximal specificity. If, however, these two rather straightforward revisions of Hempel's requirement of maximal specificity were all that is needed to fix up all of the difficulties in Hempel's account of explanation of particular events, it would have been an outrageous imposition upon the good humor of the reader to have taken so many pages to say so. The fact is quite otherwise. These rather lengthy discussions have been intended to lay the foundations for further considerations that will lead to far-reaching differences with Hempel's basic conceptions. The point that we have arrived at so far is that requirements (i) through (iv) above (see page 176) still stand, except that (iv) now contains the corrected requirement of the maximal class of maximal specificity. Requirement (i) still stands, and it demands for inductive explanation that the explanandum be made highly probable (to a degree ⩾ some chosen number r). According to this conception, an explanation is still regarded as an argument to the effect that a certain event was to be expected by virtue of certain explanatory facts. I shall maintain that we have still failed to come to grips adequately with the problem of relevance of the explanans to the explanandum.

6. Prior Weights and Degree of Inhomogeneity

One of the fundamental features of explanation according to the approach I have been suggesting is that it concerns the relations between the prior weight and the posterior weight of the explanandum event. I have attempted to indicate how a frequency theorist may get weights from probabilities, by application of the reference class rule, but so far nothing has been said about the prior probability from which the prior weight is to be taken. Let us now consider that question.

If we look at Hempel's schemata (3) and (4) for deductive and inductive explanation, respectively, we see that the form of the conclusion in each case is "x is G." However, the explanatory question to which the proffered explanation attempts to supply an answer has a more complex form. We do not ask, "Why is this thing yellow?" We ask, "Why is this Bunsen flame yellow?" We do not ask, "Why does this thing disappear?" We ask "Why does this streptococcus infection disappear?" In every case, I think, the question takes the form, "Why is this x which is A also B?" The answer then takes the form, "Because this x is also C." C must be an attribute that is statistically relevant to B within the reference class A.

The explanatory question, it seems to me, furnishes an original reference class A to which the explanandum event is referred. The probability of the attribute B within that reference class A is the *prior weight* for purposes of that explanation; normally, the class A would be epistemically homogeneous prior to the furnishing of the explanation. In cases of genuine explanation, the reference class A is not actually homogeneous with respect to the attribute B, and so a further property C is invoked to achieve a homogeneous (or more nearly homogeneous) reference class $A.C$ for purposes of establishing a *posterior weight*. If the original reference class A is homogeneous, the introduction of an irrelevant characteristic C to narrow the reference class is not only without explanatory value but is actually methodologically detrimental. In a later section of this paper, I shall offer a more formal characterization of such explanations, showing how the prior weight and posterior weight figure in the schema.

We can now introduce a quantitative measure of the degree of inhomogeneity of the reference class A which will, in a sense, indicate what has been achieved when that class is partitioned into homogeneous subclasses. Let the sequence $x_1, x_2, \ldots, x_i, \ldots$ be the ordered set A whose degree of inhomogeneity with respect to the attribute B is to be measured. Assume that there exists a unique homogeneous partition $C_1, \ldots, C_k$ of A with respect to B.[46] Let $P(A,B) = p$, and let $P(A.C_j,B) = p_j$. Assign to each element x_i of A the weight $w_i = p_j$ associated with that compartment of the partition C_j to which x_i belongs, that is, if $x_i \in A.C_j$, then $w_i = p_j = P(A.C_j,B)$. Evidently, $p = w$ is the prior weight of each x_i in A with respect to the attribute B, whereas w_i is its posterior weight in the homogeneous partition of A.

If we say, in effect, that w_i is the correct weight to assign to x_i, then we may say that $p - w_i$ is the "error" that would result from the use of the prior weight instead of the posterior weight. Squaring to make this quantity nonnegative, we may take the squared error $(p - w_i)^2$ as a measure of error. Then

$$\sum_{i=1}^{n} (p - w_i)^2$$

is the cumulative squared error involved in using the prior weight on the first n elements of A, and

$$\frac{1}{n} \sum_{i=1}^{n} (p - w_i)^2$$

is the *mean squared error*. The limit of the mean squared error as n increases without bound will be taken to represent the inhomogeneity of the reference class A with respect to the attribute B, but for reasons to be mentioned in a moment, I shall multiply this number by a constant for purposes of defining degree of inhomogeneity.

If we think of the weights w_i simply as numbers that exhibit a certain frequency distribution, then it is clear that p represents their mean value, that is,

$$\lim_{n \to \infty} \frac{1}{n} \sum_{i=1}^{n} w_i = p.$$

Then

$$\lim_{n \to \infty} \frac{1}{n} \sum_{i=1}^{n} (p - w_i)^2 = \sigma^2$$

is a measure of the dispersion of the w_i known as the *variance* (which is the square of the standard deviation σ).[47]

If the reference class A is already homogeneous, the error is identically zero for every i, and the variance is zero. The maximum degree of inhomogeneity seems to be represented by the case in which B occurs with probability $1/2$ in A, but there is a partition of A such that every element of C_1 is B and no element of C_2 is B. These conditions would obtain, for instance, if a sequence of coin tosses consisted of tosses of a two-headed coin randomly interspersed with tosses of a two-tailed coin, each of the two being used with equal frequency. In this case $p = 1/2$, and each w_i is either 0 or 1. Thus, the error on each toss is $1/2$, and the mean squared error (for each n) as well as the variance is $1/4$. In order to have a convenient measure of degree of homogeneity, let us multiply the variance by 4, so that degree of inhomogeneity ranges from 0 to 1. We can then adopt the following formal definition:

The degree of inhomogeneity of class A with respect to attribute $B = 4\sigma^2$.

It is worth noting that the transition from an inhomogeneous reference class A to a set of homogeneous subclasses $A.C_j$ provides an increase of information. In the extreme example given above, when we have only the original inhomogeneous reference class of coin tosses, we know only that the probability of a head is $1/2$, and we know nothing about whether an individual toss will result in a head or a tail. When the class is partitioned into two subclasses, one consisting of tosses with the two-headed coin and the other consisting of tosses with the two-tailed coin, we have complete information regarding the outcome of each toss; each result can be predicted with certainty.[48]

7. Causal and Statistical Relevance

The attempt to explicate explanation in terms of probability, statistical relevance, and homogeneity is almost certain to give rise to a standard objection. Consider the barometer example introduced by Michael Scriven in a discussion of the thesis of symmetry between explanation and prediction [49]—a thesis whose discussion I shall postpone until a later section. If the barometer in my house shows a sudden drop, a storm may be predicted with high reliability. But the barometric reading is only an indicator; it does not cause the storm and, according to Scriven, it does not explain its occurrence. The storm is caused by certain widespread atmospheric conditions, and the behavior of the barometer is merely symptomatic of them. "In explanation we are looking for a *cause*, an event that not only occurred earlier but stands in a *special relation* to the other event. Roughly speaking, the prediction requires only a correlation, the explanation more." [50]

The objection takes the following form. There is a correlation between the behavior of the barometer and the occurrence of storms. If we take the general reference class of days in the vicinity of my house and ask for the probability of a storm, we get a rather low prior probability. If we partition that reference class into two subclasses, namely, days on which there is a sudden drop in the barometer and days on which there is not, we have a posterior probability of a storm in the former class much higher than the prior probability. The new reference class is far more homogeneous than the old one. Thus, according to the view I am suggesting, the drop in barometric reading would seem to explain the storm.

I am willing to admit that symptomatic explanations seem to have genuine explanatory value in the absence of knowledge of causal relations, that is, as long as we do not know that we are dealing only with symptoms. Causal explanations supersede symptomatic ones when they can be given, and when we suspect we are dealing with symptoms, we look hard for a causal explanation. The reason is that a causal explanation provides a more homogeneous reference class than does a symptomatic explanation. Causal proximity increases homogeneity. The reference class of days on which there is a local drop in barometric pressure inside my house, for instance, is more homogeneous than the reference class of days on which my barometer shows a sudden drop, for my barometer may be malfunctioning. Similarly, the reference class of days on which there is a widespread sudden drop in atmospheric pressure is more homogeneous than the days on which there is a local drop, for the house may be tightly sealed or the graduate students may be playing a joke on me. [51] It is not that we obtain a large increase in the probability of a storm as we move from one of these reference classes to another; rather, each progressively better partitioning makes the preceding partitioning *statistically irrelevant*.

It will be recalled that the property C is statistically irrelevant to the attribute B in the reference class A iff $P(A,B) = P(A.C,B)$. The probability of a storm on a day when there is a sudden drop in atmospheric pressure and when my barometer executes a sudden drop is precisely the same as the probability of a storm on a day when there is a sudden widespread drop in atmospheric pressure. To borrow a useful notion from Reichenbach, we may say that the sudden widespread drop in atmospheric pressure *screens off* the drop in barometer reading from the occurrence of the storm.[52] The converse relation does not hold. The probability of a storm on a day when the reading on my barometer makes a sudden drop is not equal to the probability of a storm on a day when the reading on my barometer makes a sudden drop and there is a sudden widespread drop in the atmospheric pressure. The sudden drop in barometric reading does not screen off the sudden widespread drop in atmospheric pressure from the occurrence of the storm.

More formally, we may say that D screens off C from B in reference class A iff (if and only if)

$$P(A.C.D,B) = P(A.D,B).$$

For purposes of the foregoing example, let A = the class of days in the vicinity of my house, let B = the class of days on which there is an occurrence of a storm, let C = the class of days on which there is a sudden drop in reading on my barometer, and let D = the class of days on which there is a widespread drop in atmospheric pressure in the area in which my house is located. By means of this formal definition, we see that D screens off C from B, but C does not screen off D from B. The screening-off relation is, therefore, not symmetrical, although the relation of statistical relevance is symmetrical.[53]

When one property in terms of which a statistically relevant partition in a reference class can be effected screens off another property in terms of which another statistically relevant partition of that same reference class can be effected, then the screened-off property must give way to the property which screens it off. This is the *screening-off rule*. The screened-off property then becomes irrelevant and no longer has explanatory value. This consideration shows how we can handle the barometer example and a host of others, such as the explanation of measles in terms of spots, in terms of exposure to someone who has the disease, and in terms of the presence of the virus. The unwanted "symptomatic explanations" can be blocked by use of the screening-off concept, which is defined in terms of statistical irrelevance alone. We have not found it necesary to introduce an independent concept of causal relation in order to handle this problem. Reichenbach believed it was possible to define causal relatedness in terms of screening-off relations; but whether his program can be carried through or not, it seems that many causal relations exhibit the desired screening-off relations.[54]

8. Explanations with Low Weight

According to Hempel, the basic requirement for an inductive explanation is that the posterior weight (as I have been describing it) must be high, whereas I have been suggesting that the important characteristic is the increase of the posterior weight over the prior weight as a result of incorporating the event into a homogeneous reference class. The examples discussed thus far satisfy both of these desiderata, so they do not serve well to discriminate between the two views. I would maintain, however, that when the prior weight of an event is very low, it is not necessary that its posterior weight be made high in order to have an inductive explanation. This point is illustrated by the well-known paresis example.[55]

No one ever contracts paresis unless he has had latent syphilis which has gone untreated, but only a small percentage of victims of untreated latent syphilis develop paresis. Still, it has been claimed, the occurrence of paresis is explained by the fact that the individual has had syphilis. This example has been hotly debated because of its pertinence to the issue of symmetry between explanation and prediction—an issue I still wish to postpone. Nevertheless, the following observations are in order. The prior probability of a person contracting paresis is very low, and the reference class of people in general is inhomogeneous. We can make a statistically relevant partition into people who have untreated latent syphilis and those who do not. (Note that latent syphilis screens off primary syphilis and secondary syphilis.) The probability that a person with untreated latent syphilis will contract paresis is still low, but it is considerably higher than the prior probability of paresis among people in general. To cite untreated latent syphilis as an explanation of paresis is correct, for it does provide a partition of the general reference class which yields a more homogeneous reference class and a higher posterior weight for the explanandum event.

When the posterior weight of an event is low, it is tempting to think that we have not fully explained it. We are apt to feel that we have not yet found a completely homogeneous reference class. If only we had fuller understanding, we often believe, we could sort out causal antecedents in order to be able to say which cases of untreated latent syphilis will become paretic and which ones will not. With this knowledge we would be able to partition the reference class of victims of untreated latent syphilis into two subclasses in terms of these causal antecedents so that all (or an overwhelming majority of) members of one subclass will develop paresis whereas none (or very few) of the members of the other will become paretic. This conviction may be solidly based upon experience with medical explanation, and it may provide a sound empirical basis for the search for additional explanatory facts—more relevant properties in terms of which to improve the homogeneity of the reference class. Nevertheless, the reference class of untreated latent syphilitics is (as I understand it) epistemically homogeneous in terms of our present knowledge, so we have provided the most adequate explanation possible in view of the knowledge we possess.[56]

A parallel example could be constructed in physics where we have much greater confidence in the actual homogeneity of the reference class. Suppose we had a metallic substance in which one of the atoms experienced radioactive decay within a particular, small time period, say one minute. For purposes of the example, let us suppose that only one such decay occurred within that time period. When asked why that particular atom decayed, we might reply that the substance is actually an alloy of two metals, one radioactive (for example, uranium 238) and one stable (for example, lead 206). Since the half-life of U^{238} is 4.5×10^9 years, the probability of a given uranium atom's decaying in an interval of one minute is not large, yet there is explanatory relevance in pointing out that the atom that did decay was a U^{238} atom.[57] According to the best theoretical knowledge now available, the class of U^{238} atoms is homogeneous with respect to radioactive decay, and there is in principle no further relevant partition that can be made. Thus, it is not necessary to suppose that examples such as the paresis case derive their explanatory value solely from the conviction that they are partial explanations which can someday be made into full explanations by means of further knowledge.[58]

There is one further way to maintain that explanations of improbable events involve, nevertheless, high probabilities. If an outcome is improbable (though not impossible) on a given trial, its occurrence at least once in a sufficiently large number of trials can be highly probable. For example, the probability of getting a double six on a toss of a pair of standard dice is $1/36$; in twenty-five tosses the probability of at least one double six is over one-half. No matter how small the probability p of an event on a single trial, provided $p > 0$, and no matter how large r, provided $r < 1$, there is some n such that the probability of at least one occurrence in n trials is greater than r.[59] On this basis, it might be claimed, we explain improbable events by saying, in effect, that given enough trials the event is probable. This is a satisfactory explanation of the fact that certain types of events occur occasionally, but it still leaves open the question of how to explain the fact that this partcuilar improbable event happened on this particular occasion.

To take a somewhat more dramatic example, each time an alpha particle bombards the potential barrier of a uranium nucleus, it has a chance of about 10^{-38} of tunneling through and escaping from the nucleus; one can appreciate the magnitude of this number by noting that whereas the alpha particle bombards the potential barrier about 10^{21} times per second, the half-life of uranium is of the order of a billion years.[60] For any given uranium atom, if we wait long enough, there is an overwhelmingly large probability that it will decay, but if we ask, "Why did the alpha particle tunnel out on this particular bombardment of the potential barrier?" the only answer is that in the homogeneous reference class of approaches to the barrier, it has a 1-in-10^{38} chance of getting through. I do not regard the fact that it gets through on a particular trial inexplicable, but certainly anyone who takes explanations to be arguments showing that the event was to be expected must conclude that this fact defies all explanation.

9. Multiple Homogeneity

The paresis example illustrates another important methodological point. I have spoken so far as if one homogeneous reference class were sufficient for the explanation of a particular event. This, I think, is incorrect. When a general reference class is partitioned, we can meaningfully ask about the homogeneity of each subclass in the partition (as we did in defining degree of inhomogeneity above). To be sure, when we are attempting to provide an explanation of a particular explanandum event x, we focus primary attention upon the subclass to which x belongs. Nevertheless, I think we properly raise the question of the homogeneity of other subclasses when we evaluate our explanation. In the paresis example we may be convinced that the reference class of untreated latent syphilitics is inhomogeneous, but the complementary class is perfectly homogeneous. Since no individuals who do not have untreated latent syphilis develop paresis, no partition statistically relevant to the development of paresis can be made in the reference class of people who do not have untreated latent syphilis.

Consider the table salt example from the standpoint of the homogeneity of more than one reference class. Although the reference class of samples of table salt is completely homogeneous for water solubility, the complementary class certainly is not. It is possible to make further partitions in the class of samples of substances other than table salt which are statistically relevant to water solubility: samples of sand, wood, and gold are never water soluble; samples of baking soda, sugar, and rock salt always are.

If we explain the fact that this sample dissolves in water by observing that it is table salt and all table salt is water soluble, we may feel that the explanation is somewhat inadequate. Some theorists would say that it is an adequate explanation, but that we can equally legitimately ask for an explanation of the general fact that table salt is water soluble. Although I have great sympathy with the idea that general facts need explanation and are amenable to explanation, I think it is important to recognize the desideratum of homogeneity of the complementary reference class. I think we may rightly claim fully adequate explanation of a particular fact when (but not necessarily only when) the original reference class A, with respect to which its prior probability is assessed, can be partitioned into two subclasses $A.C$ and $A.\overline{C}$, each of which is homogeneous for the attribute in question. In the ideal case all $A.C$'s are B and no $A.\overline{C}$'s are B—that is, if x is A, then x is B if and only if x is C. However, there is not reason to believe that Nature is so accommodating as to provide in all cases even the possibility in principle of such fully deterministic explanations.

We now have further reason to reject the dissolving spell explanation for the fact that a sample of table salt dissolves in water. If we partition the general referrence class of samples of unspecified substances into hexed samples of table salt and all other samples of substances, this latter reference class is less homogeneous than the class of all samples of substances other than table salt, for we know that all unhexed samples of table salt are water soluble. To make a statistically irrelevant partition not only reduces the available statistics; it also reduces the homogeneity of the complementary reference class. This consideration also applies to such other examples as John Jones and his wife's birth control pills.

It would be a mistake to suppose that it must always be possible in principle to partition the original reference class A into two homogeneous subclasses. It may be that the best we can hope for is a partition into k subclasses $A.C_k$, each completely homogeneous, and such that $P(A.C_i, B) \neq P(A.C_j, B)$ if $i \neq j$. This is the *multiple homogeneity rule*. It expresses the fundamental condition for adequate explanation of particular events, and it will serve as a basis for the general characterization of deductive and inductive explanation.

The multiple homogeneity requirement provides, I believe, a new way to attack a type of explanation that has been extremely recalcitrant in the face of approaches similar to Hempel's, namely, the so-called *functional explanation*. We are told, for example, that it is the function of hemoglobin to transport oxygen from the lungs to the cells in the various parts of the organism. This fact is taken to provide an explanation for the blood's hemoglobin content. Trouble arises when we try to fit such explanations to schemata like (3) for deductive explanation, for we always seem to end up with a necessary condition where we want a sufficient condition. Consider the following:

(11) Hemoglobin transports oxygen from the lungs to the other parts of the body (in an animal with a normally functioning circulatory system).
This animal has hemoglobin in its blood.

This animal has oxygen transported from its lungs to the other parts of its body.

Here we have a valid deductive argument, but unfortunately the explanandum is a premise and the conclusion is part of the explanans. This will never do. Let us interchange the second premise and the conclusion, so that the explanandum becomes the conclusion and the explanans consists of all of the premises, as follows:

(12) Hemoglobin transports oxygen from the lungs to the other parts of the body.
This animal has oxygen transported from its lungs to other parts of its body.

This animal has hemoglobin in its blood.

Now we get an obviously invalid argument. In order to have a standard deductive explanation of the Hempel variety, the particular explanatory fact cited in the explanans must, in the presence of the general laws in the explanans, be a sufficient condition for the explanandum event. In the hemoglobin case the explanatory fact is a necessary condition of the explanandum event, and that is the typical situation with functional explanations.

Although the suggestion that we admit that necessary conditions can have explanatory import might seem natural at this point, it meets with an immediate difficulty. We would all agree that the fact that a boy was born does not contribute in the least to the explanation of his becoming a juvenile delinquent, but it certainly was a necessary condition of that occurrence. Thus, in general necessary conditions cannot be taken to have any explanatory function, but it is still open to us to see whether under certain special conditions necessary conditions can function in genuine explanations. The approach I am suggesting enables us to do just that.

Returning to the paresis example, suppose that medical science could discover in the victims of untreated latent syphilis some characteristic P that is present in each victim who develops paresis and is lacking in each one who escapes paresis. Then, in the reference class S of untreated latent syphilitics, we could make a partition in which every member of one subclass would develop paresis, B, whereas no member of the other subclass would fall victim to it. In such a case, within S, P would be a necessary and sufficient condition for B. If, however, medical research were slightly less successful and found a characteristic Q that is present in almost every member of S who develops paresis, but is lacking in almost every one who escapes paresis, then we would still have a highly relevant partition of the original reference class into two subclasses, and this would be the statistical analogue of the discovery of necessary and sufficient conditions. Thus, if

$$P(S,B) = p_1; \ P(S.Q,B) = p_2; \ P(S.\overline{Q},B) = p_3$$

where

$$p_1 \neq p_2 \text{ and } p_1 \neq p_3,$$

then Q is a statistically relevant statistical analogue of a necessary condition. To concern ourselves with the homogeneity of $S.\overline{Q}$ is to raise the question of further statistically relevant necessary conditions.

I have argued above that sufficient conditions have explanatory import only if they are relevant. Insertion of a sodium salt into a Bunsen flame is a relevant sufficient condition of the flame turning yellow; taking birth control pills is an irrelevant sufficient condition for nonpregnancy in a man. Similarly, being born is an irrelevant necessary condition for a boy becoming delinquent, for it makes no partition whatever in the reference class of boys, let alone a relevant one. However, as argued above, latent untreated syphilis is a relevant necessary condition of paresis, and it has genuine explanatory import.

Let us see whether such considerations are of any help in dealing with functional explanations. According to classical Freudian theory, dreams are functionally explained as wish fulfillments. For instance, one night, after a day of fairly frustrating work on a review of a book containing an example about dogs who manifest the Absolute by barking loudly, I dreamed that I was being chased by a pack of barking dogs. Presumably, my dream had the function of enabling me to fulfill my wish to escape from the problem of the previous day, and thus to help me preserve my sleep. The psychoanalytic explanation does not maintain that only one particular dream would have fulfilled this function; presumably many others would have done equally well. Thus, the fact that I had that particular wish did not entail (or even make it very probable) that I would have that particular dream, but the fact that I had the dream was, allegedly, a sufficient condition for the fulfilment of the wish. The explanatory fact, the unfulfilled wish, is therefore taken to be a necessary condition of the explanandum event—the dream—but not a sufficient condition. According to the psychoanalytic theory, the unfulfilled wish is a relevant necessary condition, however, for if we make a partition of the general reference class of nights of sleep in terms of the events of the preceding day, then the probability of this particular dream is much greater in the narrower reference class than it was in the original reference class.[61] By the way, the author of the book I was reviewing is Stephen Barker.

When we try to give a functional explanation of a particular mechanism, such as the presence of hemoglobin or the occurrence of a dream, we may feel some dissatisfaction if we are unable eventually to explain why that particular mechanism occurred rather than another which would apparently have done the same job equally well. Such a feeling is, I would think, based upon the same considerations that lead us to feel dissatisfied with the explanation of paresis. We know some necessary conditions, in all of these cases, but we believe that it is possible in principle to find sufficient conditions as well, and we will not regard the explanations as complete until we find them. How well this feeling is justified depends chiefly upon the field of science in which the explanation is sought, as the example of radioactive decay of a U^{238} atom shows.

The fact that the multiple homogeneity rule allows for the explanatory value of relevant necessary conditions and their statistical analogues does not, by itself, provide a complete account of functional explanation, for much remains to be said about the kinds of necessary conditions that enter into functional explanations and their differences from necessary conditions that are not in any sense functional. Our approach has, however, removed what has seemed to many authors the chief stumbling block in the path to a satisfactory theory of functional explanation.

10. Explanation Without Increase of Weight

It is tempting to suppose, as I have been doing so far and as all the examples have suggested, that explanation of an explanandum event somehow confers upon it a posterior weight that is greater than its prior weight. When I first enunciated this principle in section 3, however, I indicated that, though heuristically beneficial, it should not be considered fully accurate. Although most explanations may conform to this general principle, I think there may be some that do not. It is now time to consider some apparent exceptions and to see whether they are genuine exceptions.

Suppose, for instance, that a game of heads and tails is being played with two crooked pennies, and that these pennies are brought in and out of play in some irregular manner. Let one penny be biased for heads to the extent that 90 percent of the tosses with it yield heads; let the other be similarly biased for tails. Furthermore, let the two pennies be used with equal frequency in this game, so that the overall probability of heads is one-half. (Perhaps a third penny, which is fair, is tossed to decide which of the two biased pennies is to be used for any given play.) Suppose a play of this game results in a head; the prior weight of this event is one-half. The general reference class of plays can, however, be partitioned in a statistically relevant way into two homogeneous reference classes. If the toss was made with the penny biased for heads, the result is explained by that fact, and the weight of the explanandum event is raised from 0.5 to 0.9.

Suppose, however, that the toss were made with the penny biased for tails; the explanandum event is now referred to the other subclass of the original reference class, and its weight is decreased from 0.5 to 0.1. Do we want to say in this case that the event is thereby explained? Many people would want to deny it, for such cases conflict with their intuitions (which, in many cases, have been significantly conditioned by Hempel's persuasive treatment) about what an explanation ought to be. I am inclined, on the contrary, to claim that this is genuine explanation. There are, after all, improbable occurrences—such events are not explained by making them probable. Is it not a peculiar prejudice to maintain that only those events which are highly probable are capable of being explained—that improbable events are in principle inexplicable? Any event, regardless of its probability, is amenable to explanation, I believe; in the case of improbable events, the correct explanation is that they are highly improbable occurrences which happen, nevertheless, with a certain definite frequency. If the reference class is actually homogeneous, there are no other circumstances with respect to which they are probable. No further explanation can be required or can be given.[62]

There are various reasons for which my view might be rejected. In the first place, I am inclined to think that the deterministic prejudice may often be operative—namely, that x is B is not explained until x is incorporated within a reference class all of whose members are B. This is the feeling that seemed compelling in connection with the paresis example. In the discussion of that example, I indicated my reasons for suggesting that our deterministic hopes may simply be impossible to satisfy. In an attempt to undercut the hope generated in medical science that determinism would eventually triumph, I introduced the parallel example of the explanation of a radioactive decay in an alloy of lead and uranium, in which case there is strong reason to believe that the reference class of uranium 238 atoms is strictly homogeneous with respect to disintegration.

In order to avoid being victimized by the same deterministic hope in the present context, we could replace the coin-tossing example at the beginning of this section with another example from atomic physics. For instance, we could consider a mixture of uranium 238 atoms, whose half-life is 4.5×10^9 years, and polonium 214 atoms, whose half-life is 1.6×10^{-4} seconds.[63] The probability of disintegration of an unspecified atom in the mixture is between that for atoms of U^{238} and Po^{214}. Suppose that within some small specified time interval a decay occurs. There is a high probability of a polonium atom disintegrating within that interval, but a very low probability for a uranium atom. Nevertheless, a given disintegration may be of a uranium atom, so the transition from the reference class of a mixture of atoms of the two types to a reference class of atoms of U^{238} may result in a considerable lowering of the weight. Nevertheless, the latter reference class may be unqualifiedly homogeneous. When we ask why that particular atom disintegrated, the answer is that it was a U^{238} atom, and there is a small probability that such an atom will disintegrate in a short time interval.

If, in the light of modern developments in physics and philosophy, determinism no longer seems tenable as an a priori principle, we may try to salvage what we can by demanding that an explanation that does not necessitate its explanandum must at least make it highly probable. This is what Hempel's account requires. I have argued above, in the light of various examples, that even this demand is excessive and that we must accept explanations in which the explanandum event ends up with a low posterior weight. "Well, then," someone might say, "if the explanation does not show us that the event was to be expected, at least it ought to show us that the event was to be expected somewhat more than it otherwise would have been." But this attitude seems to derive in an unfortunate way from regarding explanations as arguments. At this juncture it is crucial to point out that the emphasis in the present account of explanation is upon achieving a relevant partition of an inhomogeneous reference class into homogeneous subclasses. On this conception an explanation is not an argument that is intended to produce conviction; instead, it is an attempt to assemble the factors that are relevant to the occurrence of an event. There is no more reason to suppose that such a process will increase the weight we attach to such an occurrence than there is to suppose that it will lower it. Whether the posterior weight is higher than or lower than the prior weight is really beside the point. I shall have more to say later about the function of explanations.

Before leaving this topic, I must consider one more tempting principle. It may seem evident from the examples thus far considered that an explanation must result in a change—an increase or a decrease—in the transition from the prior weight to the posterior weight. Even this need not occur. Suppose, for instance, that we change the coin-tossing game mentioned at the outset of this section by introducing a fair penny into the play. Now there are three pennies brought into play randomly: one with a probability of 0.9 for heads, one with a probability of 0.5 for heads, and one with a probability of 0.1 for heads. Overall, the probability of heads in the game is still one-half. Now, if we attempt to explain a given instance of a head coming up, we may partition the original reference class into three homogeneous subclasses, but in one of these three the probability of heads is precisely the same as it is in the entire original class. Suppose our particular head happens to belong to that subclass. Then its prior weight is exactly the same as its posterior weight, but I would claim that explanation has occurred simply by virtue of the relevant partition of the original nonhomogeneous reference class. This makes sense if one does not insist upon regarding explanations as arguments.

11. Some Paradigms of Explanation

Before attempting a general quasi-formal characterization of the explanation of particular events, I should like to follow a common practice and introduce some examples that seem to me to deserve the status of paradigms of explanation. They will, I believe, differ in many ways from the paradigms that are usually offered; they come from a set of investigations, conducted mainly by Grünbaum and Reichenbach, concerning the temporal asymmetry (or anisotropy) of the physical world.[64] Because of the close connections between causality and time on the one hand, and between causality and explanation on the other, these investigations have done a great deal to elucidate the problem of explanation.

Given a thermally isolated system, there is a small but nonvanishing probability that it will be found in a low entropy state. A permanently closed system will from time to time, but very infrequently, spontaneously move from a state of high entropy to one of low entropy, although by far the vast majority of its history is spent in states of high entropy. Let us take this small probability as the prior probability that a closed system will be in a low entropy state. Suppose, now, that we examine such a system and find that it is in a low entropy state. What is the explanation? Investigation reveals that the system, though closed at the time in question, had recently interacted with its environment. The low entropy state is explained by this interaction. The low entropy of the smaller system is purchased by an expenditure of energy in the larger environmental system that contains it and by a consequent increase in the entropy of the environment.

For example, there is a small but nonvanishing probability that an ice cube will form spontaneously in a thermos of cool water. Such an occurrence would be the result of a chance clustering of relatively nonenergetic molecules in one particular place in the container. There is a much greater chance that an ice cube will be present in a thermos of cool water to which an ice cube has just been added. Even if we had no independent knowledge about an ice cube having been added, we would confidently infer it from the mere presence of the ice cube. The low entropy state is explained by a previous interaction with the outside world, including a refrigerator that manufactures the ice cubes.

Suppose, for a further example, that we found a container with two compartments connected by an opening, one compartment containing only oxygen and the other only nitrogen. In this case, we could confidently infer that the two parts of the container had been separately filled with oxygen and nitrogen and that the connecting window had recently been opened—so recently that no diffusion had yet taken place. We would not conclude that the random motions of the molecules had chanced to separate the two gases into the two compartments, even though that event has a small but nonvanishing probability. This otherwise improbable state of affairs is likewise explained in terms of a recent interaction with the environment.

In examples of these kinds, the antecedent reference class of states of closed systems is recognized as inhomogeneous, and it is partitioned into those states which follow closely upon an interaction with the outside world and those that do not. Within the former homogeneous subclass of the original reference class, the probability of low entropy states is much higher than the probability of such states referred to the unpartitioned inhomogeneous reference class. In such cases we, therefore, have explanations of the low entropy states.

The foregoing examples involve what Reichenbach calls "branch systems." In his extended and systematic treatment of the temporal asymmetry of the physical world, he accepts it as a fundamental fact that our region of the universe has an abundant supply of these branch systems. During their existence as separate systems they may be considered closed, but they have not always been isolated, for each has a definite beginning to its history as a closed system. If we take one such system and suppose it to exist from then on as a closed system, we can consider its successive entropy states as a probability sequence, and the probability of low entropy in such a sequence is very low. This probability is a *one-system probability* referred to a *time ensemble;* as before, let us take it as a prior probability. If we examine many such systems, we find that a large percentage of them are in low entropy states shortly after their inceptions as closed systems. There is, therefore, a much higher probability that a very young branch system is in a low entropy state. This is the posterior probability of a low entropy state for a system that has recently interacted with its environment; it is a *many-system probability* referred to a *space ensemble.* The time ensemble turns out to be an inhomogeneous reference class; the space ensemble yields homogeneous reference classes.[65] Applying this consideration to our ice cube example, we see that the presence of the ice cube in the thermos of cool water is explained by replacing the prior weight it would have received from the reference class of states of this system as an indefinitely isolated container of water with the weight it receives from the reference class of postinteraction states of many such containers of water when the interaction involves the placing of an ice cube within it.

Even if we recognize that the branch systems have not always existed as closed systems, it is obviously unrealistic to suppose that they will remain forever isolated. Instead, each branch system exists for a finite time, and at each temporal end it merges with its physical environment. Each system exhibits low entropy at one temporal end and high entropy at the other. The basic fact about the branch systems is that the vast majority of them exhibit low entropy states at the same temporal end. We can state this fact without making any commitment whatever as to whether that end is the earlier or the later end. However, the fact that the systems are alike in having low entropy at the same end can be used to provide coordinating definitions of such temporal relations as *earlier* and *later*. We arrive at the usual temporal nomenclature if we say that the temporal end at which the vast majority of branch systems have low entropy is the earlier end and the end at which most have high entropy is the later end. It follows, of course, that the interaction with the environment at the low entropy end precedes the low entropy state.

Reichenbach takes this relationship between the initial interaction of the branch system with its environment and the early low entropy state to be the most fundamental sort of *producing*.[66] The interaction produces the low entropy state, and the low entropy state is the product (and sometimes the record) of the interaction. Producing is, of course, a thoroughly causal concept; thus, the relation of interaction to orderly low entropy state becomes the paradigm of causation on Reichenbach's view, and the explanation of the ice cube in the thermos of water is an example of the most fundamental kind of causal explanation. The fact that this particular interaction involves a human act is utterly immaterial, for many other examples exist in which human agency is entirely absent.

Reichenbach's account, if correct, shows why causal explanation is temporally asymmetrical. Orderly states are explained in terms of previous interactions; the interactions that are associated with low entropy states do not follow them. It appears that the temporal asymmetry of causal explanation is preserved even when the causal concepts are extended to refer to reversible mechanical processes. It is for this reason, I imagine, that we are willing to accept an explanation of an eclipse in terms of laws of motion and *antecedent* initial conditions, but we feel queasy, to say the least, about claiming that the same eclipse can be *explained* in terms of the laws of motion and *subsequent* initial conditions, even though we may *infer* the occurrence of the eclipse equally well from either set of initial conditions. I shall return to this question of temporal asymmetry of explanation in the next section.

The foregoing examples are fundamentally microstatistical. Reichenbach has urged that macrostatistical occurrences can be treated analogously.[67] For instance, a highly ordered arrangement in a deck of cards can be explained by the fact that it is newly opened and was arranged at the factory. Decks of cards occasionally get shuffled into highly ordered arrangements—just as closed thermodynamic systems occasionally develop low entropy states—but much more frequently the order arises from a deliberate arrangement. Reichenbach likewise argues that common causes macrostatistically explain improbable "coincidences." For example, suppose all the lights in a certain block go off simultaneously. There is a small probability that in every house the people decided to go to bed at precisely the same instant. There is a small probability that all the bulbs burned out simultaneously. There is a much greater probability of trouble at the power company or in the transmission lines. When investigation reveals that a line was downed by heavy winds, we have an explanation of a coincidence with an extremely low prior weight. A higher posterior weight results when the coincidence is assigned to a homogeneous reference class. The same considerations apply, but with much more force, when it is the entire eastern seaboard of the United States that suffers a blackout.

Reichenbach believed that it is possible to establish temporal asymmetry on the basis of macrostatistical relations in a manner quite analogous to the way in which he attempted to base it upon macrostatistical facts. His point of departure for this argument is the fact that certain improbable coincidences occur more often than would be expected on the basis of chance alone. Take, as an example, all members of a theatrical company falling victim to a gastrointestinal illness on the same evening. For each individual there is a certain probability (which may, of course, vary from one person to another) that he will become ill on a given day. If the illnesses of the members of the company are independent of one another, the probability of all members of the company becoming ill on the same day is simply the product of all the separate probabilities; for the general multiplication rule says (for two events) that

(1) $$P(A,B.C) = P(A,B) \times P(A.B,C),$$

but if

(2) $$P(A,C) = P(A.B,C),$$

we have the special multiplication rule,

(3) $$P(A,B.C) = P(A,B) \times P(A,C).$$

When relation (3) holds, we say that the events B and C are independent of each other; this is equivalent to relation (2), which is our definition of statistical irrelevance.

We find, as a matter of fact, that whole companies fall ill more frequently than would be indicated if the separate illnesses were all statistically independent of one another. Under these circumstances, Reichenbach maintains, we need a causal explanation for what appears to be an improbable occurrence. In other words, we need a causal explanation for the statistical relevance of the illnesses of the members of the company to one another. The causal explanation may lie in the food that all of the members of the company ate at lunch. Thus, we do not assert a direct causal relation between the illness of the leading man and that of the leading lady; it was not the fact that the leading man got sick that caused the illness of the leading lady (or the actors with minor roles, the stand-ins, the prop men, etc.). There is a causal relation, but it is via a common cause—the spoiled food. The common meal screens off the statistical relevance of the illness of the leading man to that of the leading lady. The case is parallel to that of the barometer and the storm, and the screening off works in the same way.

Reichenbach maintains that coincidences of this sort can be explained by common causes. The illness of the whole company is explained by the common meal; the blackout of an entire region is explained by trouble at the power source or in the distribution lines. If we take account of the probability of spoiled food's being served to an entire group, the frequency with which the members all become victims of a gastrointestinal illness on the same day is not excessive. If we take into account the frequency with which power lines go down or of trouble at the distribution station, etc., the frequency with which whole areas simultaneously go dark is not excessive.

Such occurrences are not explained in terms of common effects. We do not seriously claim that all members of the company became ill because events were conspiring (with or without conscious intent of some agency, natural or supernatural) to bring about a cancellation of the performance; we do not believe that the future event explains the present illness of the company. Similarly, we do not explain the blackout of the eastern seaboard in terms of events conspiring to make executives impregnate their secretaries. The situation is this: in the absence of a common cause, such as spoiled food, the cancellation of the play does not make the coincidental illness of the company any more probable than it would have been as a product of probabilities of independent events.

By contrast, the presence of a common cause such as the common meal does make the simultaneous illness more probable than it would have been as a product of independent events, and that is true whether or not the common effect occurs. The efficacy of the common cause is not affected by the question of whether the play can go on, with a sick cast or with substitutes. The probability of all the lights going out simultaneously, in the absence of a common cause, is unaffected by the question of how the men and women, finding themselves together in the dark for the entire night without spouses, behave themselves. The net result is that coincidences of the foregoing type are to be explained in terms of causal antecedents, not subsequent events. The way to achieve a relevant subdivision of the general reference class is in terms of antecedent conditions, not later ones. This seems to be a pervasive fact about our macroworld, just as it seems to characterize the microworld.

I believe that an important feature of the characterization of explanation I am offering is that it is hospitable to paradigms of the two kinds I have discussed in this section, the microstatistical examples and the macrostatistical examples. In all these cases, there is a prior probability of an event which we recognize as furnishing an inappropriate weight for the event in question. We see that the reference class for the prior probability is inhomogeneous and that we can make a relevant partition. The posterior probability which arises from the new homogeneous reference class is a suitable weight to attach to the single event. It is the introduction of statistically relevant factors for partitioning the reference class that constitutes the heart of these explanations. If these examples are, indeed, paradigms of the most fundamental types of explanation, it is a virtue of the present account that it handles them easily and naturally.

12. The Temporal Asymmetry of Explanation

It is also a virtue of the present account, I believe, that it seems to accommodate the rather puzzling temporal asymmetry of explanation already noted above—the fact that we seem to insist upon explaining events in terms of earlier rather than later initial conditions. Having noted that both the microstatistical approach and the macrostatistical approach yield a fundamental temporal asymmetry, let us now use that fact to deal with a familiar example. I shall attempt to show how Reichenbach's principle of the common cause, introduced in connection with the macrostatistical examples discussed above, helps us to establish the temporal asymmetry of explanation.

Consider Silvain Bromberger's flagpole example, which goes as follows.[68] On a sunny day a flagpole of a particular height casts a shadow of some particular length depending upon the elevation of the sun in the sky. We all agree that the position of the sun and the height of the flagpole explain the length of the shadow. Given the length of the shadow, however, and the position of the sun in the sky, we can equally infer the height of the flagpole, but we rebel at the notion that the length of the shadow explains the height of the flagpole. It seems to me that the temporal relations are crucial in this example. Although the sun, flagpole, and shadow are perhaps commonsensically regarded as simultaneous, a more sophisticated analysis shows that physical processes going on in time are involved. Photons are emitted by the sun, they travel to the vicinity of the flagpole where some are absorbed and some are not, and those which are not go on to illuminate the ground. A region of the ground is not illuminated, however, because the photons traveling toward that region were absorbed by the flagpole. Clearly the interaction between the photons and the flagpole temporally precedes the interaction between the neighboring photons and the ground. The reason that the explanation of the length of the shadow in terms of the height of the flagpole is acceptable, whereas the "explanation" of the height of the flagpole in terms of the length of the shadow is not acceptable, seems to me to hinge directly upon the fact that there are causal processes with earlier and later temporal stages. It takes only a very moderate extension of Reichenbach's terminology to conclude that the flagpole produces the shadow in a sense in which the shadow certainly does not produce the flagpole.

If we give up the notion that explanations are arguments, there is no need to be embarrassed by the fact that we can oftentimes infer earlier events from later ones by means of nomological relations, as in the cases of the eclipse and the flagpole. I have been arguing that relevance considerations are preeminent for explanations; let us see whether this approach provides a satisfactory account of the asymmetry of explanation in the flagpole case. The apparent source of difficulty is that, under the general conditions of the example, the flagpole is relevant to the shadow and the shadow is relevant to the flagpole. It does not follow, however, that the shadow explains the flagpole as well as the flagpole explains the shadow.

In order to analyze the relevance relations more carefully, I shall reexamine a simplified version of the example of the illness in the theatrical company and compare it with a simplified version of the flagpole example. In both cases the screening-off relation will be employed in an attempt to establish the temporal asymmetry of the explanation. The general strategy will be to show that a common cause screens off a common effect and, consequently, by using the screening-off rule, that the explanation must be given in terms of the common cause and not in terms of the common effect. In order to carry out this plan, I shall regard the flagpole as an orderly arrangement of parts that requires an explanation of the same sort as does the coincidental illnesses of the actors.

Consider, then, the simple case of a theatrical company consisting of only two people, the leading lady and the leading man. Let A be our general reference class of days, let M be the illness of the leading man, let L be the illness of the leading lady, let F be a meal of spoiled food that both eat, and let C be the cancellation of the performance. Our previous discussion of the example has shown that the simultaneous illness occurs more frequently than it would if the two were independent, that is,

$$P(A,L.M) > P(A,L) \times P(A,M),$$

and that the common meal is highly relevant to the concurrent illnesses, that is,

$$P(A.F,L.M) > P(A,L.M) > P(A.\bar{F},L.M).$$

It is not true, without further qualification, that the eating of spoiled food F screens off the cancellation of the performance C from the joint illness $L.M$. Although the eating of spoiled food does make it probable that the two actors will be ill, the fact that the performance has been cancelled supplies further evidence of the illness and makes the joint illness more probable. At this point experiment must be admitted. It is clearly possible to arrange things so that the play goes on, illness or no illness, by providing substitutes who never eat with the regular cast. Likewise, it is a simple matter to see to it that the performance is cancelled, whether or not anyone is ill. Under these experimental conditions we can see that the probability of the joint illness, given the common meal with spoiled food, is the same whether or not the play is performed, that is,

$$P(A.F.C,L.M) = P(A.F.\bar{C},L.M).$$

From this it follows that

$$P(A.F.C,L.M) = P(A.F,L.M).$$

If we approach the common meal of spoiled food in the same experimental fashion, we can easily establish that it is not screened off by the cancellation of the performance. We can arrange for the two leading actors to have no meals supplied from the same source and ascertain the probability $P(A.\bar{F}.C,L.M)$. This can be compared with the probability $P(A.F.C,L.M)$ which arises under the usual conditions of the two actors eating together in the same restaurants. Since they are not equal, the common spoiled food is statistically relevant to the joint illness, even in the presence of the cancellation of the performance, that is,

$$P(A.F.C,L.M) \neq P(A.C,L.M).$$

Thus, although the common cause F is relevant to the coincidence $L.M$ and the coincidence $L.M$ is relevant to C (from which it follows that C is relevant to $L.M$), it turns out that the common cause F screens off the common effect C from the coincidence to be explained. By the screening-off rule, F must be used and C must not be used to partition the reference class A. These considerations express somewhat formally the fact that tampering with the frequency of F without changing the frequency of C will affect the frequency of $L.M$, but tampering with the frequency of C without changing the frequency of F will have no effect on the frequency of $L.M$. This seems to capture the idea that we can influence events by influencing their causal antecedents, but not by influencing their causal consequents.

Let us apply the same analysis to the flagpole example. Again, let A be the general reference class, and let us for simplicity suppose that the flagpole is composed of two parts, a top and a bottom. The flagpole is in place when the two parts are in place; let T be the proper positioning of the top, and B the proper positioning of the bottom. Let M represent the flagpole makers' bringing the pieces together in the appropriate positions. Let S be the occurrence of the full shadow of the flagpole. Now, since the flagpole's existence consists in the two pieces' being put together in place and since that hardly happens except when the flagpole makers bring them together and assemble them, it is clear that

$$P(A,T.B) > P(A,T) \times P(A,B).$$

Hence, the existence of the flagpole is something to be explained. Since we know that

$$P(A.M,T.B) > P(A,T.B) > P(A.\bar{M},T.B),$$

M effects a relevant partition in the general reference class A.

The question of whether the flagpole gets put together when the flagpole makers go about putting it together is unaffected by the existence or nonexistence of a shadow—for example, the ground might be illuminated by other sources of light besides the sun, or mirrors might deflect some of the sun's rays, with the result that there is no shadow—but the flagpole is there just the same; hence,

$$P(A.M.S,T.B) = P(A.M,T.B),$$

but

$$P(A.M.S,T.B) \neq P(A.S,T.B),$$

since the shadow can easily be obliterated at will or produced by other means without affecting the existence of the flagpole. We, therefore, conclude again that the common cause screens off the common effect and that, by virtue of the screening-off rule, the causal antecedent M—and not the causal consequent S—must be invoked to effect a relevant partition in the reference class A. In this highly schematic way, I hope I have shown that the approach to explanation via statistical relevance has allowed us to establish the temporal asymmetry of explanation. Once more, the intuitive idea is that manipulating the pieces of the flagpole, without otherwise tampering with the shadow, affects the shadow; contrariwise, tampering with the shadow, without otherwise manipulating the pieces of the flagpole, has no effect upon the flagpole. The analysis of the flagpole example may, of course, require the same sort of appeal to experiment as we invoked in the preceding example.

I hope it is obvious from all that has been said that none of us is claiming that the temporal asymmetries herein discussed are anything beyond very general facts about the world. Reichenbach regards it as a matter of fact, not of logical necessity, that common causes have their particular statistical relations to the "coincidences" they explain. In a different universe it might be quite otherwise. Neither Reichenbach nor Grünbaum regards the entropic behavior of branch systems as logical necessities, so they are both prepared to deny any logical necessity to the fact that low entropy states are explained by previous interactions. Again, in a different universe, or in another part of this one, it might be different. Similarly, when I argue that explanations from earlier initial conditions are acceptable, whereas explanations from later initial conditions are inadmissible, I take this to be an expression of certain rather general facts about the world. I certainly do not take this characteristic of explanation to be a matter of logical necessity.

The general fact about the world that seems to be involved is that causal processes very frequently exhibit the following sort of structure: a process leading up to a given event E consists of a series of events earlier than E, but such that later ones screen off earlier ones. In other words, a given antecedent event A_1 will be relevant to E, but it will be screened off by a later antecedent A_2 that intervenes between A_1 and E. This situation obtains until we get to E, and then every subsequent event is screened off by some causal antecedent or other. Thus, in some deeply significant sense, the causal consequents of an event are made irrelevant to its occurrence in a way in which the causal antecedents are not. If, in Velikovsky-like cataclysm, a giant comet should disrupt the solar system, it would have enormous bearing upon subsequent eclipses but none whatever upon previous ones. The working out of the details of the eclipse example, along the lines indicated by the analysis of the flagpole example, is left as an exercise for the reader.

To say that causal antecedents can affect an event in ways in which causal consequents cannot is, of course, an utter banality. The difficulty is that causal consequents are *not* statistically irrelevant to an event, so it is rather problematic to state with any precision the kind of asymmetry of relevance we are trying to capture. Reichenbach, himself, is quite careful to point out that the notion of influencing future events is infected with a fundamental temporal ambiguity.[69] I am not at all confident that the use of the screening-off relation is sufficient for a complete account of the relevance relations we seek, but it does have a promising kind of asymmetry, and so I am led to hope that it will contribute at least part of the answer.

13. The Nature of Statistical Explanation

Let me now, at long last, offer a general characterization of explanations of particular events. As I have suggested earlier, we may think of an explanation as an answer to a question of the form, "Why does this x which is a member of A have the property B?" The answer to such a question consists of a partition of the reference class A into a number of subclasses, all of which are homogeneous with respect to B, along with the probabilities of B within each of these subclasses. In addition, we must say which of the members of the partition contains our particular x. More formally, an explanation of the fact that x, a member of A, is a member of B would go as follows:

$$P(A.C_1,B) = p_1$$
$$P(A.C_2,B) = p_2$$
$$\vdots$$
$$P(A.C_n,B) = p_n$$

where

$A.C_1, A.C_2, \ldots, A.C_n$ is a homogeneous partition of A with respect to B,

$p_i = p_j$ only if $i = j$, and

$x \in A.C_k$.

With Hempel, I regard an explanation as a linguistic entity, namely, a set of statements, but unlike him, I do not regard it as an argument. On my view, an explanation is a set of probability statements, qualified by certain provisos, plus a statement specifying the compartment to which the explanadum event belongs.

The question of whether explanations should be regarded as arguments is, I believe, closely related to the question, raised by Carnap, of whether inductive logic should be thought to contain rules of acceptance (or detachment).[70] Carnap's problem can be seen most clearly in connection with the famous lottery paradox. If inductive logic contains rules of inference which enable us to draw conclusions from premises—much as in deductive logic—then there is presumably some number r which constitutes a lower bound for acceptance. Accordingly, any hypothesis h whose probability on the total available relevant evidence is greater than or equal to r can be accepted on the basis of that evidence. (Of course, h might subsequently have to be rejected on the basis of further evidence.) The problem is to select an appropriate value for r. It seems that no value is satisfactory, for no matter how large r is, provided it is less than one, we can construct a fair lottery with a sufficient number of tickets to be able to say for each ticket that will not win, because the probability of its not winning is greater than r. From this we can conclude that no ticket will win, which contradicts the stipulation that this is a fair lottery —no lottery can be considered fair if there is *no* winning ticket.

It was an exceedingly profound insight on Carnap's part to realize that inductive logic can, to a large extent anyway, dispense entirely with rules of acceptance and inductive inferences in the ordinary sense. Instead, inductive logic attaches numbers to hypotheses, and these numbers are used to make practical decisions. In some circumstances such numbers, the degrees of confirmation, may serve as fair betting quotients to determine the odds for a fair bet on a given hypothesis. There is no rule that tells one when to accept an hypothesis or when to reject it; instead, there is a rule of practical behavior that prescribes that we so act as to maximize our expectation of utility.[71] Hence, inductive logic is simply not concerned with inductive arguments (regarded as entities composed of premises and conclusions).

Now, I do not completely agree with Carnap on the issue of acceptance rules in inductive logic; I believe that inductive logic does require some inductive inferences.[72] But when it comes to probabilities (weights) of single events, I believe that he is entirely correct. In my view, we must establish by inductive inference probability statements, which I regard as statements about limiting frequencies. But, when we come to apply this probability knowledge to single events, we procure a weight which functions just as Carnap has indicated—as a fair betting quotient or as a value to be used in computing an expectation of utility.[73] Consequently, I maintain, in the context of statistical explanation of individual events, we do not need to try to establish the explanandum as the conclusion of an inductive argument; instead, we need to establish the weights that would appropriately attach to such explanandum events for purposes of betting and other practical behavior. That is precisely what the partition of the reference class into homogeneous subclasses achieves: it establishes the correct weight to assign to *any* member of A with respect to its being a B. First, one determines to which compartment C_k it belongs, and then one adopts the value p_k as the weight. Since we adopted the *multiple homogeneity rule,* we can genuinely handle any member of A, not just those which happen to fall into one subclass of the original reference class.

One might ask on what grounds we can claim to have characterized explanation. The answer is this. When an explanation (as herein explicated) has been provided, we know exactly how to regard any A with respect to the property B. We know which ones to bet on, which to bet against, and at what odds. We know precisely what degree of expectation is rational. We know how to face uncertainty about an A's being a B in the most reasonable, practical, and efficient way. We know every factor that is relevant to an A having property B. We know exactly the weight that should have been attached to the prediction that this A will be a B. We know all of the regularities (universal or statistical) that are relevant to our original question. What more could one ask of an explanation?

There are several general remarks that should be added to the foregoing theory of explanation:

a. It is evident that explanations as herein characterized are nomological. For the frequency interpretation probability statements are statistical generalizations, and every explanation must contain at least one such generalization. Since an explanation essentially consists of a set of statistical generalizations, I shall call these explanations "statistical" without qualification, meaning thereby to distinguish them from what Hempel has recently called "inductive-statistical." [74] His inductive-statistical explanations contain statistical generalizations, but they are inductive inferences as well.

b. From the standpoint of the present theory, deductive-nomological explanations are just a special case of statistical explanation. If one takes the frequency theory of probability as literally dealing with infinite classes of events, there is a difference between the universal generalization, "All A are B," and the statistical generalization, "$P(A,B) = 1$," for the former admits no As that are not Bs, whereas the latter admits of infinitely many As that are not Bs. For this reason, if the universal generalization holds, the reference class A is homogeneous with respect to B, whereas the statistical generalization may be true even if A is not homogeneous. Once this important difference is noted, it does not seem necessary to offer a special account of deductive-nomological explanations.

c. The problem of symmetry of explanation and prediction, which is one of the most hotly debated issues in discussions of explanation, is easily answered in the present theory. To explain an event is to provide the best possible grounds we could have had for making predictions concerning it. An explanation does not show that the event was to be expected; it shows what sorts of expectations would have been reasonable and under what circumstances it was to be expected. To explain an event is to show to what degree it was to be expected, and this degree may be translated into practical predictive behavior such as wagering on it. In some cases the explanation will show that the explanandum event was not to be expected, but that does not destroy the symmetry of explanation and prediction. The symmetry consists in the fact that the explanatory facts constitute the fullest possible basis for making a prediction of whether or not the event would occur. To explain an event is not to predict it ex post facto, but a complete explanation does provide complete grounds for rational prediction concerning that event. Thus, the present account of explanation does sustain a thoroughgoing symmetry thesis, and this symmetry is not refuted by explanations having low weights.

d. In characterizing statistical explanation, I have required that the partition of the reference class yield subclasses that are, in fact, homogeneous. I have not settled for practical or epistemic homogeneity. The question of whether actual homogeneity or epistemic homogeneity is demanded is, for my view, analogous to the question of whether the premises of the explanation must be true or highly confirmed for Hempel's view.[75] I have always felt that truth was the appropriate requirement, for I believe Carnap has shown that the concept of truth is harmless enough.[76] However, for those who feel too uncomfortable with the stricter requirement, it would be possible to characterize statistical explanation in terms of epistemic homogeneity instead of actual homogeniety. No fundamental problem about the nature of explanation seems to be involved.

e. This paper has been concerned with the explanation of single events, but from the standpoint of probability theory, there is no significant distinction between a single event and any finite set of events. Thus, the kind of explanation appropriate to a single result of heads on a single toss of a coin would, in principle, be just like the kind of explanation that would be appropriate to a sequence of ten heads on ten consecutive tosses of a coin or to ten heads on ten different coins tossed simultaneously.

f. With Hempel, I believe that generalizations, both universal and statistical, are capable of being explained. Explanations invoke generalizations as parts of the explanans, but these generalizations themselves may need explanation. This does not mean that the explanation of the particular event that employed the generalization is incomplete; it only means that an additional explanation is possible and may be desirable. In some cases it may be possible to explain a statistical generalization by subsuming it under a higher level generalization; a probability may become an instance for a higher level probability. For example, Reichenbach offered an explanation for equiprobability in games of chance, by constructing, in effect, a sequence of probability sequences.[77] Each of the first level sequences is a single case with respect to the second level sequence. To explain generalizations in this manner is simply to repeat, at a higher level, the pattern of explanation we have been discussing. Whether this is or is not the only method of explaining generalizations is, of course, an entirely different question.

g. In the present account of statistical explanation, Hempel's problem of the "nonconjunctiveness of statistical systematization"[78] simply vanishes. This problem arises because in general, according to the multiplication theorem for probabilities, the probability of a conjunction is smaller than that of either conjunct taken alone. Thus, if we have chosen a value r, such that explanations are acceptable only if they confer upon the explanandum an inductive probability of at least r, it is quite possible that each of the two explananda will satisfy that condition, whereas their conjunction fails to do so. Since the characterization of explanation I am offering makes no demands whatever for high probabilities (weights), it has no problem of nonconjunctiveness.

14. Conclusion

Although I am hopeful that the foregoing analysis of statistical explanation of single events solely in terms of statistical relevance relations is of some help in understanding the nature of scientific explanation, I should like to cite, quite explicitly, several respects in which it seems to be incomplete.

First, and most obviously, whatever the merits of the present account, no reason has been offered for supposing the type of explanation under consideration to be the only legitimate kind of scientific explanation. If we make the usual distinction between empirical laws and scientific theories, we could say that the kind of explanation I have discussed is explanation by means of empirical laws. For all that has been said in this paper, theoretical explanation—explanation that makes use of scientific theories in the fullest sense of the term—may have a logical structure entirely different from that of statistical explanation. Although theoretical explanation is almost certainly the most important kind of scientific explanation, it does, nevertheless, seem useful to have a clear account of explanation by means of empirical laws, if only as a point of departure for a treatment of theoretical explanation.

Second, in remarking above that statistical explanation is nomological, I was tacitly admitting that the statistical or universal generalizations invoked in explanations should be lawlike. I have made no attempt to analyze lawlikeness, but it seems likely that an adequate analysis will involve a solution to Nelson Goodman's "grue-bleen" problem.[79]

Third, my account of statistical explanation obviously depends heavily upon the concept of *statistical relevance* and upon the *screening-off relation*, which is defined in terms of statistical relevance. In the course of the discussion, I have attempted to show how these tools enable us to capture much of the involvement of explanation with causality, but I have not attempted to provide an analysis of causation in terms of these statistical concepts alone. Reichenbach has attempted such an analysis,[80] but whether his—or any other—can succeed is a difficult question. I should be inclined to harbor serious misgivings about the adequacy of my view of statistical explanation if the statistical analysis of causation cannot be carried through successfully, for the relation between causation and explanation seems extremely intimate.

Finally, although I have presented my arguments in terms of the limiting frequency conception of probability, I do not believe that the fundamental correctness of the treatment of statistical explanation hinges upon the acceptability of that interpretation of probability. Proponents of other theories of probability, especially the personalist and the propensity interpretations, should be able to adapt this treatment of explanation to their views of probability with a minimum of effort. That, too, is left as an exercise for the reader.[81]

NOTES

1. Carl G. Hempel and Paul Oppenheim, "Studies in the Logic of Explanation," *Philosophy of Science*, XV (1948), pp. 135–75. Reprinted, with a 1964 "Postscript," in Carl G. Hempel, *Aspects of Scientific Explanation* (New York: Free Press, 1965).
2. Carl G. Hempel, "Explanation in Science and in History," in *Frontiers in Science and Philosophy*, ed. Robert G. Colodny (Pittsburgh: University of Pittsburgh Press, 1962), p. 10.
3. See also Hempel, *Aspects of Scientific Explanation*, pp. 367–68, where he offers "a general *condition of adequacy for any rationally acceptable explanation of a particular event*," namely, that "any rationally acceptable answer to the question 'why did event X occur?' must offer information which shows that X was to be expected—if not definitely, as in the case of D-N explantion, then at least with reasonable probability."
   Inductive explanations have variously been known as "statistical," "probabilistic," and "inductive-statistical." Deductive explanations have often been called "deductive-nomological." For the present I shall simply use the terms "inductive" and "deductive" to emphasize the crucial fact that the former embody inductive logical relations, whereas the latter embody deductive logical relations. Both types are nomological, for both require lawlike generalizations among their premises. Later on, I shall use the term "statistical explanation" to refer to the sort of explanation I am trying to characterize, for it is statistical in a straightforward sense, and it is noninductive in an extremely important sense.
4. Rudolf Carnap, "Preface to the Second Edition," in *Logical Foundations of Probability* (Chicago: University of Chicago Press, 1962), 2d ed., pp. xv–xx.
5. *Aspects of Scientific Explanation.*
6. "Deductive-Nomological vs. Statistical Explanation," in *Minnesota Studies in the Philosophy of Science*, III, eds. Herbert Feigl and Grover Maxwell (Minneapolis: University of Minnesota Press, 1962).
7. I called attention to this danger in "The Status of Prior Probabilities in Statistical Explanation," *Philosophy of Science*, XXXII, no. 2 (April, 1965), p. 137. Several fundamental disanalogies could be cited. First, the relation of deductive entailment is transitive, whereas the relation of inductive support is not; see my "Consistency, Transitivity, and Inductive Support," *Ratio*, VII, no. 2 (Dec. 1965), pp. 164–69. Second, on Carnap's theory of degree of confirmation, which is very close to the notion of inductive probability that Hempel uses in characterizing statistical explanation, there is no such thing as inductive inference in the sense of allowing the detachment of inductive conclusions in a manner analogous

to that in which deductive logic allows the detachment of conclusions of deductive inferences. See my contribution "Who Needs Inductive Acceptance Rules?" to the discussion of Henry E. Kyburg's "The Rule of Detachment in Inductive Logic," in *The Problem of Inductive Logic*, ed. Imre Lakatos (Amsterdam: North Holland Publishing Co., 1968), pp. 139–44, for an assessment of the bearing of this disanalogy specifically upon the problem of scientific explanation. Third, if q follows from p by a deductively valid argument, then q follows validly from p & r, regardless of what statement r is. This is the reason that the *requirement of total evidence* is automatically satisfied for deductive-nomological explanations. By contrast, even if p provides strong inductive support for q, q may not be inductively supported at all by p & r. Informally, a valid deductive argument remains valid no matter what premises are added (as long as none is taken away), but addition of premises to a strong inductive argument can destroy all of its strength. It is for this reason that the *requirement of total evidence* is not vacuous for statistical explanations.

8. See "Deductive-Nomological vs. Statistical Explanation."
9. See, for example, Hempel, "Explanation in Science and in History."
10. In the present context nothing important hinges upon the particular characterization of the parts of arguments. I shall refer to them indifferently as statements or propositions. Propositions may be regarded as classes of statements; so long as they are not regarded as facts of the world, or nonlinguistic states of affairs, no trouble should arise.
11. When no confusion is apt to occur, we may ignore the distinction between the explanandum and the explanandum event. It is essential to realize, however, that a given explanation must not purport to explain the explanandum event in all of its richness and full particularity; rather, it explains just those aspects of the explanandum event that are mentioned in the explanandum.
12. "Deductive-Nomological vs. Statistical Explanation," p. 125.
13. Ibid. See also "Inductive Inconsistencies," *Synthèse*, XII, no. 4 (Dec. 1960).
14. Hempel and Oppenheim, "Studies in the Logic of Explanation," and Hempel, "Deductive-Nomological vs. Statistical Explanation."
15. This condition has sometimes been weakened to the requirement that the premises be highly confirmed. I prefer the stronger requirement, but nothing very important hangs on the choice. See n. 76.
16. A premise occurs essentially in an argument if that argument would cease to be (deductively or inductively) correct upon deletion of that premise. Essential occurrence does not mean that the argument could not be made logically correct again by replacing the premise in question with another premise. "Essential occurrence" means that the premise plays a part in the argument as given; it does not just stand there contributing nothing to the logical correctness of the argument.
17. The requirement of total evidence demands that there should be no additional statements among our available stock of statements of evidence that would change the degree to which the conclusion is supported by the argument if they were added to the argument as premises. See Carnap, *Logical Foundations of Probability*, sec. 45B.
18. Hempel, "Explanation in Science and in History," p. 14.
19. "The Status of Prior Probabilities in Statistical Explanation," p. 145.
20. Consumer Reports, *The Medicine Show* (New York: Simon & Schuster, 1961), pp. 17–18.
21. Henry E. Kyburg, "Comments," *Philosophy of Science*, XXXII, no. 2 (April 1965), pp. 147–51.

22. Since there is no widely accepted account of inductive inference, it is difficult to say what constitutes correct logical form for inductive arguments. For purposes of the present discussion I am accepting Hempel's schema (4) as one correct inductive form.

23. The fact that the general premise in this argument refers explicitly to a particular physical object, the moon, may render this premise nonlawlike, but presumably the explanation could be reconstructed with a suitably general premise about satellites.

24. In Hempel and Oppenheim's "Studies in the Logic of Explanation," (7.5) constitutes a set of necessary but not sufficient conditions for a potential explanans, while (7.6) defines *explanans* in terms of potential explanans. However, (7.8) provides a definition of *potential explanans*, which in combination with (7.6) constitutes a set of necessary and sufficient conditions for an explanans.

25. *Aspects of Scientific Explanation*, pp. 397–403. See also Carl G. Hempel, "Maximal Specificity and Lawlikeness in Probabilistic Explanation," *Philosophy of Science*, XXXV (1968), pp. 116–33, which contains a revision of this requirement. The revision does not seem to affect the objections I shall raise.

26. Salmon, "The Status of Prior Probabilities."

27. This characterization of the logical interpretation of probability is patterned closely upon Carnap, *Logical Foundations of Probability*.

28. Salmon, "Who Needs Inductive Acceptance Rules?".

29. This point was brought out explicitly and forcefully by Richard Jeffrey in "Statistical Explanation vs. Statistical Inference" presented at the meeting of the American Association for the Advancement of Science, Section L, New York, 1967. This superb paper has since been published in *Essays in Honor of Carl G. Hempel*, ed. Nicholas Rescher (Dordrecht, Holland: Reidel Publishing Co., 1969). Hempel discussed this issue in "Deductive-Nomological vs. Statistical Explanation," pp. 156–63.

30. For an indication of some of the difficulties involved in properly circumscribing this body of total evidence, see Kyburg's comments on my paper "The Status of Prior Probabilities in Statistical Explanation," and my rejoinder. Hempel has discussed this problem at some length in "Deductive-Nomological vs. Statistical Explanation," pp. 145–49, and "Aspects of Scientific Explanation," sec. 3.4.

31. An extremely clear account of this view is given by Ward Edwards, Harold Lindman, and Leonard J. Savage in "Bayesian Statistical Inference for Psychological Research," *Psychological Review*, LXX, no. 3 (May 1963). An approach which is very similar in some ways, though it is certainly not a genuinely subjective interpretation in the Bayesian sense, is given by Carnap in "The Aim of Inductive Logic," in *Logic, Methodology and Philosophy of Science*, eds. Ernest Nagel, Patrick Suppes, and Alfred Tarski (Stanford, Calif.: Stanford University Press, 1962), pp. 303–18.

32. This treatment of the frequency interpretation takes its departure from Hans Reichenbach, *The Theory of Probability* (Berkeley and Los Angeles: University of California Press, 1949), but the discussion differs from Reichenbach's in several important respects, especially concerning the problem of the single case. My view of this matter is detailed in *The Foundations of Scientific Inference* (Pittsburgh: University of Pittsburgh Press, 1967), pp. 83–96.

33. See *Foundations of Scientific Inference*, pp. 56–96, for discussions of the various interpretations of probability.

34. See, for example, Karl Popper, "The Propensity Interpretation of the Calculus of Probability, and the Quantum Theory," in *Observation and Interpretation*, ed. S. Körner (London: Butterworth Scientific Publications, 1956), pp. 65–70, and

"The Propensity Interpretation of Probability," *British Journal for the Philosophy of Science*, X (1960), pp. 25–42.

35. See Salmon, *The Foundations of Scientific Inference*, pp. 83–96, for fuller explanations. Note that, contrary to frequent usage, the expression "$P(A,B)$" is read "the probability *from A to B*." This notation is Reichenbach's.

36. Reichenbach, *The Theory of Probability*, sec. 72. John Venn, *The Logic of Chance*, 4th ed. (New York: Chelsea Publishing Co., 1962), chap. IX, sec. 12–32. Venn was the first systematic exponent of the frequency interpretation, and he was fully aware of the problem of the single case. He provides an illuminating account, and his discussion is an excellent supplement to Reichenbach's well-known later treatment.

37. Reichenbach, *The Theory of Probability*, p. 374.

38. Carnap, *Logical Foundations of Probability*, sec. 44.

39. Richard von Mises, *Probability, Statistics and Truth*, 2d rev. ed. (London: Allen and Unwin, 1957), p. 25.

40. Also, of course, there are cases in which it would be possible in principle to make a relevant partition, but we are playing a game in which the rules prevent it. Such is the case in roulette, where the croupier prohibits additional bets after a certain point in the spin of the wheel. In these cases also we shall speak of practical homogeneity.

41. "Deductive-Nomological vs. Statistical Explanation," p. 133. Hempel also presents a résumé of these considerations in *Aspects of Scientific Explanation*, pp. 394–97.

42. "Deductive-Nomological vs. Statistical Explanation," pp. 146–47.

43. Ibid.

44. *Aspects of Scientific Explanation*, pp. 399–400. Note that in Hempel's notation "$p(G,F,)$" denotes "the probability of G, given F." This is the reverse of Reichenbach's notation.

45. Ibid., pp. 400–01. In a footnote to this passage Hempel explains why he prefers this requirement to the earlier requirement of total evidence.

46. In working out the technical details of the following definition, it is necessary to recognize that homogeneous partitions are not unique. The transfer of a finite number of elements from one compartment to another will make no difference in any probability. Similarly, even an infinite subclass C_j, if $P(A,C_j) = 0$, can fulfill the condition that $P(A.C_j,B) = p_j \neq P(A,B)$ without rendering A's degree of inhomogeneity > 0. Such problems are inherent in the frequency approach, and I shall not try to deal with them here. In technical treatment one might identify subclasses that differ by measure zero, and in some approaches the conditional probability $P(A.C_j,B)$ is not defined if $P(A,C_j) = 0$.

47. I am grateful to Prof. Douglas Stewart of the Department of Sociology at the University of Pittsburgh for pointing out that the concept of variance is of key importance in this context.

48. See James G. Greeno, "Evaluation of Statistical Hypotheses Using Information Transmitted," *Philosophy of Science*, XXXVII (1970), for a precise and quantitative discussion of this point.

49. See Michael J. Scriven, "Explanation and Prediction in Evolutionary Theory," *Science*, CXXX, no. 3374 (Aug. 28, 1959).

50. Ibid., p. 480.

51. See Adolf Grünbaum, *Philosophical Problems of Space and Time* (New York: Alfred A. Knopf, 1963), pp. 309–11.

52. Hans Reichenbach, *The Direction of Time* (Berkeley and Los Angeles: The University of California Press, 1956), p. 189.

53. Since $P(A.C.B) = P(A,B)$ entails $P(A.B,C) = P(A,C)$, provided $P(A.B,C) \neq 0$, the relevance relation is symmetrical. The screening-off relation is a three-place relation; it is nonsymmetrical in its first and second arguments, but it is symmetrical in the second and third arguments. If D screens off C from B, then D screens off B from C.

54. Reichenbach, *The Direction of Time*, sec. 22.

55. This example has received considerable attention in the recent literature on explanation. Introduced in Scriven, "Explanation and Prediction," it has been discussed by (among others) May Brodbeck, "Explanation, Prediction, and 'Imperfect Knowledge,'" in *Minnesota Studies in Philosophy of Science*, III, eds. Herbert Feigl and Grover Maxwell (Minneapolis: University of Minnesota Press, 1962); Adolf Grünbaum, *Philosophical Problems*, pp. 303–08; and Carl G. Hempel, "Explanation and Prediction by Covering Laws," *Philosophy of Science: The Delaware Seminar*, I, ed. Bernard H. Baumrin (New York: John Wiley and Sons, 1963).

56. "72 out of 100 untreated persons [with latent syphilis] go through life without the symptoms of late [tertiary] syphilis, but 28 out of 100 untreated persons were known to have developed serious outcomes [paresis and others] and there is no way to predict what will happen to an untreated infected person" (Edwin Gurney Clark, M.D., and William D. Mortimer Harris, M.D., "Venereal Diseases," *Encyclopedia Britannica*, XXIII [1961], p. 44).

57. Ralph E. Lapp and Howard L. Andrews, *Nuclear Radiation Physics*, 3rd ed. (Englewood Cliffs, N.J.: Prentice-Hall, 1963), p. 73.

58. Cf. Hempel's discussion of this example in "Aspects of Scientific Explanation," pp. 369–74.

59. Here I am assuming the separate trials to be independent events.

60. George Gamow, *The Atom and Its Nucleus* (Englewood, Cliffs, N.J.: Prentice-Hall, 1961), p. 114.

61. Cf. Hempel's discussion in "Aspects of Scientific Explanation," sec. 4.2.2, of an example of a slip of the pen taken from Freud. Hempel takes such examples to be partial explanations.

62. See Jeffrey, "Statistical Explanation vs. Statistical Inference" for a lucid and eloquent discussion of this point. In this context I am, of course, assuming that the pertinent probabilities exist.

63. Lapp and Andrews, *Nuclear Radiation Physics*, p. 73.

64. I owe an enormous intellectual debt to Reichenbach and Grünbaum in connection with the view of explanation offered in this paper. Such examples as the paradigm cases of explanation to be discussed have been subjected to careful analysis by these men. See Grünbaum, *Philosophical Problems*, chap. IX, and Reichenbach, *The Direction of Time*, chaps. III–IV, for penetrating and lucid analyses of these examples and the issues raised by them.

65. In his *Theory of Probability*, sec. 34, Reichenbach introduces what he calls the *probability lattice* as a means for dealing with sequences in which we want to say that the probabilities vary from element to element. One particular type of lattice, the *lattice of mixture* is used to describe the temporally asymmetric character of the ensemble of branch systems. The *time ensemble* and the *space ensemble* are characterized in terms of the lattice arrangement. See Reichenbach, *The Direction of Time*, sec. 14.

66. Ibid., sec. 18.

67. Ibid., chap. IV.

68. See Hempel, "Deductive-Nomological vs. Statistical Explanation," pp. 109–10; also Grünbaum, *Philosophical Problems*, pp. 307–08.

69. Reichenbach, *The Direction of Time,* sec. 6.
70. Carnap, *Logical Foundations of Probability,* sec. 44.
71. Ibid., secs. 50–51.
72. Salmon, "Who Needs Inductive Acceptance Rules?" Because of this difference with Carnap—i.e., my claim that inductive logic requires rules of acceptance for the purpose of establishing statistical generalizations—I do not have the thoroughgoing "pragmatic" or "instrumentalist" view of science Hempel attributes to Richard Jeffrey and associates with Carnap's general conception of inductive logic. Cf. Hempel, "Deductive-Nomological vs. Statistical Explanation," pp. 156–63.
73. Salmon, *Foundations of Scientific Inference,* pp. 90–95.
74. See "Aspects of Scientific Explanation," secs. 3.2–3.3. In the present essay I am not at all concerned with explanations of the type Hempel calls "deductive-statistical." For greater specificity, what I am calling "statistical explanation" might be called "statistical-relevance explanation," or "S-R explanation" as a handy abbreviation to distinguish it from Hempel's D-N, D-S, and I-S types.
75. Hempel, "Deductive-Nomological vs. Statistical Explanation," sec. 3.
76. Rudolf Carnap, "Truth and Confirmation," in *Readings in Philosophical Analysis,* eds. Herbert Feigl and Wilfrid Sellars (New York: Appleton-Century-Crofts, 1949), pp. 119–27.
77. Reichenbach, *Theory of Probability,* sec. 69.
78. Hempel, "Deductive-Nomological vs. Statistical Explanation," sec. 13, and "Aspects of Scientific Explanation," sec. 3.6. Here, Hempel says, "Nonconjunctiveness presents itself as an inevitable aspect of [inductive-statistical explanation], and thus as one of the fundamental characteristics that set I-S explanation apart from its deductive counterparts."
79. See Nelson Goodman, *Fact, Fiction, and Forecast,* 2d ed. (Indianapolis: Bobbs-Merrill Co., 1965), chap. III. I have suggested a resolution in "On Vindicating Induction," *Philosophy of Science,* XXX (July 1963), pp. 252–61, reprinted in Henry E. Kyburg and Ernest Nagel, eds., *Induction: Some Current Issues* (Middletown, Conn.: Wesleyan University Press, 1963).
80. Reichenbach, *The Direction of Time,* chap. IV.
81. The hints are provided in sec. 3.

NORWOOD RUSSELL HANSON
Formerly at Yale University

A Picture Theory
of Theory Meaning

His [Kepler's] admirable method of thinking consisted in
forming in his mind a diagrammatic or outline representation
of the entangled state of things before him, omitting all that
was accidental, observing suggestive relations between the
parts of his diagram, performing divers experiments upon it,
or upon the natural objects, and noting the results.
—C. S. Peirce
Values in a Universe of Chance

PERPLEXITIES CONCERNING SCIENTIFIC THEORIES persist because the usual
"singled valued" philosophical analyses cannot do justice to the problematic features of so complex a semantical entity. The components of
theories are like law statements and like models and hypotheses, being
conceptual entities which are used in a variety of ways—not all of these
being always compatible with the others. Thus many physicists characterize the classical laws of motion, as if they functioned in a definitional
way.[1] But sometimes these laws seem remarkably empirical.[2] Others
characterize such laws as "conventional"; they shape entire disciplines
much as the rules shape the game of chess.[3] Law statements are not
exclusively any one of these—definitions, factual claims, or conventions.
They are *all* these things.

Consider: "The sun rises in the east." It is impossible from only
hearing or seeing these words in isolation to know whether this claim is
functioning in a definitional way or in a descriptive way. Thus if tomorrow the sun parts the horizon ninety degrees from where it arose this
morning, it might still be rising in the east *if* one treats "east" as the name
of that place where the sun rises (wherever that may be). If one defines
"east" in the terms of celestial coordinates though, it will be an empirical,

233

factual, synthetic claim that the sun rises in the east. So the very meaning of "The sun rises in the east" is elusive until one comprehends this assertion's local use in a specific context. This use is quite free to change.

Much this same diversity and flexibility should mark our understanding of scientific theories. What a scientific theory is cannot be finally determined—for theories are context-dependent instruments of conceptualization. Tomorrow's inquiries can transform yesterday's scientific theories into semantical structures different from what today's philosophers pronounce them to be. (Who now reads Mach or even Schrödinger for insights into contemporary quantum theory?)

Let us look at theories in a way different from those which dominate most discussions in philosophy of science. Think of theories not as ideal deductive systems, as precise languages, or as convenient empirical shorthands. In other words, they are used sometimes as if they were definitional, analytical, calculational systems; sometimes as if they were ideal languages (well-chiseled, logicians' Esperanto); sometimes as if they were elegant compendiums of factual information. Theories are all these things, but they are more too. Explore yet another facet of scientific theories, one which disappears in the glare of the analytical spotlight.

How can theories enable us to understand a subject matter? What is the difference between a heap, or a list, of descriptive assertions and a theory, which itself is largely constituted of those same descriptions? These questions recall the contrast between a mere generalization (for example, that all white, blue-eyed, tomcats are deaf) and a law of nature (for example, that all birds' wings have a convex topside). If the generalization is imagined to be refuted, we are required only to effect a quantitative readjustment; we may have to say that 99 percent of all white, male, blue-eyed cats are deaf, rather than all of them. We will still know what cats are, however. No conceptual readjustment is forced on us by a feline counterinstance. With a law of nature, such as that all wings of birds have convex topsides, if one were to encounter a counterinstance of this, *conceptual* difficulties would ensue at once. The full concept of *bird flight* requires a wing imagined to be so shaped. Faced (*per impossible*) with a bird wing curved otherwise, one might come to doubt what a bird wing is and what role it plays in flight—doubts which do not now punctuate the thinking of aerodynamicists and ornithologists. It is as if one imagined an exception to: *All unsupported bodies in terrestrial space move toward the center of the earth.* An exception to this statement would have to be a body in a state of levitation or of "negative gravity," either of which possibilities raises doubts as to what *bodies* are in the first place.

It is sometimes said that a law of nature explains its subject matter, helps us to understand it, and makes it more intelligible and comprehensible—as against a generalization which only correlates observables via actuarial techniques; these observables may concern "unrelata" like the simultaneous occurrence of sunspots and wheat failures, where no conceptual link binds such phenomena. Analogously, a scientific theory entices philosophers because it somehow explains its subject matter; it helps us understand "interconceptions" between phenomena.

What does all this mean? What is it in such a theory that before it was formulated all the data, the descriptions, the initial conditions—however accurately recorded—did not compose into a coherent and intelligible subject matter, whereas after the theory has been generated and coupled with observations one can comprehend the subject matter?

Consider theories pro tem as conceptual entities located at the crossroads between epistemology and philosophical psychology. Think no more, for now, of the logical and the semantical aspects of scientific theories; everyone always talks about them. Let us view theories as instruments of intelligibility. Ask with me: "How does the conceptual structure of a theory make understanding possible?"

Reflect on those picture puzzles so dear to learning theorists and Gestalt psychologists. See the sheep in the tree in figure 1. Consider also figure 2. When labeled "a Mexican on a bicycle (seen from above),"

Figure 1

something happens within the perceptual field. The experience now is qualitatively different from what it had been when this drawing was a mere configuration of lines. How so?

The *how* doesn't matter (the problem is philosophical, not psychological; conceptual, not factual). That patterns affect the significance of lines, dots, shapes, and patches—which might have been in perceptual turbulence otherwise—is our fundamental datum. It has profound epistemological consequences. Knowledge is a function of how our experiences cohere. Observations made before the perceptual pattern is appreciated are epistemically distinct from the observations (and their descriptions) made after that pattern has cast them into intelligible constellations—although the observations and descriptions, those before and those after, might be "congruent."[4] The descriptive terms, the assertions, the observations themselves, when considered in terms of repeatability and what is "written on the page," might be identical both before and after the pattern is appreciated. The lines in these two drawings did not shift geometrically when the captions were assigned. Yet there is an epistemic

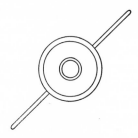

Figure 2

distinction between the earlier and later encounters, a distinction of deep importance.

Clearly, talk about *patterns* differs in type from talk of lines, shapes, and dots. Patterns do not fill the same logical space as do the lines and dots being patterned: a pattern—for example, of the Mexican atop a bicycle—is itself not detectable or visible or drawable, not as the shapes and the lines are. However, this is not to say that they are not detectable or visible at all. How else should we come to know them? Patterns are detectable and can be made visible to those who cannot see them—but not necessarily by adding more lines. Describing this encounter differs from speaking of objects of sensation as appreciated by all normal observers. Twenty-twenty vision is no guarantee of seeing the Mexican on the bicycle. Patterns are not *elements* in an epistemic configuration. Rather, the pattern is the configuration itself. By analogy, the plot of a novel is not another cluster of words; the form of a sonata is not just another cluster of notes; the architectural design of a building is not merely more bricks and beams; the aerodynamic structure of an aircraft wing—its airfoil section—is not just more ribs and skin plates; indeed, the meaning of a proposition is not only another articulated term!

Much as the level of "pattern talk" differs conceptually from that on which talk of dots, shapes, lines, and patches obtains, so also theoretical talk differs conceptually from observational and descriptive talk. The more comprehensive suggestion is this: that just as perceptual pattern recognition at once gives significance to elements perceived and yet differs from any perception of dots, shapes, and lines, so also *conceptual* pattern recognition at once gives significance to the observational elements within a theory and yet differs from any awareness of those elements vis-à-vis their primary relationship to events and objects. The way in which theories, conceptual structures, are meaningful with respect to the observation statements is qualitatively a different type of concern from that involved in discussions of how observation statements are meaningful with respect to things.

At this juncture let me inject some parenthetic autobiography. A psychological fact: there are moments when I find myself confronting a cluster of symbols, or observed anomalies, that, after having come to view through the appropriate *scientific theory*, configure, cohere, and collapse into meaningful patterns within a unified intellectual experience. This patterning, to me, seems not unrelated to what is involved when I appreciate dots and lines in a qualitatively different way after having mastered the perceptual pattern structuring those marks.[5] Consider Boyle's law as understood in 1662, then simply a stack of statistical correlations. Boyle did not extract that famous generalization himself; his followers did. That law, that correlation considered before the advent of kinetic theory and before classical statistical mechanics, resembles the dots without the pattern, the observations without the theory, the descriptions without the explanations. Boyle's law began life as the merest correlation. It explained nothing. Only when the general gas theory and the kinetic hypothesis caught up with it did Boyle's generalization come to function as laws of nature are reputed to do. Bracket with this example the historical problem concerning the anomalous motions of Saturn and Jupiter. This was a descriptive thorn in the side of astronomical explanation B.L. ("Before Laplace"). Laplace undertook to set out a conceptual framework for mechanical ideas, a stability proof in terms of which this anomaly—the apparently secular aberrations in the motions of Saturn and Jupiter—could be regarded as but local irregularities in what was really a nine-hundred-year cycle, a periodic, repetitive "aberration." This framework is a little like what one should expect in a microcinematographic film of meshing gears in a fine clock, crude and lopsided in fine scale but precise and perfectly periodic at the macrochronometric level. Descriptions of Saturn and Jupiter B.L. were independent, unrelated, and unsynchronized, whereas these same descriptions A.L. constituted almost different subjects for one's attention.

Please permit me to spell out this primitive analogy in more detail. Consider the concept of a scene. More specifically, think of a dawn seen from a hillside. There sits a landscape painter, busily conveying to his canvas a configuration like that in figure 3. Some passersby may say of this painting (figure 4) that it is "true to life," that it captures what is

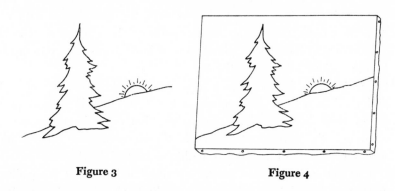

Figure 3   Figure 4

significant "out there" (figure 3). Painting is an activity of the appropriate type to capture features of the original—the tree, the hill, and the other landscape objects "out there" and "committable to" canvas. There is a structural identity between what can be seen by the painter from the hillside and what can be seen on the canvas that he has painted. And this identity is just as important for his painting's being "true to life" as is the identity at the color, shape, line level. Of course, the tree should be painted green, as it is, and not pink or silver. But no less important is the fact that it should be depicted to the left of the sun and not stretched horizontally above it. Something, which I shall designate "the scene," is "out there" for inspection; one can stand on the hillside and survey the scene to the east.

One can describe what the artist has put on canvas as "the scene he has painted." The scene on his canvas and the scene "out there" are so structurally related that it is meaningful to speak of the former as constituting a replication of the latter, something one cannot claim of sounds, textures, or tastes, no ingenious combination of which can replicate the scene at dawn; the scene as paintable eludes the powers of music, of tactile sensation, and even of cookery. Thus the term "scene," from a conceptual point of view, is specific yet Janus-faced. It alludes to an objective subject matter "out there," and it also refers to one's plastic representation of that subject matter.[6] The same scene can be both "out there" and also on canvas. That the artist has put the same scene on canvas as obtains "out there" is pertinent to whether his rendition is veridical.

I don't want to refer to the scene per se as if it were an "interim designatum." That would proliferate entities, since antiquity a philosophically suspect practice. Nonetheless, aspects of subject matters are reproducible in this way because of their possible structural identity with aspects of the reproduction—this is all I wish to assert. Many terms do similar work; "landscape" has the same mirrorlike semantical quality. The landscape is something tended by a gardener, something one can view from a distance. It is also what can be captured on canvas by a draftsman or painter. Again, the subject matter and its representations can share something of considerable conceptual importance. Were this not so, the subject matter would not be representable at all.

There are myriad such "bipartite" terms. The "planform" of a bird's wing, as referred to by ornithologists and aerodynamicists, makes reference to such geometrical relationships as the chord-span ratio, the angular sweepback of the leading edge, the relative root-to-tip rate of narrowing, and the contour shape of the wing (elliptical? rectangular? triangular?). The wing's "aspect ratio" is another such term; this is the relative thickness of the "fuselage" as against the length of the wing, as viewed directly forward or aft. The tip configuration of the wing, whether blunt or pointed or rounded, will also be part of the understood designatum of "planform." The planform and the aspect ratio of a bird or an aircraft can be drawn on a piece of drafting paper (see figure 5), and

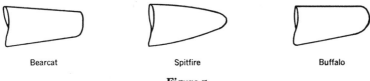

Bearcat   Spitfire   Buffalo

Figure 5

they can also be inspected in the three-dimensional wing itself, as found on the living bird or on the operational aircraft. Wittgenstein's point that the structure of the bird's song is something which is in the song itself—and also in the Gramophone recording of the song and also in the musical score which captures that song in notes (à la Delius)—is close to what I am groping for. The song, its recording, and its score share a common structure. The planforms on paper and in the actual bird wing share a common structure, as is also true of the tip configuration, the aspect ratio, the dihedral, etc. The landscape, the scene, is the common structure shared by objects in configuration "out there" and color patches in configuration on the painter's canvas.[7]

My suggestion will be that, analogously, states of affairs, that is, constellations of phenomena, are often rendered understandable and intelligible and comprehensible because some objective, structural component of those phenomena is duplicated in a corresponding structural component within some scientific theory. Scientifically understanding phenomena x, y, and z consists in perceiving what kinds of phenomena they are—how they relate one to the other within some larger epistemic context; how they are dependent upon, or interfere with, one another. Insights into such relations "out there" are generable within our perceptions of the structures of theories; these theoretical structures function vis-à-vis our linguistic references to x, y, and z in a way analogous to how the scene stands to the tree and hill "Out there" and also to the painted patches on canvas. Thus, I suggest that in contrast to the delineation of theories as "ideal languages" or "Euclidean hypothetico-deductive structures," the important function of scientific theory is to provide such structural representations of phenomena that to understand how the elements in the theoretical representation "hang together" is to discover a way in which the facts of the world "hang together." In short, scientific theories do not always argue us into the truth; they do not always demonstrate deductively and forcefully what is the case. Often they show what could be the case with perplexing phenomena, by relating representations of those phenomena in ways which are themselves possible representations of relationships obtaining "out there." Theories provide patterns for ordering phenomena. They do this, just as much as they provide inference channels through which to argue toward descriptions of phenomena.

Before proceeding, consider some classical objections to the so-called "picture theory of meaning." Clearly, if all forms of representation are taken to be fundamentally *iconic*, as would be the case in a landscape painting, then the painter will be felt to represent elements in the original three-dimensional configuration by way of iconic tokens in the copy configuration (two-dimensional). That is, his tree will share some properties of the tree out there (perspectively considered). For example, its shape, oriented with respect to the sun and the hill, will display ratios in relative height, width, and color analogous to what obtains in the original. The sun and the painter's sun will have a common geometry both internal, with respect to its discoid design and coherence, and external, with respect to its relations to tree and hill. A color transparency, for example, of the Kodak variety, could be moved from its superposition on the scene out there to superposition of the scene on canvas, and it would be logically possible for there to be shape congruence and color congruence all the way through, both in the first superposition and in the second superposition. And so that representation on canvas will stand to the original (three-dimensional) in a way which is designated as "iconic." This is proved by the congruence of the Kodak transparency with each.

Now, vis-à-vis scientific theories, where the mode of representation (if there is one) is linguistic and descriptive, it is obvious that this is not any crudely iconic representation. Theories are not simple pictures. The word "tree" has nothing iconically in common with what this word may designate, namely some actual tree. (There is nothing arboreal about "tree"!) Similarly, the word "sun" is not iconically connected with any perceptual object or any physical object. Words represent not because of property-sharing. They have no property in common with what they represent—except for onomatopoeias like "toot," "crash," "smooth," and "short." (These words seem to me relatively unimportant semantically; they certainly constitute no paradigm of word-object meaning.) It will be the *conventional correlation* of words with objects which holds our attention here. Consider "syzygy," a term well known in analytical mechanics. This word does not represent iconically any rectilinear configuration of moon, earth, and sun (which is what the word means). That this linguistic term means what it does is not due to any iconic relationship with objects in the solar system, although one can perceive that there is something about these designations ("y", "y", "y") which seems to tie in with the three-bodied problem involved—sun, earth, and moon. Nonetheless, "syzygy" is not related to planets as a painting may be related to trees. To hear it for the first time is not to know (simply from the configuration of the sounds and symbols) that it connects semantically with moon, earth, and sun—in the way in which "toot" might connect semantically with a passing train or "buzz" with a passing saw. In other words, statements paradigmatically designate; then they characterize their designata as being of this or that type, or having these or those properties.

Thus we have: "The moon is a pocked sphere," where "The moon" is the designation of an astronomical object and "is a pocked sphere" characterizes that object, that designatum. Pictures represent in a nondesignatory way; they are nonspecific with respect to the attention-directing they may stimulate. Does figure 6 designate the moon, its sphericity, its discoidity, its pockmarks, its yellow color, or what? Statements place one's attention precisely on particular designata, and then they discriminate between and select from the appropriate alternative characterizations of that designatum. Thus, of all the things that may be true of the moon as depicted in figure 6—for example, that it is spherical, that it

Figure 6

appears as discoid, that it is pockmarked—the statement "The moon is a pocked sphere" selects one of these specific data as its unique and direct message and articulates it pointedly. That is why it is true that a picture is worth a thousand words; a picture is a thousand times less specific than a short, sharp statement. But, by the same token, one word is worth a thousand pictures; a statement can supply a focus for the attention that is different in type from anything generable via confrontation with a picture.

These objections to the picture theory are well known, and yet I am going to suggest something sometimes suggested by others—that all this critical carping on the distinctions between originals and icons, as against originals and statements, really misses the profound point of the picture theory of meaning. Objections concerning the noniconic ways in which words and statements represent really deal with the hyperfine structure of discourse versus pictorial representation. They are directed to the ways in which words like "moon" are, or are not, correlatable in function with line configurations such as those in figure 7. Aside from

Figure 7

such hyperfine structural differences, statements and drawings remain deeply analogous vis-à-vis representational features to be discussed shortly. Thus the objections to the picture theory advanced by such people as Edna Daitz and Irving Copi concern just the minute superficialities of word tokens and claim tokens. What else could be the point of noting that "cat" does not look feline and that "moon" sheds no light?

However, let us attend to the structure of discursive knowledge in more general terms and not restrict our interest to the indivisible tokens through which that structure is conveyed. Consider the structure of discourse itself and the corresponding structure of representational knowledge. These different kinds of structures can perhaps convey insight and information about the structures of the originals in much the same way—so much so that there is yet more to be said for the classical picture theory of meaning with respect to how it helps us understand linguistic meaning. A claim such as "The sun rises to the right of the juniper" does indeed have something in common structurally with figure 8, a configuration that an artist might put on canvas. The structure

Figure 8

common to both the claim and the sketch makes it possible to learn from both to what extent they might be veridical. Certainly the claim and the sketch, because of some common structure, stand or fall together; either both of them have the structure of the original (that is, the "sun-to-the-right-of-juniper" structure), or neither of them has. The picture theory, then, as articulated at high speed in the *Tractatus*[8] and at very slow speed in Wisdom's "Logical Constructions," may be articulated improperly in both contexts; for these celebrated expositions dwell excessively on language token, physical-object correspondences, and not enough on structural correspondences. The latter will constitute the philosophical burden of this essay, a burden relieved somewhat, I hope, by special illustrations from fluid mechanics which will serve as a typical scientific theory in the analysis to follow.

Our paradigm should not be the interconnections and resemblances between paintings and their subject matter (or between photos and their subject matter), but rather between such a thing as a map and its subject matter. Even better, for our purposes, will be the logical linkages between a chart and its subject matter, or between a highly schematic diagram and its subject matter. Let us begin with maps, our first approximation. To be useful at all, a map must share some structure with the original terrain. This much stresses the iconic and might even cause one to think of maps as if they were stylized paintings from above (high above) ground or even vast aerial photographs; they are neither of these things. Still, maps do resemble paintings and photos in that they must share some representational structure with the original terrain mapped or else they would not be at all reliable or informative. A map has got to indicate to us, for example, that the Greater Pittsburgh Airport is northwest by west of the Golden Triangle Point, that the Allegheny County Airport is roughly south, and that the Monroeville Airport is roughly east. These must be fairly stable and veridical representations; if the representation fails in this respect, then it simply will not be doing for us what a map is expected to do, namely, to provide us with such a representation of the geographical structure of Pittsburgh that if we can "locate ourselves within" the representation, we have located ourselves within Pittsburgh. Still, noting this requirement of cartographic verisimilitude is compatible with recognizing that there is an extensive conventional vocabulary within any map—so much so that a painting or a photo of Pittsburgh from above just cannot serve as a map of Pittsburgh. Thus one has in the map a legend, and by appreciating how one designates tall towers, state capitols, railway lines, airports, parks, etc., one can make one's way through the urban jungle with the aid of this graphic, but necessarily stylized, representation of that jungle.

Such reference points as those in figure 9 cannot be expected in an aerial photograph, of course, anymore than one can expect them to stand forth when viewing a large metropolitan area from a high-altitude reconnaissance aircraft. After all, aviators often get lost flying from Baltimore to Boston in the clearest weather—even though the terrain is stretched below them in a most detailed dioramic display. They still require a map to "clarify" what is before their eyes, although no visible detail of the megalopolis below escapes their view. It is this conventional vocabulary of maps which helps us learn to "read" them, something which never happens in the confrontation with representational art. This conventional symbolization is much more extensive in en route aircraft navigational charts, especially when one actually begins to work out things like distance from the ground and one's distance from appropriate objects in the air space and terrain ahead, by extraordinarily stylized and conventional blobs and shapes of color which, for a novice, would simply be unintelligible. (Notice how in figure 10, an en route aircraft navigational map, the outlines of famous land masses have almost disappeared, in favor of the much more stylized depiction of "airways"—so much so that "map" almost seems less appropriate a word for this rendition than does "chart.")

These features of aircraft maps and navigational charts are valuable not because they represent iconically, but rather because they have a

Landmarks (with appropriate note). . . . . (Numerals indicate elevation above sea level of top)	■ Factory ■ Stack 875'
Oil Tanks.	● ● ● ●
Oil Fields.	Я Я
Dams.	
Rapids and Falls.	
Elevations (IN FEET) {Highest on chart (devoid of tint). . . . . Highest in a general area. . . . . Spot.	● **1115** ● **1085** ● 950
Mines and Quarries.	⚒
Mountain Passes.	)(
Lookout Stations (Elevation is base of tower).	▲ 75 (Site) 1025 (ELEV)
Ranger Stations.	♪
Coast Guard Stations	⚓ CG 79
Pipe Lines.	*PIPE LINE*
Race Tracks or Stadiums.	⬯ RT ■
Open-Air Theaters	▽ ■ Open-air

Figure 9

fixed and widely applicable significance; this fact makes any attempt to comprehend the geographic complexities of Pittsburgh in principle like the comprehending of any other urban geographic subject, such as Harrisburg, Philadelphia, New York, or Boston. Maps must be read, as pictures need not be. They require training in reading them and legends, glossaries, and vocabularies. In just these aspects they differ from country to country. An aviator in Great Britain finds that local maps contrast markedly with American versions of the same locale—not in the iconic details, of course, but in all those other features which make it possible for an aerial chart to serve as his guide. He soon realizes that there is much to learn and to comprehend in order merely to understand what the "message" is in these maps; for cities, towers, airports, train tracks, bridges, monuments, etc., are symbolized in different ways on United Kingdom aviation maps than on those from the United States. Both are useful only to the extent that there is structural verisimilitude between the map representation and what is being represented (the terrain below). But how can one "read" the chart so as to be sure of this? After all, maps are not photographs. The map and the original do have

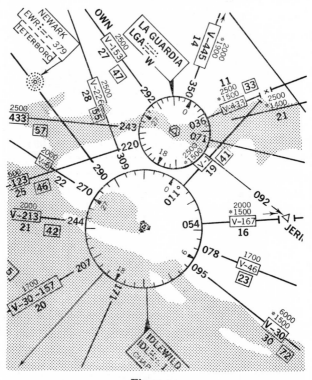

Figure 10

geometrical structure in common. Still, the terrain is marked neither with As, Bs, and Cs nor with standardized representations of cathedrals and canyons. Moreover, there are also other types of representation—other than the iconic, that is.

Consider charts even more abstract than our example of the aircraft navigational map. The correlations of structures are really quite different in some of these explicitly nonrepresentational line-clusters. The sort of chart seen in the economic and financial sections of the *New York Times,* for certain purposes, correlates within a single representation the ages, family sizes, income brackets, parental employment, major subjects, high school standings, hometowns, and career objectives of, say, all Pittsburgh undergraduates in residence. Imagine how a dean's office might have a whole battery of wall charts to indicate something of what constitutes the cross-sectional makeup of the undergraduate student body. Or consider the jungle of Detroit automotive products; to understand fully the present "state of the art" would virtually require charts—charts which would in some manner represent and correlate, say, the respective weights, speeds and powers, payload, mileage and durability, operational ease, instrumentation, and reliability—of our present zoo of compacts, wagons, sedans, and sports cars. Thus one would have to be enabled to contrast the ability in acceleration or braking performance of a Detroit compact with a foreign super car like the Saab, Citroen, Bentley, and Jaguar. The graphic display of these data will be an intricate rococo pattern at best, and it certainly will not look like a picture of an automobile.

Charts of considerable complexity might be needed to understand the performance properties of some small internal combustion engine. One would have to relate within the same curvilinear configuration a potpourri of parameters concerning things like compression differentials, average fuel flow, the brake mean effective pressure, the revolutions per minute of the crankshaft, the generator shaft, the drive shaft, the oil temperature after five minutes of idle running, the mixture ratios of fuel and air, the lubricant's viscosity, and the coolant's efficiency—all these things as considered in this one small engine at a moment, perhaps five minutes after it has been fired up. From such a graphic display of parameters one could then delineate with accuracy the power plant's total performance after having run for ten minutes, fifteen minutes, or any time whatever. These lines and their interrelationships would indicate changes in the numerical values of variables whose functions really constitute what is the essence of the machine in question. The R2800-30W Pratt and Whitney aircraft engine *is* what appears in figure 11, which is all of its constituent dynamic parameters—their waxings, wanings, and influxes—and a representational description of that power plant's performance. (This chart and the engine have much in common —structurally, that is!)

The concept of the chart gets quite complex and abstract as we address a modern, high-performance aircraft wing. A wing *is* its coefficient of lift, plus its coefficient of drag, its frontal resistance, and its skin friction; it is the eddying turbulence at the trailing edge, the vorticity, and the wing-tip configuration; it is the stalling point, where the bound-

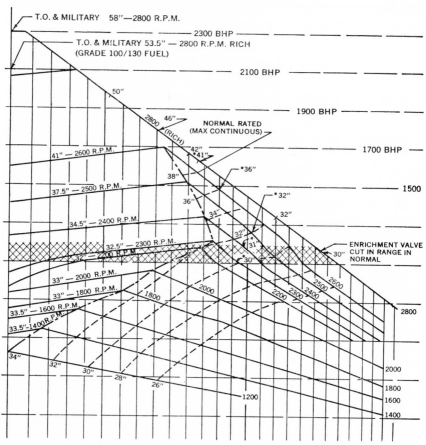

Figure 11

ary layer lifts off the upper surface, and the starting vortex. Understanding a wing, appreciating what it does and what it is, *is* just being aware of how such lines would slope and intersect when representing the aerodynamics of that shape. (See figure 12.) To contrast the S-shaped airfoil section envisaged by Richard von Mises and a quite distinct airfoil section known as NACA 2412 is to appreciate intuitively how the appropriate graph representations of the aerodynamic parameters of these wings will differ or will be the same. (See figure 13.) Aerodynamicists are really discussing the properties of such *shapes* when they discuss stagnation points, laminarity at given angles of attack, the stability of burble zones in the boundary layer, the vorticity of the downwash, and the induced drag and form drag of a particular wing section's shape—none of these being photographable (or at least not primarily photo-

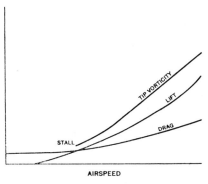

Figure 12

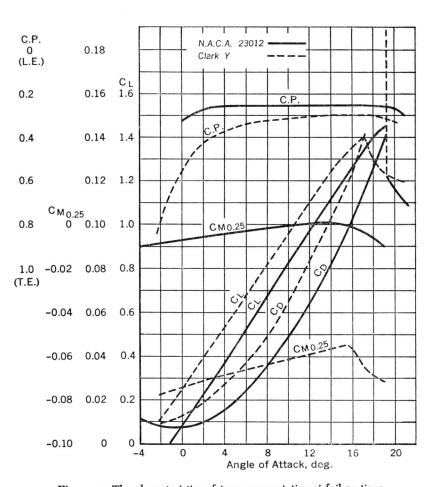

Figure 13. The characteristics of two representative airfoil sections

graphable). These locutions really concern the structural interrelationships between those functions which describe the slopes of the
performance lines on aerodynamic airfoil charts. No two airfoils have the
same characteristic charts; when two charts are identical, they designate
one and the same airfoil shape—something known without actually
introducing the shape as a third bit of evidence. The charts are the
wings; they are everything aerodynamically significant about the wings.
Yet they are certainly not pictures. They are structural representations of
aspects of the wing, however.

Such charts, diagrams, and graphs as I have described are veridical in
a large number of instances. They are informative because they share
structures with the actual wings in question—dynamical structures, not
geometrical structures. They provide a pattern through which the multi-
form and chaotic manifestations of the original appear as correlated
parameters. These patterns provide conceptual *gestalts* which allow inferences from one parameter to another parameter throughout a charted
system of data lines. Thus from knowledge of what in fact is the
numerical value of x (the respective angle of attack) supposed to obtain
with NACA 2412, plus a knowledge of the airspeed, one can infer to a
value for the 2412's coefficient of lift, coefficient of drag, and its trailing
edge turbulence. These data tumble right out of this representational
approach, since to know the shape of the general parameter configuration and the value of one parameter is to be positioned for inference to
all the other parameters which describe the 2412 (the parameter configuration as graphed describes NACA 2412 just as a compass describes a
circle; the 2412 is completely delineated via the drawn parametric interactions).

Once a constellation of parameters is captured in a graph, simple
Cartesian transformations can render them algebraic, and less obviously
pictorial and geometrical. Any cluster of crisscrossed lines in a plane
becomes totally comprehensible when transformed into a cluster of algebraic formulas, each one of which "programs" where every point on a data
line will fall. Thus coordinate geometers easily transform a horizontal
straight line into $y = k$; a vertical one into $x = k$; a sloping straight line
(through x,y, with slope m) into $y - y_1 = m(x - x_1)$; a circle (radius r,
center at a,b) into $(x - a)^2 + (y - b)^2 = r^2$; a parabola (vertex at o,
focus at a,o) into $y^2 = 4ax$; a hyperbola (center at o, focuses at c,o and
$- c - o$, transverse axis $2a$) into $x^2/a^2 - y^2/(c^2 - a^2) = 1$, etc. Familiar
exponential and logarithmic curves are $y = e^x, y = e^{-x}, y = log_e x$, etc. The
Gaussian distribution so important in laboratory work is $y = e^{-kx^2}$. In

general, one can find an algebraic descriptive equation for any locus of points defined geometrically. Hence all lines, or segments thereof, are completely described algebraically. For physical curves like projectile paths, light rays, celestial orbits, etc., and for physical surfaces like ship hulls, airfoil sections, gas flow lines, etc., this analytical technique is obviously very powerful, for the full battery of algebraic procedures is at once made available to the physical understanding of any two-dimensional curve or three-dimensional curve-set. The situation is identical for the analysis of data graphs, the observation-point lines on which are interrelated simply by algebraically interlinking the equations of those lines. The result is a very much fuller understanding of the dynamical properties of the subject matter partially described by each observation line; this is signally true in studies of airfoils, engines, markets, and societies. The sequence is always from observations, to numerical descriptions (of those observations), to point location (on a graph), to curve construction (out of the observation points), to algebraic description (of the parametric curves), to functional interrelation of the algebraic descriptions. The end result is an algebraic structure which *is* (or at least is analogous to) the dynamical structure of the actual airfoil, the engine, the market, or the society. (As the algebra exfoliates in logical space, so does the airfoil, or the engine, perform in actual space-time.)

This natural development from discrete measurements to algebraic formula clusters is open to an insight advanced by Wittgenstein, in which he states that "language sprang from hieroglyphics, but the essence of representation remains." Analogously, I shall argue that after curvilinear data graphs are rendered into algebraic and function-theoretic forms, the essence of structure remains as between the processes in the subject matter and the processes in the algebra. If the graphs were informative as structural representations of the aerodynamics of a wing or the thermodynamics of an internal combustion engine—and if there is no difference that makes any difference between the graphed lines and the algebraically symbolized lines—then the algebra is in and of itself informative as a structural representation of the original subject matter, wing, or engine. Physical theory, I submit, can be thought of as a result of compiling, meshing, and unifying many such charts. I do not mean this statement genetically or factually, of course; it is not a historical remark. But reflect on the conceptual possibilities here. If a physical theory is construed pro tem as a result of data-chart compiling, if each graph is structurally related to complexes of phenomenal processes, and if the data-graph compilation grows toward theorizing by being transformed into generalized algebra from which one then undertakes to detect still higher orders of formal and inferential connection—if all this is allowed to stand, it will suggest how phenomena and theories relate, how measurements and algebra connect, how the world and our ideas of it are linked. And in this view of the rise of theory, the essence of representation is not lost. The graph lines represent data structure; when the data ascend or descend numerically, when they oscillate or spread randomly, the graph lines do the same. They share structures to that extent. But the algebra does this to no less extent; thus the algebraic descriptions are also representative of processes in the phenomena. But complex and inelegant algebraic expressions can be traced to simple, powerful, and elegant higher order algebraic claims. Equations fuse, collapse into one another, and reveal themselves as but special cases of much wider and abstract mathematics. Most sets of equations can be shown to be functions of other sets, given sufficient time and ingenuity (and computers). In fact, for any particular body of descriptive equations, it can usually be shown that there is an indefinitely large number of functional relationships which bind the euqations. It is the special office of theoretical insight to opt for *one* of these functional reticula as against others equally faithful to the data. Thus, that a theory should square with the facts is a necessary condition for its acceptance. But it is not a sufficient condition; that is, after threading data into alternative abstract functional relationships—there always will be alternatives—further considerations must be weighed in the court of scientific theory.

The sophisticated complexities of theoretical language should never obscure the fact that much of the understanding such language systems provide issues from structural insights they afford into the three-dimensional behavior of the original phenomena. Before developing an example or two of what I am getting at (and, as usual, the examples may prove richer than the analysis), let me discuss some advantages of our analogy. Remember, the analogy concerns theories considered as structural patterns, or perhaps "structuring patternings" would be more felicitous. Theories as blueprints are our primary interest; theories as narrative will barely concern us. One immediate fallout from this way of viewing theories is that it becomes difficult to speak of them as being descriptively true or false in the manner that observation statements are descriptively true or false. Indeed, it becomes hard to think of them as being true or false in any simple sense. This is compatible with many arguments by many philosophers of science: "of theories ask not whether they are true or false, but only whether they apply or do not apply";

"theories are not simple conjunctions of observation statements"; etc. So, just as patterns suggested within a pictorial configuration are not true or false, but rather are effective or ineffective vis-à-vis the color patches they are intended to relate, so also patterns set out within a theoretical representation are not true or false, but rather are such that they do function effectively vis-à-vis the observed phenomena or do not function effectively vis-à-vis those phenomena. For theoretical patterns to "function effectively" vis-à-vis phenomena, they must provide conceptual structures which permit one to move inferentially from descriptions of one phenomenon to descriptions of other phenomena, much as a pictorial

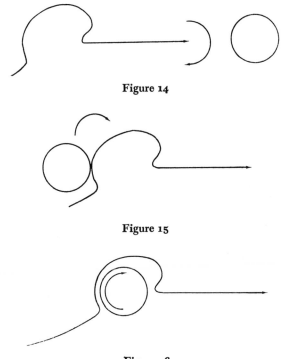

Figure 14

Figure 15

Figure 16

pattern permits the eye to move smoothly from one color path to another within a larger picture. The patches in figure 14 are "given meaning" in one way through the pictorial pattern in figure 15 and are "given meaning" in quite a different way in the pictorial pattern in figure 16.

It is the same with theories, wherein the inference patterns of one will structure observation statements in *this* way, whereas the inference patterns of another will structure them in *that* way. *It was not the observed data, but the patterns* which distinguished geocentrism from heliocentrism in sixteenth-century astronomy, which distinguished wave theory from corpuscle theory in seventeenth-century optics, which distinguished phlogistication from oxidation in eighteenth-century combustion theory, which distinguished vitalism from mechanism in nineteenth-century biology, and which distinguishes the Copenhagen interpretation from those of its critics in twentieth-century microphysics.

If someone, for example, were to look at figure 17 and identify it as an x-ray tube, that designation would invoke a kind of pattern. So also

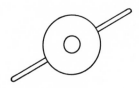

Figure 17

would designating it "a donut with toothpicks." But these do not happen to be patterns which make one feel the patches and lines pulling together when it is dubbed "a Mexican on a bicycle." Your spontaneous laughter, then, was a clear behavioral signal that the marks did knit together for you in a dramatic way. This is part of what Ernest Nagel had in mind when he urged that in his view "theoretical-predicates-as-variables" constitutes a healthy philosophical attitude. That it is not at all clear just how theories can be represented as true or false is a part of what he is gesturing at and a part of what my remarks so far are meant to embrace.

So much for the first point concerning the sense in which theories are true or false. The second one concerns something Hempel made reference to in 1951, namely, determining the meaning of an observation statement or an observational term. This cannot even be assessed except in terms of the theoretical structure within which such statements and terms figure. It reminds one at once of comparable difficulties in determining the meaning of such a claim as "It is close in here"; one does not know whether such words are uttered in a small, smoky room or in a large hall filled with university officials or in an embarrassing situation or what. Similarly, were "The sun rises in the east" spoken by one for whom "east" is the name of where the sun rises, no matter where, the claim could not be false. However, when "east" designates a direction, as determined, for example, by celestial coordinates, the claim could be false. One must know the context of utterance in order to know the meaning of the claim uttered. Observation statements within a theory derive much of their semantical content from the structural framework within which they do figure.

Consider a pronouncement like "red now" in an astrophysical context, wherein it may be a highly technical reference to such things as photo-spectrometric tabulations of the red shift and consequent confirmation or disconfirmation of rarified cosmological theories. (This is not unrelated to Hoyle's recent abandonment of the steady state theory.) In an ophthalmological context, of course, "red now" would be semantically charged in a different way. In a philosophical (phenomenalistic) context it would be different again. In a chemical context (involving titrations) it may be different yet again, etc. This point connects with one that Wilfrid Sellars makes concerning how it is that theories indicate why particulars fall under the empirical laws they do fall under. It seems to me remarkably analogous to understanding why it is that certain dots and shapes and lines cluster within a pictorial configuration as they do, given a "significance pattern." Of course, this point also ties in with the view that theories, in general, stand or fall en bloc. Observation statements that do not fit into the overall theoretical conception are treated as anomalous, and one's reactions to the anomaly vis-à-vis the theory's structure are of the greatest importance here. The advance of the perihelion of Mercury wrecked classical celestial mechanics in toto; [9] the nonconservation of parity did not wreck quantum mechanics. Different responses to anomalies indicate that different meanings are attached to the structural principles of the theory in question.

Understanding how phenomena are sometimes felt to become comprehensible when viewed through a particular theory is somewhat analogous to the "Gestalt click," and this itself is instantiated via the pictorial illustrations which support the analogy I have been shaping. It also relates to another point that Sellars has been zealous in making, namely, that theoretical terms themselves might well be construed as meaningful in the restricted sense that they are *structurally effective*. This point pierces the classical positivistic and hypothetico-deductive positions which urge that the real meaning to be found in theoretical structures is all imported "upward" from the observational statement level (Braithwaite's semantical zip-fastener); the rest is syntactical veneer—all form and no content. But significance—and meaning—is often purely a matter of form and not of content. One may have theoretical terms like ψ in quantum mechanics or ι in classical thermodynamics, "structural terms" which are important for forming the framework in terms of which the observational details hang together; these terms, and operations on them, are what make that structure apparent and effective; and as such they constitute an important part of "the meaning" dimension of these terms. Yet they are in no obvious sense connected with factual observations. The meaning of ι is not "zipped" up into it from the ostensive correlation level of observation statements, and ψ is not a semantic composite of all the special coordinating definitions which strap quantum mechanics to laboratory facts. Rather, a theory and its constituent abstract terms can be considered a sort of conceptual gestalt for observation statements. There are, in the history of science, several examples of mere unstructured inference, lacking any appropriate accompanying gestalt.

This was the criticism of Leibniz, Poleni, and Bernoulli of Newton's *Philosophia naturalis principia mathematica* (1687) when it first appeared. They treated it as being but a mere algorithm—one which failed altogether to link up the sort of causal explanations and speculative understanding so important in their kind of natural philosophy. Leibniz regarded Newton's theory simply as an algorithm which, almost per accidens, seemed incredibly effective at grinding out numbers. But this was not really natural philosophy in Leibniz's sense—thus their criticisms. Parts of quantum field theory today are describable in much this same way; consider the mathematical divergencies which result immediately from the technique of renormalization. Non-Hermitian S-matrices, negative probabilities, "ghost" states, etc., are the ghastly issues of this inelegant manner of forcing calculation even after understanding has departed. One's comprehension of the subject matter of microphysical radiation is almost to be distinguished here, from the degree to which numbers can be successfully churned out. In the seventeenth century analytical mechanics required no stacks of statistics, successive approximations, confirmed predictions, etc., because the full understanding of classical physics was embodied within a terse, powerful, rich, and elegant system of inferential patterns. Nothing like that obtains in microphysics today, and undigested statistics have taken the place of natural philosophy—but this replacement may not matter since computers have taken the place of natural philosophers.

A further example of this business of drawing mere inferences sans any appropriate gestalt can be seen in the many aerodynamic recipes used in the late nineteenth century by aeronauts and by physicists, in order even crudely to approximate to this incredibly multiparametric, turbulent subject matter. This example is to be contrasted with theories which, at rosy moments in the history of science, are felt to be imbued with an intrinsic gestalt, a key to the intelligibility of the phenomenon in question. This was the attitude of Euler and Helmholtz toward Newtonian mechanics itself. At one stage Euler asked: "What is it to explain any phenomenon in nature? It is simply this, to 'reduce' that perplexing macrophenomenon to a mechanical analysis of its microconstituents, their energies, motions and positions." This was carrying the argument of the *Principia* all the way. To explain was to "apply Newton."

Many moments in the history of analytical mechanics, especially those involving planetary theory, fall within the plot just delineated. The contrast between geocentric and heliocentric theory is clearly representable in these structural terms: the facts were equally accessible to protagonist and antagonist. The *arrangement* of those facts, however, is what generates the dramatis personae within the history of astronomy. The arguments which terminate in an hypothesis's positing the existence of some trans-Uranic object, the planet Neptune, and the structurally identical arguments which forced Leverrier to urge the existence of an intra-Mercurial planet, the planet "Vulcan," to explain the precessional aberrations of our "innermost" solar system neighbor are formally one and the same. They run: (1) Newtonian mechanics is true; (2) Newtonian mechanics requires planet P to move in exactly this manner, x, y, z, . . . ; (3) but P does not move à la x, y, z; (4) so either (a) there exists some as-yet-unobserved object, o, or (b) Newtonian mechanics is false. (5) 4b) contradicts 1) so 4a) is true—there exists some as-yet-undetected body which will put everything right again between observations and theory. The variable "o" took the value "Neptune" in the former case; it took the value "Vulcan" in the latter case. And these insertions constituted the zenith and the nadir of classical celestial mechanics, for Neptune *does* exist, whereas Vulcan does not.[10] Some of Kepler's arguments leading to his first law are also relevant here; Tycho Brahe had the data, and Kepler divined their structures. And there are many other historical instances of this recognizing-of-patterns-in-data which highlight the evolution of science.

Now, Book Two of Newton's *Principia* was designed to represent a difficult and turbulent subject matter, namely, the dynamics of fluids. Newton makes the important suggestion that all undulatory behavior within a fluid subject matter can be represented in punctiform terms. He urges that the laminar flow of fluids, their turbulence, their vorticities, their currents, viscosities, resistances, eddies, densities, etc., all can be treated simply as a mathematical manifestation-in-the-large of what is fundamentally appreciable in terms of those punctiform interactions articulated in Book One of the *Principia*. By the time Newton composed his *Principiate*, certain empirical facts that were well known had been bypassed. Galileo had concerned himself with the resistance a fluid exerts against a given object, as had Leonardo before him. Both reckoned this resistance to be directly proportional to the velocity of the fluid itself or at least to the relative velocity of the fluid as against a moving object. Huygens perceived the matter more clearly and set up a little experiment, the end result of which was his discovery that the fluid's resistance to any object moving through it is proportional to the square of that relative velocity. This is the figure we use today. Thus, whether it be a fish swimming through a still stream or rapids rushing against a rock, the resistance felt by fish or rock is proportional to V^2, that is, the square of the fluid relative to the motion of the fish, or rock, through it.

Now, one of the interesting moves made in Book Two of the *Principia*, a move which had a remarkable effect on subsequent history of science, consists precisely in a pictorial type of representation (again, where "pictorial" is in inverted commas; Newton's is not a presentation of pictures in any graphic sense). Instead, his is a presentation of structures

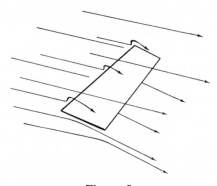

Figure 18

in the sense more difficult to comprehend—more difficult but essential to our understanding of what *this* scientific theory (fluid mechanics) does by way of "explaining" its subject matter. Newton concerns himself with the resistance that will obtain between a fluid flow which is assumed to be laminar (without internal rotations or turbulence) and a given flat plate. The latter can be thought of as being simply two-dimensional (of zero thickness) for the purposes of this inquiry. (See figure 18. It has been drawn tilted so it can be viewed from the port quarter.) Newton construes the force exerted upon this particular plane plate by this fluid, or, rather, by the indefinitely large number of particles of which this fluid is constituted as being proportionally related to a number of things. One of these, of course, will be the *density* of the fluid. In liquid mercury the force upon the plate will be greater for a given relative velocity than it will in ordinary water. Moreover, this force will vary with the relative velocity with which the fluid moves against the underside of the plate; when it passes slowly, the force manifested will be different from when it is moving very quickly (as every water-skier knows). This force is also a function of the area of this particular plane plate; a postcard will not receive the hydrodynamic shock experienced by a barn door, other things being equal.

Finally, and here is the joker in Newton's deck, the force that the liquid exerts upon the plate will be a function of the sine squared of the angle of inclination of the plate to the liquid. That is the representation Newton gives. (See figure 19.) In short, it should be pointed out that

$$F \propto \rho V^2 S \sin \alpha^2$$

is within the "mechanics of perfect fluids"; this law is a simple deduction from having assumed our fluid to be *inviscid* (that is, there are no in-

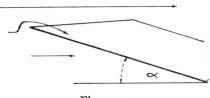

Figure 19

ternal interactions, or frictions, between the particles of the fluid) and *irrotational* (that is, the particles of this fluid are restricted vis-à-vis their degrees of freedom in motion; they do not rotate; no vortices are genera-ble within the fluid). And finally, this ideal fluid is *incompressible* (that is, calculations do not depend on the *springiness* or "squeeze" of the fluid). These assumptions, although profoundly counterfactual, make the "per-fect fluid" game go beautifully. The developed algorithm of Euler and Bernoulli is a magnificent and elegant symbol system to comprehend. The mere fact that it applies to nothing should not be taken too seriously, perhaps. The same is true of many "idealized" physical theories. And, of course, *this* theory is fundamentally punctiform.

We were discussing one of the consequences of this particularity when one represents the fluid resistance on the underside of a flat plate in terms of the sine squared of the angle of incidence. This particulate model of such an ideal fluid comports well with such a sine-squared law of resistance. Applied to *practical* cases within fluid mechanics, however, the model and its associated "laws" are less than wholly satisfactory. How does the air support a bird—like an albatross or even a humming-bird—which always displaces a volume of air which weighs but a small fraction of its own total weight? The condor seemingly soars eternally, supported upon a substance, the air, which (volume for volume) weighs less than one-thousandth of what the bird weighs. Were a similar situation to obtain with water and ships, the oceans' floors would be strewn with wreckage in less than five minutes. Archimedes's law, then, conveys little for our understanding of flight. Nor do the analyses of Leonardo and the influential Newton. Suppose, à la Leonardo and Newton, that the only thing that holds birds aloft is the pressure upon the underside of the wing. Then a bird such as an albatross (moving at an airspeed of fifteen miles an hour), in order to achieve the value for F compatible with these other known parameters, must angle its wing incidence up to something like sixty degrees! A Boeing 707, to move as it does at take off or cruising speed, would also have to tilt up its wings to about sixty-five degrees, given the sine-squared law! This would force calculations for the associated drag which would be totally out of the question. Birds and planes could not soar and glide as they do with wings tilted up like snow plow blades! The sine-squared law requires that we would have to minimize the angle of incidence (α) in order to keep the drag factor within the bounds of conceivability. This would also necessitate enlarging the value for the wing area, S, so that the actual area of the 707's wing would have to be about the size of two football fields—either that or boost the value of V^2 by fantastic increases in propulsive power, the result of which would be fantastic increases in gross weight.

Reflections such as these forced many to conclude that, by mechanical means alone, birds could not fly. The only effective manner of calculating such parameters derived from Newton's *Principia* is what constituted "analysis of natural phenomena" in the eighteenth century. But this analysis (the sine-squared law in particular) was quite incompatible with what one observed in the case of bird flight. Therefore, there must be something extraphysical or extranatural about bird flight—something beyond Newton, which came to mean "beyond science" and was simply attributed by many to divine intervention, occult qualities, miracles, etc. The extant literature thus makes the special problem of flight extraordinarily interesting from the point of view of the history of science.

As *we* know, there is another natural effect primarily responsible for flight. Newton did not realize (how could he?) that this further effect contributes to the flight of a bird at least five times more upward force, or "lift," than is generated by way of direct impact of the fluid molecules on the underside of a wing. Newton's picture of material collisions, although not negligible in the phenomenon of flight, is minor indeed when compared with the "suction" operative on the topside of the wing.[11] The magnitude of this lifting force is a function of the shape of the wing, but in some cases it can be fifty times greater than anything calculable via the classical sine squared law.

Consider the Magnus effect (so-called because a Professor Magnus, in Germany, worked with it in the early nineteenth century); it is actually stated quite clearly in Newton's *Principia*. See figure 20. Here a cylindri-

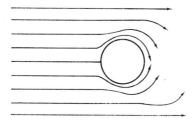

Figure 20

cal rod is seen on end. Laminar (nonturbulent) airflow comes from left to right. The d'Alembert paradox states in effect that on either side of the stagnation point air will flow symmetrically around the rod, curling in behind it and impinging on the aftermost point (diametrically opposite the stagnation point). How like Aristotle and the *antiperistasis* theory articulated in Book IV of the *Physica!*[12] Indeed, in a perfect fluid the force exerted on the trailing surface of the rod would be exactly equal to that exerted on the leading edge. The effective resistance of the rod to the fluid flow would then be zero, a calculational result which is paradoxical because, although consistent with the analysis that Newton provides within ideal fluid mechanics, it is obviously counterfactual. Submerged rods resist streamflow redoubtably. (See figure 21.) At this juncture Newton entertains that a rotation is given to the cylinder itself (clockwise as we view it). (See figure 22.) This rotation of the cylindrical rod affects the liquid flow across the rod as shown in figure 22. Indeed, the rod itself now moves "upward" through the flow, a phenomenon for which there was no reasonable expectation in eighteenth-century fluid mechanics. Today, an appeal to boundary-layer theory is required for

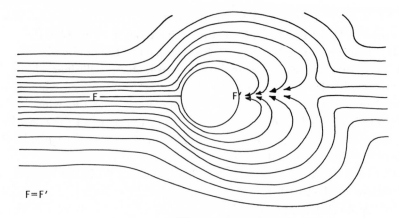

Figure 21

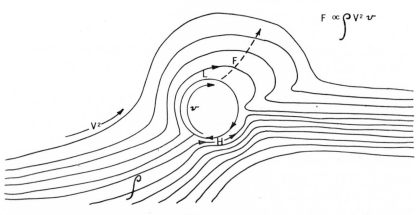

Figure 22

any complete understanding of the Magnus effect. Thus, at the top of the cylinder the particles constituting the rod's surface move in the same direction (left to right) as the particles of the fluid. The reverse obtains at the cylinder's underside; there the rod particles and the fluid particles move in opposite directions, against each other, with resultant friction and turbulence. These latter produce a pressure against the bottom of the cylinder at the same time as laminarity increases above the rod, with consequent decrease in turbulence and pressure against the cylinder's upper surface. Hence the rod moves upward, from high pressure toward low pressure.[13]

At the topside, indeed, a further effect also operates—the Bernoulli effect. This instructs us that, in laminar liquid flow, where velocity of flow increases relative pressure decreases and where velocity decreases pressure increases; also, where pressure decreases velocity increases and where pressure increases velocity decreases. $\rho V^2 + P = \text{cons.}$ At the rod's topside V^2 increases, so P decreases and the rod lifts. What is actually operating is (1) a positive pressure beneath the rod and (2) a negative pressure above (due to the Bernoulli effect). Newton notices the translation of the rod across the streamline, and he remarks the force (F) of that translation. In other words, the Magnus translation will be a function of the velocity of this fluid's motion across the rod (V^2_{fl}), and of the angular velocity of the rod's rotation (V^2_r); $V^2_{fl} \pm V^2_r = V^2$. When fl and r have the same sense of motion, their velocities are added; when their senses are opposed, the rotatory velocity is subtracted from the flow velocity. The resultant variation in V^2 will determine a corresponding variation in P (since they both add up to a constant figure for a given fluid of a given density ρ). From knowledge of this variation in topside pressure, one can calculate the degree of translation of the rotating cyclindrical object across the laminar liquid flow. Increase this rotatory velocity and you increase the displacement across the laminar flow.

It became clear, with the failure of the sine-squared law, that some other force was operative—some force beyond the positive pressure exerted on the underside of an object inclined to a fluid flow. It was the ornithologists (not the physicists, incidentally) who first tumbled to this conclusion. Natural philosophers had really "let the side down" when it came to fluid mechanics. Every time they talked about hydrodynamics they talked about *perfect* fluids, irrotational, inviscid, and incompressible fluids. Plumbers, engineers, and aeronauts could not do a thing with theories of perfect fluids because there just were no such things. Every fluid they had anything to do with was highly viscous, very compressible under certain conditions, and certainly rotational. These characteristics made the entire structure of the Euler-Bernoulli calculations virtually useless for any practical study of fluids. And when ornithologists began to study the physics of the flight of birds, dissatisfaction with the analytical fluid mechanics of the physicists reached its zenith. After all, contra the sine-squared law, birds could fly—they do! Some professional physicists appreciated the situation and joined the ornithologists pro tem. Thus Lord Rayleigh made a number of serious contributions to our understanding of what it was in the *actual* motion of the swan's wing which made it aerodynamically effective during take off and landing. It slowly dawned on biologists, fledgling aeronauts, and finally upon physicists that there was *something* about the convex camber on the upper surface of the airfoil section of a swan's wing—or that of a condor, an albatross, or a gull—which was definitely responsible for the dramatic lifting effect operative in, and responsible for, bird flight.

The leading practical question for aeronauts and aerodynamicists then became one of determining precisely the upward swerve—the lift—of actual airfoil shapes (and not just rotating rods) across the flowing fluid. Each new airfoil, each new bird wing, had its own geometrical and physical properties. Were these to be calculated anew for each different wing? How wonderful it would be to have a computational technique that would allow generalized airfoil theory to determine lift with just the precision possible for the description of rods, rotating within streams! After all, in the Magnus effect the rod's cross section is circular, the streamflow is laminar, and the entire calculation is structured by symmetry considerations. Vary the angular velocity of rotation, or the laminar velocity of streamflow, or the roughness of the cylindrical surface, or the density of the fluid envelope and the upward swerve will vary accordingly, predictably and exactly. How then to construct a theory of the airfoil so that the same parameters could be interrelated with comparable precision—as against the early exposures to aerodynamic phenomena, wherein the physical properties of each airfoil had to be determined separately through observation?

It was a triumph for Kutta and Joukowsky when they discovered precisely such a calculational technique. Their researches paved the way for a computational aid which makes the determination of an airfoil's lift no more difficult for us today than was the determination of a rotating cylinder's upward swerve through a fluid difficult for latter-day Newtonians. I will delineate this final example in some detail, since it is clearly a case of perplexing phenomena being rendered intelligible through a theory which shows the structure of those phenomena in a representational way.

Let us review: an ideal fluid impinging broadside upon a cylindrical rod will curl around behind the rod in symmetrical fashion, turning in against the trailing edge with a force of impact exactly equal to the force of impact at the stagnation point forward. The paradoxical result will be that such a rod will experience, and exert, no effective resistance within such an ideal fluid. Now rotate the rod rapidly. Because of friction and turbulence below, as against the lack of such effects above—plus an increased fluid velocity (V^2) above (and further decreased pressure)—the rod will lift across the flow lines with a force (F) that is proportional to the density of the fluid (ρ), its laminar velocity (V^2), and the velocity of the rod's rotation (v).

When this much was known about cylindrical rods and their rotatory behavior within nonideal fluids, airfoil theory was still in a primitive experimental state. To learn how a given airfoil shape would "lift" across laminar flow lines, one could not analyze and calculate; it was necessary to experiment and observe. To that extent, full understanding of the physical phenomenon involved was lacking.

Bird wings lift, but no one knew how or why. Rotating cylindrical rods lift, and this seemed to be fairly well understood. F. W. Lanchester had *the* physical insight within unnumbered millennia of wonderment about flight. Just as the rod's rotation itself is a kind of *circulation* within a laminar flowstream, so also an airfoil may also be a circulation—indeed, a simple deformation of the kind of circulation one sees operative in the Magnus effect. Now, the Magnus effect is equally well instantiated in the complete absence of a physical rod. If, across a flowstream one can induce a vortical motion within the fluid (as, for example, by rotating

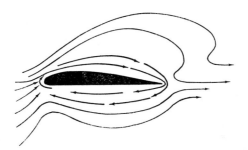

Figure 23

vanes on the walls of the tank or tunnel), that very vortex will itself swerve upward just as calculated in the case of the rod. And the magnitude of that "lift" across the flow itself will be determinable as a function of the fluid flow velocity, the fluid density, and the angular velocity of the vortex rotation. (See figure 23.)

Suppose now, muses Lanchester, that an airfoil—for example, a bird's wing—is uniquely suited to induce vortical circulation within the fluid flowstream. Suppose, that is, that the main function of a wing were not physically to support a flying object, but rather to induce around itself a vortex, a circulation, whose behavior within the flowstream is in principle identical to the behavior of a vortex manifesting the Magnus effect. In this view the first function of an airfoil shape would thus be to generate vortical circulations around itself, the properties of which circulations will differ in accordance with differences in the airfoil shape itself. Thus some airfoil shapes will more sharply decrease pressure above for slower flowstreams than others will, but they will pay for the slower flowstreams by generating greater trailing-edge turbulence at increased relative velocities. Other shapes will move and lift effectively through flowstreams of higher velocity, but will be almost like flat plates when the flowstream is more leisurely.

The primitive representation of an airfoil-induced circulation, then, would be something like figure 24. Lanchester than argued that there

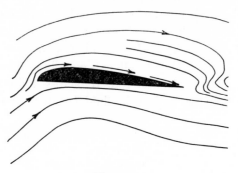

Figure 24

would be a tendency for the flowstream to curl beneath the trailing edge of the wing and to proceed forward; this tendency would be obliterated by increased velocity in the flowstream itself, which would "wash" this turbulence far astern. This physical insight immediately collapses a number of striking observations and forms one impressive pattern of "explanation."

What makes the airfoil start to lift? What is it about the shape of an airfoil that induces within the flow around it an effect comparable to what is generated by the rotation of a submerged rod? Lanchester argues that the air flows around a symmetrical shape in a way analogous to a fluid flowing around a nonrotating cylinder. Particles of air separated at the leading edge will arrive at the trailing edge at the same time and, with but a modicum of turbulence, will proceed astern with no observable effect on the airfoil itself. If, however, the upper surface is cambered much more radically than the lower surface, then two particles of air separated at the leading edge will not reach the trailing edge at the same time. And, because of the slight angle of inclination of the shape to the streamflow, there will be greater positive pressure below the wing than above it, and, hence, the particle arriving at the trailing edge before its "twin" which traveled over the upper surface will tend to move upward immediately after passing the trailing edge. Indeed, a vortex will be generated at the trailing edge of the wing, the immediate effect of which is to institute a small Magnus generator within the flowstream above. This "starting vortex" will pull the particles of fluid across the upper surface of the airfoil much more rapidly. This action will further increase the velocity of airflow above, which will in turn further decrease the pressure within the fluid above the wing. As the relative velocity of the wing through the flowstream increases, Lanchester argues, this "starting vortex" will suffer the same fate as did the earlier particles which sought to curl around over the trailing edge and proceed forward again, that is, they will be swept astern. But what has happened above the upper surface of the airfoil shape is, in every physical way, analogous to what happened at the upper surface of the rotating rod in the Magnus effect example. In other words, if there is *any* relative velocity of the flowstream over the airfoil, a primary circulation is instituted. (See figure 25.) This immediately gives way to the starting vortex astern, which in turn draws the fluid over the wing so as to reduce most dramatically the effective pressure above and to increase most dramatically the difference between the effective pressure below and that above.

How then, by way of this model, does the whole aircraft (bird, airplane) fly? Figure 26 clearly portrays Lanchester's representation. In short, by analogy with the kind of circulation one witnesses in the Magnus effect, there is instituted around a wing an initial "vortex loop." Increase in relative velocity of airfoil through flowstream washes the starting vortex astern, leaving a "vortex hoop." So, around any aircraft moving at constant speed in constant attitude, there is found a *vortex horseshoe*. As soon as any parameter is varied—that is, if forward speed is increased, or the angle of attack is increased, or the density of the fluid is increased—a new starting vortex appears, with its resultant vortex hoop. And again, this at once gives way to the vortex horseshoe—all by close analogy with what obtains in the Magnus effect.

Every other complex feature of aeronautical design can be at once related to this fundamental insight within aerodynamic theory, for it was Lanchester who perceived in the phenomenon of circulation the key to the ancient problem of mechanical flight. In fact, were it not technologically so formidable a thing to achieve, a wing with a surface which moved à la the cylindrical rod of our Magnus effect case would be very efficient indeed. Boundary-layer problems on the topside would be drastically reduced, because the air molecules adhering to the wing surface might move almost with the velocity of the free molecules in the airflow. Beneath the wing they would be moving in the opposite way, with the already discussed friction and turbulence thereby generated creating

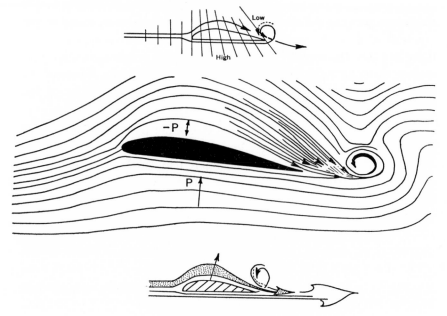

Figure 25

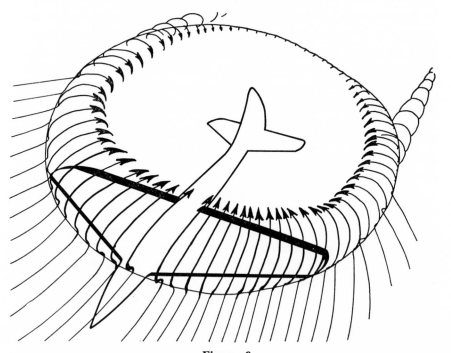

Figure 26

enormous pressure differentials across the airfoil. Indeed, many recent practical advances in boundary-layer control have tried to achieve something like this—namely, by sucking air molecules adhering to the upper surface down into the body of the wing itself. Literally thousands of related phenomena all can be organized and aligned in terms of this monumental structural insight of F. W. Lanchester.

Still, although this much constitutes a qualitative insight into the physical processes attending the lifting airfoil, it still provides no calculus in terms of which one might *calculate* the lift on any given airfoil shape. Enter Kutta and Joukowsky. If, to follow on with Lanchester's insight, the airfoil is just a solid object whose main function is to generate a circulation with certain characteristics—just as the cylindrical rod is a solid object whose only function is to generate a circulation with characteristics such as those noted under the heading "Magnus Effect" —then the geometrical differences between airfoils and cylinders must correspond to the fluid mechanical differences between the "lift" of flight and the "lift" of the Magnus effect. Suppose that there were a transformation technique which could allow one to "reduce" the complex shapes of airfoils to the simple shapes of cylindrical rods. This would, in effect, be tantamount to reducing the complex circulation around a wing to the simple circulation around a rotating cylinder. And, vice versa. Thus, from the simple determination of the fluid mechanical properties of a rotating cylinder, one could (by way of such a transformational technique) calculate the correspondingly more complex parameters associated with particular airfoil shapes. And the specific deformations required in the geometrical transformation would give the clue to the numerical adjustment required in noting how, for example, the lift on a given airfoil at a given angle of attack will compare with the lift on a given circulating cylinder.

The Kutta-Joukowsky transformation technique is intuitively quite simple. Begin with a cylinder imaged to be made of a perfectly plastic substance. We imagine it to be deformed as shown in figure 27. What has happened here is that there has been no deformation along the vertical axis, whereas units of measurement have doubled along the horizontal axis. And one would expect of such a resultant shape that the characteristics of a circulation around such an oval rod might be directly calculable in some similar way. Thus, if the total underside turbulence of the rod were x, the total underside turbulence beneath the oval would then be $2x$.[14]

Infinite variations are possible, of course. A circle can be deformed (according to rules) in an indefinitely large number of ways. Thus suppose we place our original cylindrical section eccentrically upon our reference lines. And suppose that we opt for a trebling of each of the four areas now apparent, under the restriction that the boundary line remain

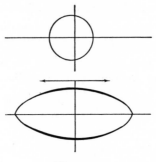

Figure 27

smooth, that is, still described by a continuous function. Figure 28 shows such a deformation. Associated with this deformation would be an identically shaped vortical circulation, the properties of which would be to the original "circular" circulation as is the new shape to the original cylinder.

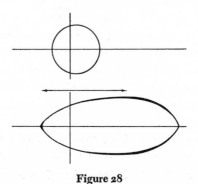

Figure 28

The way is now clear for the generation of almost any shape one can envisage. Thus, given figure 29 and the fundamental intuition of Lanchester to the effect that it is not the shape of such a object but rather the fluid mechanical characteristics of a vortex circulating around such a shape, it is this which generates the specific dynamical characteristics that one should be able now to correlate with any shape within actual tests.

Thus take any bird wing. Determine the plane geometry of a representative airfoil section of that wing, and then determine to what degree this shape is a deformation of a corresponding circle. This deformation indirectly determines what will be the corresponding aerodynamical properties of this airfoil section. All one must do is to calculate the fluid mechanical behavior of a cylindrical vortex as undistorted. Then one takes the numerical descriptions of that behavior (for given velocities, densities, relative laminarities, etc.) and simply makes distortions of those numbers analogous to the geometrical distortions that the given airfoil section constitutes vis-à-vis the original cylinder. The reader can already appreciate that this procedure will require a subtle and advanced analytical technique. But the basic principle is simple. Indeed, in the hands of von Mises and von Karman, this transformational technique has become one of the algorithmic glories of contemporary aerodynamic

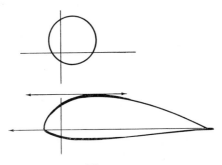

Figure 29

theory. And it is all based on the fundamental insight of Lanchester. To the degree that this is so, one could say that all modern airfoil theory is effective to the understanding of flight and to the construction of aircraft, to precisely the degree that the formal operations within the theory are the structural analogue of the dynamical operations within the phenomena themselves.

Within no time all the drawings and the graphs and the geometrical representations disappear, and their Cartesian algebraic equivalents take their place. Then these equivalents are interrelated through infinitely more subtle "internal relations," well known to the theory of functions. However, notwithstanding the unbelievable advances in calculation and theoretical transformation made available to physicists and aerodynamicists by this Cartesian development, this approach is effective precisely because of the *original* structural felicity of the basic representational intuition, for example, that of Lanchester and the de facto phenomenon itself.

* * *

We have come a long way 'round to nirvana in this exposition of *a picture theory of theory meaning*. After our initial feinting with the picture theory of the *Tractatus* and Wisdom's "Logical Constructions," we considered how observational data—measurement numbers, position-time event registrations, process description, etc.—all were map-able on data graphs, the parametric representations of which have a structural identity with the dynamical features of the phenomena themselves. Thus the identifying structure of a physical process can be represented on a graph, just as the "essence" of a given power plant can be completely encapsulated on a chart or graph appropriate to it. So there is something in section 3 of the *Tractatus* which remains not fully explored by philosophers to date. *Data charts are representations of physical processes in virtue of the fact that they have the same structure as the process itself. They are structural pictures of the dynamical reality.*

Although this much may not seem to constitute the familiar treatment of the nature of scientific theories, I submit that it is but a small step from the structural representations of data graphs to the most sophisticated algebraic treatment of complex physical processes, complete with some insight into what we mean by "explanations via theory"; for a curve of any slope on a data graph is representable (in principle) algebraically by way of a Cartesian transformation. "And the essence of representation is not lost thereby!" [15] Whatever structure the curves on the data graph had, the corresponding algebraic formulations have precisely the same structure. Therefore, if the charts and the phemonena were related by structural identities, the algebra and the phenomena are also related in this way. Indeed, the algebra is the structure of the physical process.

At this point the algebraist assumes command. Any list of algebraic statements will be such that there may be an indefinitely large number of ways of interconnecting them, by inventing functions powerful enough so that one can infer from any one expression of the list to any other one. The result, an exploration within the theory of functions, permits the perception of, and creation of, ever higher-order conceptual patterns. And it is within these that our paradigm examples of physical *explanation* are to be traced. This is also the point where scientific creativity and theoretical insight assume command; for, in general, there will be an indefinite and large number of possible connections between algebraized data—from which the scientist selects.

That all this constitutes a much more plausible account of the nature of scientific theories than can be provided by the "ideal language" approach is made clear within a contemporary discipline which is as yet a long way from being "ideal"; its problems are largely unresolved and one's efforts can hardly henceforth be devoted to axiomatic elegance, as is the case in our present reflections concerning classical mechanics (the discipline which provides the model for the "ideal language" approach to theory). Fluid mechanics—and its exciting contemporary offspring aerodynamics—is a multiparametric nightmare. All major advances seem to have been made through the uses of the theorist's imagination; it is there that intuitions of structural interconnections within the phenomena are instantiated within a structural model which becomes no less representational after all the initial diagrams are destroyed and replaced by the most sophisticated mathematical analyses, transformations, and higher order theoretical connections.

Scientific theories enable us to understand perplexing phenomena precisely because they enable us to see on the page some of the same structures which are there in the phenomena themselves. The theory allows us to comprehend what makes things "go"—and to work our ways into the phenomena, along the dynamical structures (as it were) by way of inferences through the algebra which itself has the same structure as the phenomena, or at least *a* structure compatible with the phenomena.

Perhaps the best name for this function of theories vis-à-vis their subject matters should make no mention of "pictures" at all. Perhaps our title might more justifiably have been "A Structural Representation Theory of What Theories Do." This approach, as recommended within the foregoing, will not only let us make sense of the advancing frontier of science, with all its brow-breaking perplexities, but it will also shed light on the perennial philosophical problems concerning the meaning of theories, the semantical status of theoretical terms, the interrelationship between laws and generalization and measurements and observation— and, finally, it will be *scientific understanding* itself of which we may yet have some satisfactory explanation.

NOTES

1. Kolin, and sometimes Poincaré, for example.
2. Mach and Broad frame the second law as fundamentally a factual statement based on experience.
3. Reichenbach and Poincaré are cases in point.

4. The temporal references "before" and "after" are inessential. This exposition would not suffer if "independently of" and "dependent upon" were interjected.
5. Cf. the earlier illustrations.
6. Yet the designation is specific in that it excludes myriad other kinds of representations of the world. The real steak's possession of properties which can induce gustatory delight in me is no part of any *scene* of that steak. The nightingale's song is replicable, but not because it is part of a scene. The scrape I endure may be due to the icy rough surface of the granite I clamber upon, but the scrape is not a replication of any part of the granite, whereas my visual memory of the granite may, indeed, have properties of the granite block itself—such as those an artist could commit to canvas in a painting of that block.
7. *Facts* are the common structure shared by events "out there" (as when they are "hard" and "stubborn" and must "be faced") and by the truth as stated about those events (as when we "state the facts," "list" them, and base theories on them).
8. L. Wittgenstein, *Tractatus Logico-Philosophicus*, trans. D. F. Pears and B. F. McGuiness (New York: Humanities Press).
9. At least this is true of the assumption that classical celestial mechanics could be applied in principle to *any* astronomical phenomena, anywhere and anyhow in the universe.
10. Cf. N. R. Hanson, "Leverrier: The Zenith and Nadir of Newtonian Mechanics," *Isis*, 53 (1962), pt. 3.
11. "Suction" is just a quick way of designating "pressure differential"—which moves the wing from the high pressure region below it toward the low pressure region above. So this is, in effect, the same phenomenon Newton had in mind with his sine-squared law, only the actual difference in pressure is many times greater than could be calculated by that Law.
12. Cf. N. R. Hanson, "Aristotle (and Others) on Motion Through Air," *Review of Metaphysics* (1965).
13. This much alone is quite compatible with the analysis of Leonardo. At least it adds but little to the main point he made—wherein it was increased pressure below the surface which forced the plate upward. However, the microphysical friction and turbulence beneath the rod, and the lack of same above, of course, eluded him.
14. Of course, the increase will not always be linear in this way. In most cases the relationship will be very complex from a functional point of view. But there will be *some* functional connection between the geometrical deformation and the corresponding fluid mechanical deformation.
15. Wittgenstein, *Tractatus*, sec. 4.016.

PAUL K. FEYERABEND
University of California

Problems of Empiricism, Part II

> It seems to me that [Galileo] suffers greatly from continual
> digressions, and that he does not stop to explain all that is
> relevant at each point; which shows that he has not examined
> them in order, and that he has merely sought reasons for
> particular effects, without having considered . . . first
> causes . . . ; and thus that he has built without a foundation.
> —Descartes

> I am [indeed] unwilling to compress philosophical doctrines
> into the most narrow kind of space and to adopt that stiff,
> concise, and graceless manner, that manner bare of any
> adornment which pure geometricians call their own, not
> uttering a single word that has not been given to them by
> strict necessity. . . . I do not regard it as a fault to talk about
> many and diverse things, even in those treatises which have
> only a single topic . . . for I believe that what gives gran-
> deur, nobility, and excellence to our deeds and inventions
> does not lie in what is necessary—though the absence of it
> would be a great mistake—but in what is not.
> —Galileo [1]

1. Introduction

IN AN EARLIER PAPER [2] I have argued that it is not only possible but also
desirable to introduce and elaborate hypotheses which are inconsistent
with highly confirmed theories and with the evidence. The reasons that I
gave were partly abstract, partly historical. In the abstract part I tried to
show that evidence which refutes a theory can often be found only with

This essay is a continuation of "Problems of Empiricism," which appeared in
Beyond the Edge of Certainty (New York: Prentice-Hall, 1965), and is derived
from a series of lectures I gave at the London School of Economics in the fall of
1967 in which I tried to explain my dissatisfaction with the method of conjectures
and refutations as described by K. R. Popper. It owes much to Professor Imre Lakatos

the help of an alternative so that the advice to postpone alternatives until the first refutation has occurred puts the cart before the horse. The historical part introduced examples which explained how this works in practice.

The present essay continues the argument in a somewhat different style and with a much more chaotic result. It is mainly historical. Abstract considerations are used only sparingly, in the form of comments on the historical material. This material, I think, shows that there is something seriously amiss with the professional philosophy of science of today (it may be different with the amateurs [3]). Contemporary philosophy of science not only fails to describe adequately some of the most exciting episodes in the history of thought; it would also have given extremely bad advice to the participants.

The change of style and the prevalence of history mirror the influence of political analysis and of the philosophy of historical and dialectical materialism. Political analysis is wider and much more realistic than the philosophy of science in that it considers historical conditions including the peculiarities of individual human thought. It also realizes that the historical conditions always contain layers of different age and of different sophistication so that a progressive idea may be impeded, not by any intrinsic disadvantage, but by the fact that it arises in backward surroundings.[4] Conversely, we may be forced to defend absurd and unrealistic views, for it is only in this way that the impeding conditions (which provide the criteria of absurdity) can be overcome. *Historical and dialectical materialism* on the other hand has developed a scheme—the dialectics—for dealing reasonably with chaotic conditions of this kind. The application to our topic consists in pointing out that science, too, contains layers of different age and sophistication and that a new theory may be in trouble not because of any intrinsic disadvantage, but because of the antediluvian character of the views which appear at the observational level.[5] We may, therefore, again be forced to defend absurd,

whose amused interest was one of the main motives for pursuing the matter further and whose pithy formulations helped me considerably in putting it to paper. When writing the paper, I frequently remembered the quizzical look on the pretty face of Mrs. Helena Sheiham during my lectures, and I only hope that the paper is clearer than what she heard half a year ago. I also hope that the paper comes up to Dr. Mary Hesse's call for a "more detailed investigation" which she raised in her review of *Problems of Empiricism*, p. I, in *Ratio*, IX (June 1967), p. 92. I am sure it will please Prof. Paul Meehl who years ago tried to persuade me, by words and by drinks, to adopt the point of view expressed here only to be rewarded with a barrage of Popperian arguments. For support of research I am again indebted to the National Science Foundation.

unrealistic, refuted views, for it is only in this way that the impeding conditions (the faults of the observational ideology) can be overcome.[6]

More specifically, I am indebted to T. S. Kuhn's *Structure of Scientific Revolutions*, to I. Lakatos's *Proofs and Refutations*, to Hanson's *Discovery of the Positron*, and, in a very special way, to the work of Vasco Ronchi (see note 57). Kuhn has made it clear to me that history cannot be dismissed out of hand as being irrelevant to the methodologist. The methodologist deals with a certain sticky material—theories—and he wants to change it. Knowledge of the shape in which the material is available and of the circumstances under which the change has to be carried out is essential to his task. Lakatos has convinced me that Kuhn's approach is still too abstract and that there is only one way to acquire the needed information: case studies. But case studies—and this seems to be the main point of his essay—do not need to degenerate into gossip (as does much specialist history) but can be truly philosophical, being guided by abstract principles and giving rise to them in turn. Hanson's magnificent study of an important episode in the history of modern physics, which I once criticized rather severely,[7] exhibits the chaotic character of science and thereby shows the tremendous abyss that exists between a certain philosophical picture of science and the real thing. My greatest debt, however, is to Vasco Ronchi. Indeed, the first part of this essay, up to and including section 6, is but a faithful repetition of his views with only a few comments added here and there. Considering the contributions of these authors, I now set myself the task of widening this abyss so that the mechanisms which underlie the actual development of knowledge will stand out and be recognized more easily. In this I have simply followed Bohr who in his early work tried to emphasize "the extent to which [the new quantum theory] conflicts with the admirably coherent group of conceptions which have been rightly called the classical theory of electrodynamics. On the other hand," Bohr continues, "I have tried to convey . . . the impression that just by emphazising this conflict it may also be possible in the course of time to discover a certain coherence in the new ideas."[8] The aim of the present essay is exactly the same: to progress by emphasizing the contrast between the customary methodologies and certain important episodes in the history of thought.

A methodology that emerges from an analysis of this kind differs from the existent systems by its lack of dogmatism and by its openness. Each rule, each demand, that it contains is asserted only conditionally, like a rule of thumb, and it can be overthrown, or replaced by its opposite, as the result of an examination of concrete cases (this applies even to such "fundamental" demands as the demand for consistency, for falsification, for agreement with observational results, for maximal content in given conditions, and so on). It is not claimed that rules or prescriptions are now derived from facts (though there would not be anything wrong even with this procedure). It is only asserted that inspection of facts educates a sensible person and makes him aware of demands whose existence, urgency, and relevance he had not realized before. Nor is it possible to say in advance under what circumstance a rule may be suspended and how one will behave then. The future cannot be foreseen and our actions under new and as yet untried circumstances cannot be predicted (after all, we do learn, and we do have new ideas). Also, looking back into history, we find that for every rule one might want to defend, there exist circumstances where progress was made by breaking the rule. All this means that methodology can at most offer a somewhat chaotic list of rules of thumb and that the only principle we can trust under all circumstances is that *anything goes.*[9]

One peculiar and gratifying result of the move away from methodology may be mentioned at once, for it will not be taken up again in the present essay.[10] The inventions and tricks which help a clever man through the jungle of facts, a priori principles, theories, mathematical formulas, methodological rules, pressure from the general public and his "professional peers" and which enable him to form a coherent picture out of apparent chaos are much more closely related to the spirit of poetry than one would be inclined to think. Indeed, one has the suspicion that the only difference between poets and scientists is that the latter, having lost their sense of style now try to comfort themselves with the pleasant fiction that they are following rules of a quite different kind which produce a much grander and much more important result, namely, the Truth.

2. Galileo's View of the Copernican Revolution

Replying to an interlocutor (in the *Dialogue Concerning the Two Chief World Systems*[11]) who expresses his astonishment at the small number of Copernicans, Salviati, who "act[s] the part of Copernicus,"[12] explains the situation as follows:

You wonder that there are so few followers of the Pythagorean opinion [that the earth moves] while I am astonished that there have been any up to this day who have embraced and followed it. Nor can I ever sufficiently admire the outstanding acumen of those who have taken hold of this opinion and accepted it as true; they have through sheer force of intellect done such violence to their own senses as to prefer what reason told them over that which sensible experience plainly showed them to be to the contrary. For the arguments against the whirling of the earth which we have already examined are very plausible, as we have seen; and the fact that the Ptolemaics and the Aristotelians and all their disciples took them to be conclusive is indeed a strong argument of their effectiveness. But the experiences which overtly contradict the annual movement are indeed so much greater in their apparent force that, I repeat, there is no limit to my astonishment when I reflect that Aristarchus and Copernicus were able to make reason so conquer sense that, in defiance of the latter, the former became mistress of their belief.[13]

A little later Galileo notes that "they [the Copernicans] were confident of what their reason told them." [14] And he concludes his brief account of the origins of Copernicanism by saying that "with reason as his guide he [Copernicus] resolutely continued to affirm what sensible experience seemed to contradict." Then Galileo repeats: "I cannot get over my amazement that he was constantly willing to persist in saying that Venus might go around the sun and might be more than six times as far from us at one time as at another, and still look always equal, when it should have appeared forty times larger." [15]

The "arguments against the whirling of the earth" to which Galileo refers in this brief account of the origins of Copernicanism are presented by him as follows:

First, whether the earth is moved either in itself, being placed in the center, or in a circle, being removed from the center, it must be moved with such motion by force, for this is not its natural motion. Because, if it were, it would belong also to all its particles. But every one of them is moved along a straight line towards the center. Being thus forced and preternatural, it cannot be everlasting. But the world order is eternal; therefore, etc.—Second, it appears that all other bodies which move circularly lag behind, and are moved with more than one motion, except the *primum mobile*. Hence, it would be necessary that the earth be also moved with two motions; and if that were so, there would have to be variations in the fixed stars [parallax].[16] But such are not to be seen; rather, the same stars always rise and set in the same place without any variations.—Third, the natural motion of the parts and of the whole is toward the center of the universe, and for that reason it also rests therein.[17]— Finally . . . , as the strongest reason of all . . . heavy bodies, . . . falling down from on high, go by a straight and vertical line to the surface of the earth. This is considered an irrefutable argument for the earth being motionless.[18] For if it made the diurnal rotation, a tower from whose top a rock was let fall, being carried by the whirling of the earth, would travel many hundreds of yards to the east in the time the rock would consume in its fall, and the rock ought to strike the earth that distance away from the base of the tower. . . . This argument is fortified with the experiment of a projectile sent a very great distance upward; [19] this might be a ball shot from a cannon aimed perpendicular at the horizon. In its flight and return this consumes so much time that in our latitude the cannon and we would be carried together many miles eastwards by the earth, so that the ball, falling, could never come back near the gun. . . . They add moreover a third and very effective experiment of shooting a cannon ball point blank to the east, and then another one with equal charge at the same elevation to the west; the shot towards the west ought to range a great deal farther out than the one to the east. For when the ball goes towards the west, and the cannon, carried by the earth, goes east, the ball ought to strike the earth at a distance from the cannon equal to the sum of the two motions, one made by itself to the west, and the other by the gun, carried by the earth, towards the east. On the other hand, from the trip made by the ball shot towards the east it would be necessary to subtract that which was made by the cannon following it . . . Now experiment shows the shots to fall equally; therefore the cannon is motionless, and consequently the earth is, too. Not only this, but shots to the south or north likewise confirm the stability of the earth; for they would never hit the mark that one had aimed at, but would always slant towards the west because of the travel that would be made towards the east by the target, carried by the earth while the ball was in the air. And not merely shots along the meridians, but even those made to the east or west would not range truly; for the easterly shots would carry high and the westerly low whenever they were aimed point blank. . . . Hence, in no direction would shooting ever be accurate; and since experience is contrary to this, it must be said that the earth is immovable.[20]

[The] experiences which overtly contradict the annual movement [and which] are . . . much greater in their apparent force [than even the dynamical arguments which were just presented consist in the fact that] Mars, when it is close to us . . . would have to look sixty times as large as when it is most distant. Yet no such difference is to be seen. Rather, when it is in opposition to the sun and close to us it shows itself only four or five times as large as when, at conjunction, it becomes hidden behind the rays of the sun.[21]

Another and greater difficulty is made for us by Venus which, if it circulates around the sun, as Copernicus says, would be now beyond it and now on this side of it, receding from and approaching towards us by as much as the diameter of the circle it describes. Then, when it is beneath the sun and very close to us, its disc ought to appear to us a little less than forty times as large as when it is beyond the sun and near conjunction. Yet the difference is almost imperceptible.[22]

In an earlier essay, *The Assayer,* Galileo expresses himself still more bluntly. Replying to an adversary who has raised the issue of Copernicanism, Galileo remarks that *"neither Tycho, nor other astronomers, nor even Copernicus could clearly refute [Ptolemy]"* inasmuch as a most important argument taken from the movement of Mars and Venus stood always in their way" (my italics; this "argument" is mentioned again in the *Dialogue,* which has just been quoted). And he concludes that "the two systems" (the Copernican and the Ptolemaic) are *"surely false."* [23]

We see that Galileo's view of the origin of Copernicanism differs markedly from the more familiar historical accounts. He neither points to new facts which offer inductive support to the idea of the moving earth, nor mentions any observations that would refute the geocentric point of view but be accounted for by Copernicianism. Quite the contrary, he emphasizes that not only Ptolemy but Copernicus, too, is refuted by the facts; [24] and he praises Aristarchus and Copernicus for not having given up in the face of such tremendous difficulties. He praises them for having proceeded *counterinductively*.

This, however, is not yet the whole story; for although it may be admitted that Copernicus has acted simply on faith, it may also be said that Galileo has found himself in an entirely different position. Galileo, after all, invented a new dynamics. And he invented the telescope. The new dynamics, one might want to point out, removes the inconsistency between the motion of the earth and the "conditions affecting ourselves and those in the air above us." [25] And the telescope removes the "even more glaring" clash between the changes in the apparent brightness of Mars and Venus as predicted on the basis of the Copernican scheme and as seen with the naked eye. This, incidentally, is also Galileo's own view. He admits that "were it not for the existence of a superior and better sense than natural and commonsense to join forces with reason" he would have been "much more recalcitrant towards the Copernican system." [26] The "superior and better sense" is of course the telescope, and one is inclined to remark that the apparently counterinductive procedure was as a matter of fact induction (or conjecture plus refutation plus new conjecture), *but based on a better type of experience* than was available to Galileo's Aristotelian predecessors.[27] This matter must now be examined in some detail.

3. The Telescope, Invention and Terrestrial Tests

The telescope is "a superior and better sense" that gives new and more reliable evidence for judging astronomical matters. How is this hypothesis examined, and what arguments are presented in its favor?

In the *Sidereus Nuncius*, the publication which contains his first telescopic observations and which was also the first important contribution to his fame, Galileo writes that he "succeeded [in building the telescope] through a deep study of the theory of refraction." [28] This statement suggests that he had *theoretical reasons* for preferring the results of telescopic observations to observations with the naked eye. But the particular reason he gives—his insight into the theory of refraction —is not correct; nor is it sufficient (see note 46).

The reason is not correct, for there exist serious doubts as to Galileo's knowledge of contemporary physical optics. In a letter to Giuliano de Medici of October 1, 1610, more than half a year after the publication of the *Sidereus Nuncius*,[29] Galileo asks for a copy of Kepler's "Optics of 1604," [30] pointing out that he has not yet been able to obtain it in Italy.[31]

Jean Tarde, who asks Galileo in 1614 about the construction of telescopes of preassigned magnification, reports in his diary that Galileo regards the matter a difficult one and that he has found Kepler's optics of 1611 [32] so obscure "that perhaps its own author had not understood it." [33] In a letter to Liceti of June 23, 1640, two years before his death,[34] Galileo remarks that as far as he is concerned, the nature of light is still in darkness. Even if we look at such utterances with the care that is needed in the case of a whimsical author like Galileo, we must yet admit that his knowledge of optics was inferior by far to that of Kepler. This conclusion has also been drawn by Prof. E. Hoppe, who sums up the situation as follows:

Galileo's assertion that, having heard of the Dutch telescope, he reconstructed the apparatus by mathematical calculation must of course be understood with a grain of salt; for in his writings we do not find any calculations and the report, by letter, which he gives on his first effort says that no better lenses had been available; six days later we find him on the way to Venice with a better piece to hand it as a gift to the Doge Leonardo Donati. This does not look like calculation; it rather looks like trial and error. The calculation may well have been of a different kind, and here it succeeded, for on Aug. 25, 1609, his salary was increased by a factor of three.[35]

"Trial and error"—this means that "in the case of the telescope it was *experience* and not mathematics that led Galileo to a serene faith in the reliability of his device." [36] This second hypothesis on the origin of the telescope is also supported by the testimony of Galileo, who writes on other occasions that he has tested the telescope "a hundred thousand times on a hundred thousands stars and other objects." [37] Such tests produced great and surprising successes. The contemporary literature— letters, books, gossip columns—testifies to the extraordinary impression which the telescope made as a means for improving *terrestrial vision.* Julius Caesar Lagalla, professor of philosophy in Rome, describes a meeting of April 16, 1611, at which Galileo demonstrated his device:

"We were on top of the Janiculum, near the city gate named after the Holy Ghost, where once is said to have stood the villa of the poet Martial, now the property of the Most Reverend Malvasia. By means of this instrument, we saw the palace of the most illustrious Duke Altemps on the Tuscan hills so distinctly that we readily counted its each and every window, even the smallest; and the distance is sixteen Italian miles. From the same place we read the letters on the gallery, which Sixtus erected in the Lateran for the benedictions, so clearly, that we distinguished even the periods carved between the letters, at a distance of at least two miles.[38]

Other reports confirm this and similar events.[39] Galileo himself points to the "number and importance of the benefits which the instrument may be expected to confer, when used by land or sea." [40] The terrestrial success of the telescope was, therefore, assured. Its application to the stars, however, was an entirely different matter.

4. The Telescope, Extrapolation to the Sky

To start with, there is the problem of telescopic vision. This problem *is* different for celestial and terrestrial objects, and it was also known to be different (or thought to be different) in the two cases.[41]

It was thought to be different because of the contemporary idea (which was supported by an impressive amount of evidence, extending over more than two thousand years [42]) that celestial objects and terrestrial objects are formed from different material and obey different laws. This idea entails that the results of an interaction of light (which connects both domains and has special properties) with terrestrial objects cannot without further discussion be extended to the sky. To this physical idea was added, entirely in accordance with the Aristotelian theory of knowledge [43] (and also with the present views on the matter), the hypothesis that the senses are acquainted with the close appearance of terrestrial objects and are, therefore, able to perceive them distinctly, even if the telescopic image should be vastly distorted or disfigured by colored fringes. The stars are not known from close by. Hence, we cannot in their case use our memory for separating the contributions of the telescope and those which come from the object itself.[44] Moreover, all the familiar cues (such as background, overlap, and knowledge of nearby size) which constitute and aid our vision on the surface of the earth are absent in the sky so that new and surprising phenomena are bound to occur.[45] Only a new theory of vision, containing both hypotheses concerning the behavior of light in the telescope and hypotheses concerning the reaction of the eye under exceptional circumstances, could have bridged the gulf between the heavens and the earth that was and still is such an obvious fact of physics and of astronomical observation.[46] I shall soon comment on the theories that were available at the time, and we shall see that they were unfit for the task and were refuted by plain and obvious facts. For the moment I want to stay with the observations themselves in order to comment on the contradictions and difficulties which arise when one tries to take the celestial results of the telescope at their face value, as indicating stable, objective properties of the things seen.

5. The Problem of Celestial Data

Some of these difficulties have already appeared in the above report of the *Avvisi* (see note 38), which ends with the remark that "some of them [the participants of the gathering] went out expressly to perform this observation [of "four more stars or planets, which are satellites of Jupiter . . . , as well as two companions of Saturn" [47]], and even though they stayed until one o'clock in the morning, they still did not reach an agreement in their views." Another meeting that became notorious all over Europe makes the situation even clearer. About a year earlier, on April 24 and 25, 1610, Galileo brought his telescope to the house of his opponent Magini in Bologna to demonstrate it to twenty-four professors of all faculties. Horky, Kepler's overly excited pupil wrote of this occasion:

I never slept on the 24th or 25th of April, day or night, but I tested the instrument of Galileo's in a thousand ways,[48] both on things here below and on those above. *Below it works wonderfully;* in the heavens it deceives one as some fixed stars [Spica Virginis, for example, is mentioned, as well as a terrestrial flame [49]] are seen double. I have as witnesses most excellent men and noble doctors . . . and all have admitted the instrument to deceive. . . . This silenced Galileo and on the 26th he sadly left quite early in the morning . . . not even thanking Magini for his splendid meal.[50]

Magini wrote to Kepler on May 26, "He has achieved nothing, for more than 20 learned men were present; yet nobody has seen the new planets distinctly [*nemo perfecte vidit*]; he will hardly be able to keep them." [51] A few months later (in a letter signed by Ruffini) he repeats, "Only some with sharp vision were convinced to some extent." [52] After these and other negative reports had reached Kepler from all sides, like a paper avalanche, he asks Galileo for witnesses:

I do not want to hide it from you that quite a few Italians have sent letters to Prague asserting that they could not see those stars [the moons of Jupiter] with your own telescope. I ask myself how it can be that so many deny the phenomenon, including those who use a telescope. Now if I consider what occasionally happens to me, then I do not at all regard it as impossible that a single person may see what thousands are unable to see.[53] . . . Yet I regret that the confirmation by others should take so long in turning up. . . . Therefore I beseech you, Galileo, give me witnesses as soon as possible.[54]

In his reply of August 19 Galileo refers to himself, the Duke of Toscana, and Guliano, "as well as many others in Pisa, Florence, Bologna, Venice, Padova who however remain silent and hesitate. Most of them are entirely unable to distinguish Jupiter, or Mars, or even the Moon as a planet" [55]—not a very reassuring state of affairs, to say the least.

Today we understand a little better why the direct appeal to telescopic vision was bound to lead to disappointment, especially in the initial stages. The main reason, one already foreseen by Aristotle, was, of course, the fact that the senses applied under abnormal conditions are liable to give an abnormal response. Some of the older historians had an inkling of the situation, but they speak *negatively;* they try to explain the absence of satisfactory observational reports, the poverty of what is seen in the telescope.[56] They are unaware of the possibility that the observers might have been disturbed by *strong positive illusions* also. The extent of such illusions was not realized until recently, mainly as the result of the work of Ronchi and his school.[57] Here the greatest variations are reported in the placement of the telescopic image and, correspondingly, in the observed magnification. Some observers put the image right inside the telescope,[58] making it change its lateral position with the lateral position of the eye, exactly as would be the case with an afterimage or a reflex inside the telescope—an excellent proof that one must be dealing with an "illusion." [59] Others place the image in a manner that leads to no magnification at all, although a linear magnification of over thirty may have been promised.[60] Even a doubling of images can be explained as the result of a lack of proper focusing.[61] Adding to these psychological difficulties the many imperfections of the contemporary telescopes,[62] one not only can well understand the scarcity of satisfactory reports, but is astonished at the speed with which the reality of the new phenomena was accepted and, as was the custom, publicly acknowledged.[63] This development becomes even more puzzling when we consider that many reports of even the best observers were either plainly false, and capable of being shown as such at the time, or else self-contradictory.

Thus Galileo reports unevennesses, "vast protuberances, deep chasms, and sinuosities" [64] at the inner boundary of the lighted part of the moon whereas the outer boundary "appear[s] not uneven, rugged, and irregular, but perfectly round and circular, as sharply defined as if marked out with a pair of compasses, and without the indentations of any protuberances and cavaties." [65] The moon, then, seemed to be full of mountains at the inside, but perfectly smooth on its periphery and this despite the fact that the periphery changed as the result of the slight librations of the lunar body. The moon and some of the planets, for example, Jupiter, seemed to be enlarged whereas the apparent diameter of the fixed stars decreased: the former were brought nearer whereas the latter were pushed away. "The stars," writes Galileo, "fixed as well as erratic, when seen with the telescope, by no means appear to be increased in magnitude in the same proportion as other objects, and the Moon itself, gain increase of size; but in the case of the stars such increase appears much less, so that you may consider that a telescope, which (for the sake of illustration) is powerful enough to magnify other objects a hundred times, will scarcely render the stars magnified four or five times." [66]

The strangest features of the early history of the telescope emerge, however, when we take a closer look at Galileo's pictures of the moon. It takes only a brief look at Galileo's drawings and at photographs of similar phases to convince the reader that "none of the features recorded . . . can be safely identified with any known markings of the lunar landscape." [67] (See figures 1 and 2 in this essay.) Looking at such evidence, it is easy to think that "Galileo was not a great astronomical observer; or else that the excitement of so many telescopic discoveries made by him at that time had temporarily blurred his skill or critical sense." [68]

Now this assertion may well be true (though I rather doubt it in view of the extraordinary observational skill which Galileo exhibits on other occasions [69]). However, it is poor in content and, I submit, not very interesting. No new suggestions emerge for additional research and the possibility of a test is rather remote also. [70] There are, however, other hypotheses which do lead to new suggestions and which show us how complex the situation was at the time of Galileo. Let us consider the following two:

Hypothesis i. Galileo recorded faithfully what he saw and in this way left us evidence of the shortcomings of the first telescopes as well as of the peculiarities of contemporary telescopic vision. Interpreted in this way, Galileo's drawings are reports of exactly the same kind as the reports emerging from the experiments of Stratton, Ehrisman, and Kohler [71]—except that the characteristics of the physical apparatus and the unfamiliarity of the observed objects must also be taken into account. [72] We must remember, too, the many conflicting views which were held even at Galileo's time about the surface of the moon [73] and which may have influenced what observers saw. [74] What is needed in order to shed more light on the matter is an empirical collection of all the early telescopic results, preferably in parallel columns, including whatever pictorial representations have survived. [75] Subtracting instrumental peculiarities, such a collection adds fascinating material to a yet-to-be-written history of perception (and of science). [76] This is the content of hypothesis i.

Hypothesis ii is more specific than hypothesis i and develops it in a certain direction. I have been considering it, with varying degrees of enthusiasm, for the last two or three years, and my interest in it has been revived by a recent letter of Professor Stephen Toulmin to whom I am grateful for his clear and simple presentation of the view. It seems to me, however, that the hypothesis is confronted by many difficulties and must perhaps be given up.

Hypothesis ii, just like hypothesis i, approaches telescopic reports from the point of view of the theory of perception, but it adds that the practice of telescopic observation and the acquaintance with the new telescopic reports changed not only what was seen through the telescope but also what was seen with the naked eye. It obviously is of importance for our evaluation of the contemporary attitude toward Galileo's reports.

That the appearance of the stars and the moon may at some time have been much more indefinite than it is today was originally suggested to me by the existence of various theories about the moon which are incompatible with what everyone can plainly see with his own eyes. Anaximander's theory of partial stoppage (to explain the phases of the moon), Xenophanes's belief in the existence of different suns and different moons for different zones of the earth, Heraclitus's assumption that eclipses and phases are caused by the turning of the basins which for him represent the sun and the moon [77]—all these views run counter to the existence of a stable and plainly visible surface, a "face" such as we "know" the moon to possess. The same is true of the theory of Berossos, which occurs as late as Lucretius [78] and even later in Alhazen (see note 85).

Now, such disregard for phenomena, which for us are quite obvious, may be due either to a certain indifference toward the existing evidence which, however, was as clear and as detailed as it is today or else to a difference in the evidence itself. It is not easy to choose between these alternatives. Having been influenced by Wittgenstein, Hanson, and others, I was for some time inclined toward the second alternative, but it now seems to me that this version is ruled out both by physiology (psychology) [79] and by historical information. Thus we need only remember (see notes 15, 22, and accompanying text) how Copernicus disregarded the difficulties arising from the variations in the brightness of Mars and Venus, which were well known at the time, to come to a more realistic evaluation, for example, of Polemarchus's remark that such variations are "imperceptible." [80] And regarding the face of the moon, we see that Aristotle refers to it quite clearly when observing that "the stars do not *roll*. For rolling involves rotation: but the 'face,' *as it is called*, of the moon is always seen" (my italics).[81]

We may infer, then, that the occasional disregard for the stability of the face was due not to a lack of clear impressions, but to some widely held views about the unrealiability of the senses. This inference is supported by Plutarch's discussion of the matter which plainly deals not with what is seen (except as evidence for or against certain views) but with certain explanations of phenomena otherwise assumed to be well known: "To begin with," he says,

it is absurd to call the figure seen in the moon [φαινόμενον εἶδος ἐν τῇ σελήνῃ] an affection of vision, . . . a condition which we call bedazzlement [glare]. Anyone who asserts this does not observe that this phenomenon should rather have occurred in relation to the sun, since the sun lights upon us keen and violent, and moreover does not explain why dull and weak eyes discern no distinction of shape in the moon but her orb for them has an even and full light whereas those of keen and robust vision make out more precisely and distinctly the pattern of facial features and more clearly perceive the variations. . . .[82] The unevenness also entirely refutes the hypothesis, for the shadow that one sees is not continuous and confused, but is not badly depictured by the words of Agesianax: She gleams with fire encircled, but within / Bluer than lapis show a maiden's eye / and dainty brow, a visage manifest.—In truth, the dark patches submerge beneath the bright ones which they encompass . . . and they are thoroughly entwined with each other so as to make the delineation of the figure resemble a painting.[83]

Later on, the stability of the face is used as an argument against theories which regard the moon as being made of fire or of air, for "air is tenuous and without configuration, and so it naturally slips and does not stay in place." [84] The appearance of the moon, then, seemed to be a well-known and distinct phenomenon. What was in question was the relevance of this phenomenon for astronomical theory.[85]

We can safely assume that the same was true at the time of Galileo.[86]

But then we must admit that Galileo's observations could be checked with the naked eye and could in this way be exposed as illusory.

Thus the circular monster below the center of the disk of the moon [87] is well above the threshold of observation with the naked eye (its diameter is larger than three and one half minutes of arc) while a single glance convinces us that the face of the moon is not anywhere disfigured by a blemish of this kind. It would be interesting to see what contemporary observers had to say on the matter [88] or, if they were artists, what they had to draw on the matter.

To summarize what has emerged so far, Galileo had only slight acquaintance with contemporary optical theory. His telescope gave surprising results on the earth and these results were duly praised. However, trouble was to be expected in the sky, as we know now. Trouble promptly arose: the telescope produced spurious and contradictory phenomena, and some of its results could be refuted by a simple look with the unaided eye. Only a new theory of telescopic vision could possibly bring order into the chaos (which may have been still larger, due to the different phenomena seen at the time even with the naked eye) and separate appearance from reality. Such a theory was developed by Kepler, first in 1604 and then again in 1611.[89]

6. Kepler's Theory

According to Kepler the place of the image of a punctiform object is found by first tracing the path of the rays emerging from it according to the laws of (reflection and) refraction until they reach the eye and by then using the principle (which lies at the basis of all of optics as it is taught today) that "the image will be seen in the point determined by the backward intersection of the rays of vision from both eyes" [90] or, in the case of monocular vision, from the two sides of the pupil.[91] This rule, which proceeds from the assumption that "the image is the work of the act of vision" ("cum imago sit visus opus"),[92] is partly empirical,[93] partly geometrical. It bases the position of the image on a "metrical triangle" ("triangulum distantiae mensorium"),[94] or a "telemetric triangle" as Ronchi calls it,[95] that is constructed out of the rays which finally arrive at the eye and is used by the eye and the mind to place the image in the proper distance. Whatever the optical system, whatever the total path of the light rays from object to observer, the mind of the observer utilizes its very last part only and bases its visual judgment, the perception, on it.

It is clear that this rule constituted a considerable advance over and above all previous thought. However, only a second is needed to show that it is entirely false: take a magnifying glass, determine its focus, and look at an object close to it. The telemetric triangle now reaches beyond the object to infinity. A slight change of distance brings the Keplerian image from infinity to close by and back to infinity. No such phenomenon is ever observed. We see the image, slightly enlarged, in a distance that is usually identical with the actual distance between the object and the lens. The visual distance of the image remains constant however much we vary the distance between lens and object and even when the image becomes distorted and, finally, diffused.[96]

This situation, then, existed in 1610 when Galileo published his telescopic findings. How did Galileo react to it? The answer has already been given: he raised the telescope to the state of a "superior and better sense." [97] What were his reasons for doing so? This question brings us back to the problems raised by the evidence (against Copernicus) that was reported in section 2.

7. The "Two Unknowns"

According to the Copernican theory, Mars and Venus approach to and recede from the earth by a factor of 1:6 and 1:8 respectively (these are approximate numbers). Their change of brightness should be 1:40 and 1:60 respectively (these are Galileo's values [see notes 21, 22]). Yet Mars changes very little, and the variation in the brightness of Venus "is almost imperceptible." [98] These experiences "overtly [contradict] the annual movement [of the earth]." [99] The telescope, on the other hand, produces new and strange phenomena, some of them exposable as illusory by observation with the naked eye (see notes 65, 67), some contradictory, some having even the appearance of being illusory (see notes 57, 58), whereas the only theory that could have brought order into this chaos, Kepler's theory of vision, is refuted by evidence of the plainest kind possible. But—and with this we come to what I think is the central feature of Galileo's procedure—*there are telescopic phenomena*, namely, the telescopic variations of the brightness of the planets *which agree more closely with Copernicus than do the results of observation with the naked eye.* Seen through the telescope, Mars does indeed change as it should according to the Copernican view. Compared with the total performance of the telescope, this change is still quite puzzling (see notes 66, 97). It is just as puzzling as is the Copernican theory when compared with the pretelescopic evidence. But the change is in harmony with the predictions of Copernicus. It is this harmony, rather than any deep understanding of cosmology and optics, which for Galileo proves Copernicus and the veracity of the telescope in terrestrial as well as in celestial matters. And it is this harmony on which he builds an entirely new view of the universe.

"Galileo," writes Ludovico Geymonat, referring to this aspect of the situation,

was not the first to turn the telescope upon the heavens, but . . . he was the first to grasp the enormous interest of the things thus seen. And he understood at once that these things fitted in perfectly with the Copernican theory whereas they contradicted the old astronomy. Galileo had believed for years in the truth of Copernicanism, but he had never been able to demonstrate it, despite his exceedingly optimistic statements to friends and colleagues [he had not even been able to remove the refuting instances, as we have seen and as he says himself (see note 23)]. Should direct proof [should even mere agreement with the evidence] be at last sought here? The more this conviction took root in his mind, the clearer to him became the importance of the new instrument. In Galileo's own mind, faith in the reliability of the telescope and recognition of its importance were not *two separate acts;* . . . rather, they were *two aspects of the same process* (my italics).[100]

Can the absence of independent evidence be expressed more clearly? "The *Nuncius*," writes Franz Hammer in the most concise account I have seen of the matter, "contains two unknowns, the one being solved with the help of the other." [101] This is entirely correct except that the "unknowns" were not quite that unknown but were known to be false, as Galileo says himself (see note 23). It is this rather peculiar situation, this harmony between two interesting but refuted ideas, which Galileo exploits in order to prevent the elimination of either.

Exactly the same procedure is used to preserve his new dynamics. We shall see that this science, too, was endangered by observable events. To eliminate the danger, Galileo introduces friction and other disturbances with the help of ad hoc hypotheses, treating them as tendencies *defined* by the obvious discrepancy between fact and theory rather than as physical events *explained* by a theory of friction for which new and independent evidence might someday become available (such a theory arose only much later, in the eighteenth century, see note 167). Yet the agreement between the new dynamics and the idea of the motion of the earth which Galileo successfully increases with the help of his method of *anamnesis* (see section 14) again had the effect that both became more acceptable.

The reader will realize that a more detailed study of historical phenomena like these creates considerable difficulties for the view that the transition from the pre-Copernican cosmology to the philosophy of the seventeenth century consisted in the replacement of a refuted theory by a more general conjecture which explained the refuting instances, made new predictions, and was corroborated by the observations carried out to test these new predictions. And one will perhaps see the merits of a different view which asserts that whereas the pre-Copernican astronomy was in trouble (was confronted by a series of refuting instances), the Copernican theory was in even greater trouble (was confronted by even more drastic refuting instances),[102] but that being in harmony with still further inadequate theories, it gained strength and was retained, the refutations being made ineffective by ad hoc hypotheses and clever techniques of persuasion.[103] This would seem to be a much more adequate description of the developments at the time of Galileo.

I shall now interrupt the historical narrative in order to show that the account just given is not only factually adequate but also perfectly reasonable, and that the attempt to enforce some of the more familiar methodologies of the twentieth century (such as the method of conjectures and refutations) would have had disastrous consequences.[104]

8. Methodological Considerations

A prevalent tendency in contemporary methodological discussions is to approach the problem of knowledge *sub specie aeternitatis,* as it were.[105] One asks, given initial conditions and accepted observations, What conclusions can we draw with respect to our theories? The answers vary considerably. Some would say that it is possible to determine degrees of confirmation and that theories can be evaluated by them. Others would reject any logic of confirmation and would judge theories by their content and the actual falsifications which have taken place. But all assume that observations which have been assembled for the purpose of testing a theory and which satisfy the current standards of precision and relevance are already decisive, and that they can and perhaps even must be used here and now either to eliminate the theory or to enthrone it to the exclusion of all alternatives.

This assumption overlooks that our knowledge has layers of different age and different sophistication and that an acceptable cosmology may be in trouble because it is compared with observations flowing from an outmoded theory of cognition. The criteria for good observations and good measurements which are taken for granted at a certain time may be far behind in maturity and sophistication the theory that they are supposed to judge (they usually are behind a new *cosmology,* for physiology is a much more complex and much more difficult subject than astronomy). But if there is such a difference of maturity between theories and observations, then a direct judgment of the former by the latter is as sensible as is the attempt to judge modern optics by an Aristotelian experience (this, incidentally, was the basis of Goethe's theory of colors). One must then wait until one's views on the nature of knowledge have had time to catch up, or, to express it in more general terms, one must realize that knowledge is not a collection of timeless ideas, but the temporal result of a complex temporal process in which the right judge very often turns up hundreds of years after the things to be judged.

Let me explain this situation in somewhat greater detail. I shall start by pointing out that what counts and what does not count as relevant evidence (in matters of cosmology) depends not only on the theory tested but also on the current views concerning the process of cognition, on the medium in which the tests are taking place (the terrestrial atmosphere, for example), on the motion and other properties of the observation post (the earth, for example), and on many other supplementary assumptions. Some of these assumptions are not even explicit, but are part of the meaning of our observational terms. (The idea that there cannot be any nonoperative motion [106] which has no effect whatever on the processes in the moved object is exactly of this kind, as we shall see in section 13. The same applies to the common distinction between veridical perceptions and illusory perceptions which was used with such great effect and, as an empiricist is bound to say, with perfectly good reason against the telescope).

Now, it is clear that a cosmology may be highly corroborated not because it is correct, but because it is connected with inadequate *auxiliary sciences*, as we shall call disciplines of the kind just enumerated. The harmony between the cosmology and such inadequate sciences may be so great that opposing facts have no chance of making themselves felt. Conversely, a change of cosmology—from Ptolemy to Copernicus, for example—will disrupt the harmony, and refuting instances will arise in great numbers even if the change should happen to be an improvement. (Who indeed would expect the idea of the motion of the earth to be compatible with observations which are interpreted by Aristotelian dynamics or by the more familiar idea of the "operative character" of all motion?) In such circumstances the existence of refuting instances only shows that the new cosmology is at cross-purposes with the traditional auxiliary subjects; it does not show that the cosmology is inherently unsound. Thus the Aristotelian view concerning the process of cognition (which perfectly fitted Aristotelian dynamics and Aristotelian cosmology) contained the assumption that a normal observer who is in possession of his faculties is adapted to the universe and capable of giving adequate reports on all aspects of it. "The traditional conception of nature," writes Hans Blumenberg, commenting on this feature,

was connected with a kind of *postulate of visibility* which corresponds both to the finite extension of the universe and to the idea that it was related to man as its center. That there should be things in the world which are inaccessible to man not only now, or for the time being, but in principle, and because of his natural endowment and which could therefore be never seen by him—this was quite inconceivable for later antiquity as well as for the Middle Ages.[107]

Precisely such a view is implied by the new Copernican cosmology, especially in the radical form given to it by Bruno: the observer is no longer situated at the center; he is separated from the true laws both by the special physical conditions of his observation post, the moving earth, and by the idiosyncrasies of his main instrument of observation, the human eye. Therefore, what is needed for a test of the Copernican cosmology is not just a simpleminded and direct comparison of its predictions with what is seen, but the interpolation, between the laws of the new cosmology and the data of observation, of a well-developed *meteorology* (in the good old sense of the word, as dealing with things below the moon), of an equally well-developed science of *physiological optics* dealing with both the subjective (brain) and the objective (light, structure of eye, lenses) aspects of vision, as well as of a new *dynamics* stating the manner in which the motion of the earth might influence the physical processes on its surface. One can say at once that the evidence obtained in accordance with the older Aristotelian views is bound to clash with the new astronomy, and one is also perfectly justified, for the reasons just given, to regard the clash as irrelevant.

The most important element in this peculiar and complex situation is, however, the time factor. The Copernican view conflicts with the evidence that has been assembled in accordance with the older ideas of cognition and is directly compared with it. This is not a relevant test. Relevant tests must interpolate meteorological, dynamical, and physiological disturbances between the basic laws and the perceptions of the observer. Now, there is no guarantee that the supplementary sciences providing these disturbances and their laws will be available at once, after the discovery of the new astronomy. Quite on the contrary—such a sequence of events is extremely unlikely because of the vast complexity of the phenomena one has to consider. One simply cannot demand of a single man, or even of a single generation, to provide at one stroke not only a new theory of the planetary system, but also a new psychology, a new dynamics, and perhaps even a new theology. (Impatient critics always seem to overlook that man can do only so many things at one time, and they seem to forget that their own pet theories did not arise in a few years either.)

But what is a natural philosopher to do under these circumstances? What is he to do when he realizes that it may take hundreds of years before the first reasonable auxiliary hypothesis appears? It is clear that he must preserve the new astronomy and the new laws it contains. He must preserve them, and he must even try to develop them in order to keep alive and to articulate further the motive for the invention and the improvement of the supplementary sciences. Strictly speaking, to fulfill these demands, he must develop methods which permit him to retain his theories in the face of plain and unambiguous refuting facts, even if testable explanations for the clash are not immediately forthcoming. This is the first thing he must do. In developing such methods, he must be aware of the fact that our total knowledge contains parts of different ages which are adapted to different types of evidence (which in turn are interpreted by different layers of auxiliary science) and which cannot be used indiscriminately in the judgment of new and revolutionary theories. A careful selection of data must be made, even before the arrival of hypotheses which would give reasons for the selection. All this means, of course, that the new theory is quite intentionally cut off from some of the data which had supported its predecessor, it is made more metaphysical, and the support it does obtain is arranged with the help of ad hoc hypotheses.

A new period in the history of science, therefore, commences with a backward movement that returns us to an earlier stage when theories were vaguer and had smaller empirical content. This backward movement is not just an accident; it has a definite function and is essential if we want to overtake the status quo, for it gives us the time and the breathing space that is needed for the development of the new auxiliary sciences. A natural philosopher who propounds a new and revolutionary theory must also realize that such sciences are not bound to arrive, that their invention will often be the result of accidents, and that it is his task not only to proceed and construct, but also occasionally to wait until the right accident occurs. There is no provision for this in our methodologies where judgments of increase of content, falsification, corroboration, and confirmation are made here and now, taking it for granted that the right kind of effort (experiment, logical analysis) must yield the right kind of result (discovery of refuting evidence and of ad hoc hypotheses), the latter of which can, therefore, be used at once to condemn and to remove a newly proposed view. Such an impatient procedure, I repeat, overlooks that the elements which play a role in the comparison between theory and experiment—namely, the theory to be judged; the data; the auxiliary sciences which are used in the interpretation and the arrangement of the data; the epistemologies which we use for assembling, interpreting, and often even for constituting our data—often arrive in the wrong temporal order, causing a perfectly good and "modern" cosmology to be combined with an atrocious and antediluvian theory of cognition and to be endangered thereby.[108] Of course, nobody can know in advance just where the scandals lie, but one can make a guess and, by so doing, proceed in the way just described: remove data from the theory, connect other data with it in an ad hoc manner, and hope that auxiliary sciences giving reasons for such arbitrariness will eventually arise (one may even try to create the impression that they have already arrived). This, I said, is the first step to be carried out by the inventors of a new cosmology.

We have seen that Galileo resolutely adopts this method.[109] What is not so well known is that the transition from classical physics to relativity and to the quantum theory has precisely the same structure except that the missing data are taken over from the preceding science, classical physics, and that the existence of ad hoc connections is no longer regarded as a vice. Seen from the point of view of the present paper (and also from the point of view of Bohr's philosophy [110] [see also note 133]), the quantum theory and the theory of relativity are intermediate stages, comparable to Copernican science at the time of Galileo, between a wrong cosmology (classical physics) and the correct, empirically adequate, and conceptually uniform cosmology of the future. The intermediate and provisional character of these two disciplines becomes very clear from the way in which they use the older classical notions.

9. Survival of Classical Ideas, A: The Quantum Theory

To start with, there are excellent reasons to suppose that the *quantum theory* does not "[reduce] to classical mechanics as a special case" [111] unless it is supplemented both by classical concepts and by classical assumptions. To see this, consider the following two questions:

> i) Is it true that all problems which are successfully dealt with by (nonrelativistic) classical physics (classical mechanics) can also be treated by the elementary quantum theory without help from additional assumptions which (a) do, and (b) do not involve classical ideas?
>
> ii) Is it true that the quantum theory does not make any assertion that disagrees with assertions made by classical physics (classical mechanics) in its own proper domain and not contradicted by experience? (The latter qualification is added in order to exclude such cases as superfluidity, superconductivity, the existence of stable material objects, visual fluctuations under weak illumination, and so on.)

A negative answer to (i) would force us to supplement the quantum theory with principles obtained from inspecting the macrodomain and expressed, in the case of (ia), in classical terms.

A negative answer to (ii) would force us to eliminate certain consequences of the quantum theory—and this again on the basis of known macroscopic facts.

It seems that neither (i) nor (ii) admits an affirmative answer.

Concerning (i), we can start by pointing out that a complete subsumption of the macrolevel under the laws and concepts of the quantum theory assumes the existence of quantum theoretical descriptions for every macroscopic situation or else for every situation that can be described in classical terms. For example, there must be Hermitian operators corresponding to the Eulerian angles and the temporal changes of the Eulerian angles of a rigid body so that the essential theorems of rigid body mechanics can be stated in quantum mechanical terms. Also —and this point is often overlooked in the discussion of problems such as the one just introduced—the quantum theory of measurement must be extended to macroscopic situations, and it must be explained how the complementary features of microobjects manage to disappear on the macrolevel.

This program runs into trouble at the very beginning. The problems which arise are connected partly with the quantum theory of measurement proper, where "for some observables, in fact for the majority of them (such as xyp_z), nobody seriously believes that a measuring apparatus exists"[112] (the reference here is to *microscopic* magnitudes). And concerning *macroscopic* magnitudes, we encounter difficulties of definition already in the simplest cases, such in the case, mentioned above, of the Eulerian angles of a rigid body.[113] The attempt at a general characterization of macroscopic variables has more recently been taken up by various authors, such as Ludwig[114] and the members of the Italian school,[115] but this very attempt has revealed the need for new principles, which one has then tried to discover with the help of an "inverted principle of correspondence which, starting from the quantum theory, allows us to guess at the limits of very large systems, the main features of macrophysics acting as guides in our guesswork."[116] Now, whereas such additional principles (which have been compared to the exclusion principle[117]) might be accepted as legitimate additions which do not transcend the framework of the quantum theory, this can no longer be said about the additions and modifications required in connection with the second question.

Concerning this second question, it suffices to point out that the quantum theory is a *linear* theory, whereas classical mechanics (with such idealized exceptions as the theory of small vibrations) is not. And as all those approximations which are based on quantum theoretical hypotheses about the macrolevel preserve linearity, we have good reason to expect deviations.[118] Drastic examples of such deviations have been given by Schrödinger[119] and Einstein.[120] And exactly these drastic deviations are produced in every quantum mechanical measurement: the measurement of a magnitude in a system that is not in one of its eigenstates separates these eigenstates (for example, by spreading them out in space, as is done in the Stern-Gerlach experiment), but it does not destroy the interference terms (this is true even if the original state of both the system and the measuring apparatus should happen to be a mixture rather than a pure state [121]). Thus the state function of a single electron that has passed a device showing some kind of periodicity and now starts interacting with the molecules on a photographic plate is broken up into many tiny packets with complex phase relations between them. And as long as there are these phase relations, the electron cannot be said to be in any particular packet but must be described as being "in all of them at once." [122] But, then, the quantum theory as we have described it so far (which is not yet the quantum theory of Bohr and of the Copenhagen school) does not allow us to say that the electron has interacted with a particular molecule and has left its mark. This indefiniteness will spread as more objects are allowed to participate in the interaction, and it may even reach the mind of the conscious observer making it impossible for him to say that he has received definite information, no matter how certain he himself may feel about it.[123]

Now, at this point one might be tempted to say that the difficulty is purely imaginary. One might wish to emphasize that although there are these phase relations between different wave packets, even on the macroscopic level, which may be difficult to deal with mathematically, the logic of the situation is quite simple. Owing to their complexity, and owing also to the crudity of our observations on the macrolevel (looking, etc.), the phase relations will forever escape detection.[124] But as Professor Putnam pointed out in conversations at the Minnesota Center for the Philosophy of Science (Summer 1957) [125] and as Bohm and Bub have explained more recently,[126] the removal of the interference terms does not yet return us to the classical level. Such removal changes an assembly of interfering wave packets which are jointly occupied by the electron into an assembly of isolated wave packets which, however, are still jointly occupied by the electron. And as the classical level is reached only when we are allowed to assign the electron to a single wave packet, we need a further transition which cannot in any way be regarded as an approximation.[127] Now, it was a discovery of the first order to find that interpreting the terminal assembly—the assembly that emerges in our natural surroundings and can be described in the usual macroscopic terms—as a collective in the sense of classical statistics is a consistent procedure which also leads to correct predictions.[128] It was this discovery which allowed physicists to combine wave mechanics with Born's interpretation and which was adopted by the founders of the quantum theory. Let us repeat the assumptions which are involved in it:

1) Interference terms are disregarded—but the approximation is excellent and the differences undiscoverable on the macrolevel.

2) Macroobjects are assumed to have well-defined properties which inhere in them independently of measurement and do not spread.

3) These properties are identified with the properties ascribed to the objects by classical physics (such as the position of a pointer or of a mark on a photographic plate).

Now, before going any further, it is extremely important to see that (2) and (3) are not necessary for turning the formalism of the quantum theory into a physical theory.[129] We can connect the formalism with "the facts" without using either assumption. As a matter of fact, we do connect it in this manner at the beginning of each measurement, when we represent both the object and the apparatus by wave functions without paying attention to the demands of the classical level (use of plane waves in scattering theory as well as in the above thought experiment with electrons). But if the theory resulting from such a use of the formalism is to satisfy the demands of the correspondence principle, then its predictions must be drastically modified. It is in this connection that assumptions (1) to (3) become relevant. Of course, (1) is quite harmless. It is an approximation of the exact same kind as we use in classical physics. However, (2) and (3) are very different. They eliminate not some minute and perhaps negligible quantitative consequences of the principle of superposition but some quite noticeable qualitative effects. It is only after the application of (2) and (3) that we obtain agreement with the principle of correspondence and can guarantee the universal (macroscopic) validity of Born's interpretation. At the same time we must deny the universal validity of the superposition principle and must admit that it is but a (very useful) instrument of prediction. "In fact," says Bohr, commenting on this feature of the situation, "wave mechanics, just as the matrix theory, on this view represent a symbolic transcription of the problem of motion of classical mechanics adapted to the requirements of the quantum theory and only to be interpreted by an explicit use of the quantum postulate," [130] that is, "with the aid of the concept of free particles." [131]

We see that the relation of the quantum theory to classical physics is, indeed, very different from that of a comprehensive point of view to a special case of it, using as it does essential parts of the classical framework in the prediction of phenomena.[132] Bohr has long ago expressed this situation by calling both classical physics and the quantum theory "caricatures . . . which allow us, so to speak, to asymptotically represent actual events in two extreme regions of phenomena." [133] But we also see that the difference is but an expression of the fact that quantum mechanics is still dwelling in the intermediate stage where the relation to the data is established not with the help of the proper auxiliary sciences but directly and in an ad hoc fashion. Such a stage is certainly necessary, as I tried to show in the last section. However, it is equally clear that it cannot be regarded as the last word.[134]

10. Survival of Classical Ideas, B: Relativity

Now it is surprising to see that the theory of relativity shows a very similar incompleteness and, partly at least, for the very same reasons. To start with, the adequacy of the *general theory of relativity* is now seriously endangered by Professor Dicke's results, which use the resources of classical physics in an essential way.[135] And concerning its generality, we must remember that any actual calculation employs in its premises both the theory and the classical mechanics. First, classical ideas turn up in the calculation of planetary orbits which rests on classical perturbation theory, adding but a small relativistic correction.[136] We have no clue what a purely relativistic calculation would yield (the problem of stability, for example, is entirely open). This is a practical difficulty, but in view of certain features of relativity, one wonders whether it does not reflect matters of principle also. Secondly, we need the classical ideas when dealing with extended solid objects. The reason is that we do not possess a concise relativistic description of those phenomena which were described (approximately) in classical physics with the help of the notion of a rigid body.[137] Classical ideas, then, occur essentially in all those predictions which today are regarded as decisive tests of the theory of relativity. It is, therefore, a mistake to assign such predictions to the content of relativity exclusively, as most philosophers seem to be inclined to do, and to declare that an increase in content and in explanatory power has taken place.[138]

So much for the actual situation. However, I suspect that the general theory of relativity is also unable in principle to produce many empiricial predictions. My reasons are, roughly, as follows: the theory is very simple. It is also coherent to a high degree. This means that distant points of any curve expressing a law of the theory are strongly dependent upon one another. Hence, if large parts of the curve move in an arbitrarily selected domain with certain simple topological properties (such as the domain of theoretical entities), then it is very unlikely that intermediate parts will leave the domain (and move, say, into the domain of observable entities).

This last consideration is, of course, highly conjectural. However, taken together with the preceding description of the actual situation, it shows that there may be something seriously wrong with the customary philosophical evaluation of the general theory of relativity.[139] (Einstein's own ideas are untouched by these considerations, for he was interested in "unification" and not in "verification by little effects." "This however," he continues, "is properly appreciated only by few."[140])

11. Methodological Considerations Continued

The first step on the way to a new cosmology is, we have said, a step back: apparently relevant evidence is pushed aside, new data are brought in by ad hoc connections, and the empirical content of science is drastically decreased. Now, the cosmology that happens to be in the center of attention and whose adoption causes the changes just described differs from other views in only one respect: it has features which at the time in question seem attractive to some people. But there exists hardly any idea that is totally without merit and that might not also become the starting point of concentrated effort (this, if I understand it correctly, is an important corollary of the *philosophy of possibilism* as it has been developed by Arne Naess[141]). No invention is ever made in isolation, and no idea is, therefore, completely without abstract or empirical support. Now if partial support and partial plausibility suffice to start a new trend (and I have suggested that they do), if starting a new trend means taking a step back from the evidence, if any idea can become plausible and can receive partial support—then the step back is, as a matter of fact, a step forward, away from the tyranny of tightly knit, highly corroborated, and gracelessly presented theoretical systems (for the last item, gracelessness, cf. the quote by Galileo preceding this essay). "Another different error," writes Bacon on precisely this point, "is the . . . peremptory reduction of knowledge into arts and methods, from which time the sciences are seldom improved; for as young men rarely grow in stature after their shape and limbs are fully formed, so knowledge, whilst it lies in aphorisms and observations, remains in a growing state; but when once fashioned into methods, though it may be further polished, illustrated, and fitted for use, it no longer increases in bulk and substance." [142] The similarity with the arts arises at exactly this point. As soon as it is realized that close empirical fit is no virtue and must be relaxed at times of progress, in the very same moment style, elegance of expression, simplicity of presentation, and tension of plot and of narrative again become important features of our writing (one wonders why one should ever have thought them to be irrelevant [see note 103]).

For—and with this we come to the second step in the defense of a new cosmology—we now must create and maintain interest in a theory that has partly been removed from the observational plane and is inferior to its rivals (see note 24), and, lacking the appropriate auxiliary sciences, we must do this by means other than observation and content-increasing argument. It is in this context that much of Galileo's work should be seen. This work has often been likened to propaganda [143]—and propaganda it certainly is. But propaganda of this kind is not a marginal affair that may or may not be added to allegedly more substantial means of defense and that should perhaps be dropped by a "professionally honest scientist" (whatever animal *he* is).[144] In the circumstances which we are considering now, propaganda is of the essence. It is of the essence because interest must be created at a time when the usual methodological reasons have no point of attack and because this interest must be maintained, perhaps for centuries, until the reasons arrive.

Thirdly, one must admit that the reasons, that is, the appropriate auxiliary sciences, need not at once turn up in full formal splendor but may at first be quite inarticulate and may conflict with the existing evidence. Agreement, or partial agreement, with the cosmology is all that is needed in the beginning. The agreement shows that the reasons are at least relevant and that they may someday produce full-fledged positive evidence. Thus the idea that the telescope shows the world as it really is leads into many difficulties. But the support it lends to and receives from Copernicus is a hint that we might be moving in the right direction. We have here an extremely interesting relationship between a general view and the particular hypotheses which constitute its evidence. It is almost universally assumed that general views do not mean much unless the relevant evidence can be fully specified. Carnap, for example, asserts that "there is no independent interpretation for L_T [the language in terms of which a certain theory or world view is formulated]. The system T [the axioms of the theory and the rules of derivation] is itself an uninterpreted postulate system. [Its] terms obtain only an indirect and incomplete interpretation by the fact that some of them are connected by correspondence rules with observational terms." [145] "There is no independent interpretation," says Carnap—and yet an idea such as the idea of the motion of the earth which is inconsistent (or even incommensurable) with contemporary evidence, which is upheld by declaring this evidence to be irrelevant, and which is, therefore, cut off from the most important facts of contemporary astronomy manages to become a nucleus, a point of crystallization for the aggregation of other inadequate views which gradually increase in articulation and finally fuse into a new cosmology including new kinds of evidence.

There is no better account of this process than the description which John Stuart Mill has left of the vicissitudes of his education. Referring to the explanations which his father gave him on logical matters, he writes: "The explanations did not make the matter at all clear to me at the time; but they were not therefore useless; they remained as a nucleus for my observations and reflections to crystallize upon; the import of his general remarks being interpreted to me, by the particular instances which came under my notice afterwards." [146] In exactly the same manner the Copernican view, though incomprehensible from the point of view of a strict empiricism, and even refuted, was needed in the construction of the supplementary sciences even before it became testable with their help and even before it in turn provided them with supporting evidence of the most forceful kind. Is it not clear that our beautiful and shining methodologies which demand from us that we concentrate on theories of high empirical content and which implore us to take risks and to take refutations seriously would have provided extremely bad advice in the circumstances? (The advice to test his theories would have been quite useless for Galileo, who was at any rate faced by an embarrassing amount of prima facie refuting instances, who was unable to explain them for he lacked the necessary knowledge [though not the necessary intuitions] [see note 167], and who had, therefore, to explain them away so that a possibly valuable hypothesis might be saved from premature extinction). And is it not equally clear that we must become more realistic, that we must cease to gape at the imaginary shapes of an ideal philosophical heaven (of a "third world" as Popper has expressed it) and must start considering what will help us in this *material world*, given our erring brains, our imperfect measuring instruments, and our faulty theories? One can only marvel at how reluctant philosophers and scientists are to adapt their general views to an activity in which the latter already participate (and which, if asked, they would not want to give up).

It is this reluctance, this psychological resistance, which makes it necessary to combine abstract argument with the sledgehammer of history. Abstract argument is necessary, because it gives our thoughts direction. History is necessary, at least in the present state of philosophy, because it gives our thoughts force. We now return to history. More specifically, we shall deal with the dynamical "arguments against the whirling of the earth" to which Galileo refers in his brief account of the origins of Copernicanism (see note 20) and which constitute a second major difficulty for the Copernican cosmology.

12. Natural Interpretations

Whereas the telescope replaces certain *sense impressions* by others that are in greater harmony with the Copernican idea, the dynamical arguments and counterarguments leave sense impressions untouched and deal with *theoretical notions* instead. More specifically, they consider theoretical ideas which are so closely connected with our perceptions that it is difficult to isolate and criticize them. They deal with none other than those "operation[s] of the mind which [follow] close upon the senses" [147] which Bacon wanted to examine with special care. Such operations, such notions, will be called *natural interpretations*. It is by an analysis of natural interpretations that Galileo tries to remove the "arguments against the whirling of the earth" which have been quoted in section 2 (see note 18). Let us take another look at his analysis:

Heavy bodies, . . . falling down from on high, go by a straight and vertical line to the surface of the earth. This is considered an irrefutable argument for the earth being motionless. For if it made the diurnal rotation, a tower from whose top a rock was let fall, being carried by the whirling of the earth, would travel many hundreds of yards to the east in the time the rock would consume in its fall, and the rock ought to strike the earth that distance away from the base of the tower. (126) [148]

In considering the argument, Galileo at once admits the correctness of the sensory content of the observation made, namely, that "heavy bodies, falling from a height, go perpendicular to the surface of the earth" (125). Reflecting on an author (Chiaramonti) who sets out to convert Copernicans by repeatedly mentioning this fact, Galileo says:

I wish that this author . . . would not put himself to such trouble trying to have us understand from our senses that this motion of falling bodies is simple straight motion and no other kind, nor get angry and complain because such a clear, obvious, and manifest thing should be called into question. For in this way he hints at believing that to those who say such motion is not straight at all, but rather circular, it seems that they see the stone move visibly in an arc since he calls upon their senses rather than their reason to clarify the effect. This is not the case, Simplicio, for just as I . . . have never seen nor ever

Figure 2. Moon, age seven days (first quarter)

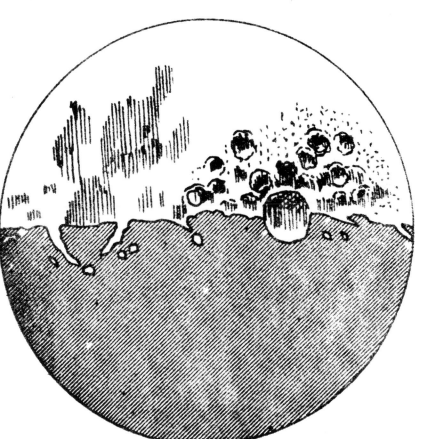

Figure 1. The shape of a lunar mountain and a walled plain, from Galileo, *Sidereus Nuncius* (Venice, 1610)

expect to see the rock fall any way but perpendicularly, just so do I believe that it appears to the eyes of everyone else. It is therefore better to put aside the appearance, on which we all agree, and to use the power of reason either to confirm its reality, or to reveal its fallacy. (256)

The correctness of the observation is not in doubt. What *is* in question is its "reality" or "fallacy." What is meant by these expressions?

This question is answered by an example

from which one may learn how easily one may be deceived by simple appearances, or, let us say, by the impressions of one's senses. This event is the appearance to those who travel along a street by night of being followed by the moon, with steps equal to theirs, when they see it gliding along the eaves of the roofs. Then it looks to them just as would a cat really running along the tiles and putting them behind it; an appearance which, if reason did not intervene, would only too obviously deceive the senses. (257)

In this example we start with a sensory impression and consider a statement that is forcefully suggested by it (the suggestion is so strong that it has led to entire systems of belief and rituals, as becomes clear from a closer study of the lunar aspects of witchcraft and of other religions). Now "reason intervenes"; the statement suggested by the impression is examined, and one considers other statements in its place. The nature of the impression is not changed a bit by this activity (this is only approximately true but we can omit, for our present purpose, the complications arising from the interaction of impression and proposition), but it enters new observation statements and plays new—better or worse—roles in our knowledge. What are the reasons and methods which regulate such exchange?

To start with, we must become clear about the nature of the total phenomenon: appearance plus statement. There are not two acts—the one, noticing a phenomenon; the other, expressing it with the help of the appropriate statement—*but only one,* namely, saying, in a certain observational situation, "The moon is following me" or, "The stone is falling down." A separation is noted only when the language used for expressing one's observations is either not well known, that is, when it is not spoken fluently, or when it is a theoretical language which is for the first time confronted with the results of observation. Under such circumstances we may indeed "look for the right word" so that statement and phenomenon are experienced as being apart and waiting to be related. But in everyday life such a division does not occur, and describing a familiar situation is for the speaker an event in which statement and phenomenon are firmly glued together.

This unity is the result of a process of learning that starts in one's childhood. From very early years we learn to react to situations with the appropriate (linguistic or other) movements. The teaching procedures both shape the appearance or the phenomenon and establish a firm connection with words so that finally the phenomena seem to speak for themselves, without the assistance of extraneous knowledge. They just *are* what the associated statement asserts them to be. The language they "speak" is, of course, influenced by the beliefs of earlier generations, beliefs which have been held for such a long time that they no longer appear as separate principles but enter the terms of everyday discourse and, after the described training, seem to emerge from the things themselves.

Now, at this point we may wish to compare, in our imagination and quite abstractly, the results of the teaching of different languages, incorporating different ideologies. We may even want to change some of these ideologies consciously and adapt them to more "modern" points of view. It is very difficult to say how this activity will change our situation unless we make the further assumption that the quality and structure of sensations (perceptions), or at least the quality and structure of those sensations and perceptions which enter science, are independent of their linguistic expression. I am doubtful about even the approximate validity of this assumption (which can be refuted by simple examples), and I am sure that we are depriving ourselves of new and surprising discoveries as long as we remain within the limits defined by it. Yet the present essay will consciously remain within these limits.

Making our additional simplifying assumption, we can distinguish between (a) sensations and (b) those "operation[s] of the mind which [follow so] closely upon the senses" [147] and are so firmly connected with their reactions that a separation is difficult to achieve. These operations we have already mentioned, and we have called them natural interpretations. We see now that the idea of a natural interpretation is an approximation which neglects the interaction between thinking and perceiving.

In the history of thought, which entirely moves within the limits of this approximation, natural interpretations have been regarded either as a priori presuppositions of science or else as prejudices which must be removed before a serious examination can start. The first assumption is made by Kant and, in a very different manner and on the basis of very different talents, by some contemporary linguistic philosophers. The second assumption is due to Bacon (there are, however, predecessors, such as the Greek skeptics). Galileo is one of those rare thinkers who neither wants to retain natural interpretations nor wants to eliminate them altogether. Wholesale judgments of this kind are alien to his ways of thinking. He insists on a critical discussion to decide which natural interpretations can be retained and which must be replaced. This does not always become clear from his writings. Quite the contrary, the methods of reminiscence to which he appeals so frequently are designed to create the impression that nothing has changed and that we continue describing our observations by means of the old and familiar terms. Yet his attitude is relatively easy to ascertain: natural interpretations are necessary. The senses alone cannot give us a true account of nature. What is needed for arriving at a true account are "the . . . senses *accompanied by reasoning*" (255, my italics). Moreover, in the arguments dealing with the motion of the earth, it is this reasoning and not the message of the senses, the appearance, that makes trouble. "It is therefore better to put aside the appearances on which we all agree, and to use the power of reason either to confirm [their] reality, or to reveal [their] fallacy" (256). "To confirm the reality, or reveal the fallacy of appearances" means, however, to examine the truth of those natural interpretations which in our natural thinking are so intimately connected with appearances that we no longer regard them as separate assumptions.

Galileo carries out the examination first by making the interpretations explicit. Considering the unity, within common sense (and established scientific practice), of appearance and interpretation, this is not at all a simple matter. He then replaces them by different assumptions which may not be quite so evident and which must, therefore, be smuggled in so that the "natural language" of the appearances which is, of course, the language of the original interpretations cannot be mobilized against them. And he finally shows that the replacement allows us to give a better account of phenomena. The replacement is facilitated by insinuating that no change has taken place and that one has only used well-known, plausible, but temporarily forgotten ideas. Let us now take a look at the first natural interpretation implicit in the argument from falling stones.

13. Galilean Relativity

According to Copernicus, the motion of a falling stone should be "mixed straight-and-circular" (248). And by the "motion of the stone" is meant not just its motion relative to some visible mark in the visual field of the observer, or its observed motion, but rather its motion in the solar system, or in (absolute) space, or its real motion.[149] The familiar facts that are appealed to in the argument seem to assert a different kind of motion; they seem to assert a simple vertical motion. This assertion refutes the Copernican hypothesis only if the concept of motion that occurs in the observation statement is the same as the concept of motion that occurs in the Copernican prediction. Therefore, the observation statement "The stone is falling straight down" must likewise refer to a movement in absolute space. The senses must "speak the language" of absolute motion. Is this condition satisfied?

It is satisfied in the context of seventeenth-century everyday thought —or at least this is what Galileo tells us. He tells us that the everyday thinking of the time contains the idea of the "operative" character of all motion (171) or, to use well-known philosophical terms, it entails a naïve realism with respect to motion: except for occasional and avoidable illusions, seen motion is identified with real (absolute) motion. According to what has been said in the last section, the distinction between real motion and seen motion is not explicitly drawn. One does not first distinguish the seen motion from the real motion and then connect the two by a correspondence rule. Quite the contrary, one describes, perceives, and acts toward the seen motion as if it were already the real thing. We must conclude, then, either that the Copernican hypothesis is not in agreement with the facts—and this is indeed the conclusion drawn by Galileo's contemporaries—or that our senses mislead us even in simple matters. In the latter case "the criterion of science itself [namely, experience] will be badly shaken if not completely overturned" (248).[150] Seen from the point of view of seventeenth-century thought, the argument is impeccable. How can it be overcome?

The argument from falling stones seems to refute the Copernican view. This may be due to an inherent disadvantage of Copernicanism, but it may also be due to the existence of natural interpretations which are in need of improvement. The first task, then, is to discover and to isolate these unexamined obstacles to progress.

It was Bacon's belief that natural interpretations could be discovered by a method of analysis that peels them off, one after another, until the sensory core of every observation is laid bare. This method has serious drawbacks: (i) natural interpretations of the kind considered by Bacon are not just added to a previously existing field of sensations; they are instrumental in constituting the field. Eliminate all natural interpretations and you also eliminate the ability to think and to perceive. (ii) Disregarding this fundamental function of natural interpretations, it should still be clear that a person who faces a perceptual field without a single natural interpretation at his disposal would be completely disoriented. He could not even start the business of science. (3) The fact that we do start, even after some Baconian analysis, shows that the analysis has stopped prematurely. And it has stopped at precisely those natural interpretations of which we are not aware and without which we cannot proceed. It follows that the intention to start from scratch, after a complete removal of all natural interpretations has taken place, is self-defeating.

Nor is it possible to unravel even partly the cluster of natural interpretations. At first sight the task would seem to be simple enough: one takes observation statements, one after the other, and analyzes their content. However, concepts which are hidden in observation statements are not likely to reveal themselves in the more abstract parts of language. And if they do, it will still be difficult to nail them down (concepts just as percepts are ambiguous and dependent on background). Moreover, the content of a concept is determined also by the way in which it is related to perception. Yet, how can this way be discovered without circularity? Perception must be identified, and the identifying mechanism will contain some of the very same elements which govern the use of the concept to be investigated. We never penetrate this concept completely, for we always use part of it in the attempt to find its constituents.[151] There is only one way to get out of the circle, and it consists in using an external measure of comparison including new ways of relating concepts and percepts. Removed from the domain of natural discourse and from all those principles, habits, and attitudes which constitute its "form of life," such an external measure will look strange indeed. This, however, is not an argument against its use. Quite the contrary, such an impression of strangeness reveals that natural interpretations are at work, and it is a first step toward their discovery. Let us explain this situation with the help of our example.

The argument is intended to show that the Copernican view is not in accordance with "the facts." Seen from the point of view of these "facts," the idea of the motion of the earth appears to be outlandish, absurd, obviously false, to mention only some of the expressions which were frequently used at the time (and which are still heard wherever professional squares confront a new and counterfactual theory). This knowledge makes us suspect that the Copernican view is an external measuring stick of the kind just described. We can now turn the argument around and use it as a detecting device that helps us to discover what precisely it is that excludes the motion of the earth. In doing so, we first assert the motion of the earth and then inquire what changes will remove the contradiction. Such an inquiry may take considerable time, and there is a good sense in which we can say that it is not yet finished, not even today. The contradiction, therefore, may stay with us for decades and even centuries. Still, it must be upheld until we have finished our examination or else the examination—the attempt to discover the archaic components of our knowledge—cannot even start. (This, incidentally, is one of the reasons one can give for retaining and for inventing counterfactual theories.) We conclude, then, that ideological ingredients of our knowledge and, more especially, of our observations are discovered with the help of theories which are refuted by them. They are discovered counterinductively.

Let us repeat what has been asserted so far. Theories are tested and, possibly, refuted by facts. Facts contain ideological components, older views, natural interpretations, which have vanished from sight or perhaps never were formulated in an explicit manner. These components are highly suspicious, first, because of their age, because of their archaic origin, and, secondly, because their very nature protects them from critical examination and always has protected them from such an examination. Considering a contradiction between a new and intelligent theory and a collection of "firmly established facts," the best procedure is, therefore, not to abandon the theory but to use it for the discovery of hidden principles which are responsible for the contradiction. Counterinduction is an essential part of this process of discovery. (Excellent historical example: the arguments of Parmenides and Zeno against motion and atomicity. Diogenes of Sinope, the Cynic, took the simple course that would be taken by many contemporary scientists and by all contemporary philosophers—he refuted the arguments by rising and walking up and down. The opposite course, recommended here, led to much more interesting results, as is witnessed by the history of the argument. One should not be too hard on Diogenes, however, for it is also reported that he beat up a pupil who was content with his refutation, exclaiming that he had given reasons which the pupil should not accept without additional reasons of his own.)

Having discovered a particular natural interpretation, the next question is how it is to be examined and tested. Obviously we cannot proceed in the usual way: derive predictions and compare them with "the results of observation." These results are no longer available. The idea that the senses, employed under normal circumstances, produce correct reports of real events, real absolute motions included, has been found to be an essential part of the anti-Copernican argument and has now been removed from it and from all observational statements. But without it our sensory reactions cease to be relevant for tests. This conclusion has been generalized by some rationalists who have decided to build their science on reason only and who have ascribed to observation only an insignificant auxiliary function. Galileo does not adopt this procedure. If one natural interpretation causes trouble for an attractive view, if its elimination removes the view from the domain of observation, then the only way out of trouble is to use other interpretations and to see what happens. The interpretation which Galileo uses restores the senses to their position as instruments of exploration but only in respect to the reality of relative motion. Motion "among things which share it in common" is "non operative," that is, "it remains insensible, imperceptible, and without any effect whatever" (171). Therefore, Galileo's first step in the joint examination of the Copernican doctrine and of a familiar but hidden assumption of the older point of view consists in replacing the latter by a different assumption or, to use modern terminology, he introduces a new observation language.

This is, of course, an entirely legitimate move. The observation language which enters the argument has been in use for a long time and is familiar. Considering the structure of common idioms, on the one hand, and of the Aristotelian philosophy, on the other, neither this use nor the familiarity can be regarded as a test of its underlying principles. These principles, these natural interpretations, occur in every description, and extraordinary cases that might create difficulties are defused with the help of "adjuster words" [152] such as "like" or "analogous" which divert them so that the basic ontology remains unchallenged. However, a test is urgently needed. It is needed especially in those cases where the principles seem to threaten a new theory. It is then quite reasonable to introduce alternative observation languages and to compare them with the original idiom and with the theory under examination. Proceeding in this way, we must make sure that the comparison is fair, that is, we must not criticize an idiom that is supposed to function as an observation language because it is not yet well known and is, therefore, less strongly connected with our sensory reactions and less plausible than is another, more common idiom. Superficial criticisms of this kind which have been elevated into an entire new "philosophy" abound in discussions of the mind-body problem. Philosophers who want to introduce and to test new views thus find themselves faced not with arguments which they could most likely answer but with an impenetrable stone wall of well-entrenched reactions. (These reactions are not at all different from the attitude of people ignorant of foreign languages who feel that a certain color is much better described by "red" than by "rosso.") As opposed to such attempts to win by appeal to familiarity ("I know what pains are, and I also know, from introspection, that they have nothing whatever to do with material processes!") we must emphasize that a comparative judgment of observation languages (materialistic observation language, phenomenalistic observation language, objective-idealistic observation language, theological observation language, etc.) can start only when all of them are spoken equally fluently. Assuming this condition to be satisfied, how will the judgment be carried out?

Let me assert at this point that although it is possible to consider, and to actively apply, various rules of thumb and although we may in this way arrive at a satisfactory judgment, it is not at all wise to go farther and to turn these rules of thumb into necessary conditions of science. For example, one might be inclined to say, following Neurath, that an observation language A is preferable to an observation language B if it is at least as useful as B in our everyday life (one always seems to assume that observation languages should be employed not only in laboratories but also at home and in the natural surroundings of the scientist) and if more comprehensive theories are compatible with it than are with B. Such a criterion takes into account that both our perceptions (natural interpretations included) and our theories are fallible, and it also pays attention to our desire for a harmonious and comprehensive point of view. However, we must not forget that we find and improve the assumptions hidden in our observational reports by a method that makes use of inconsistencies. Hence, we might prefer B to A as a starting point of analysis and we might in this way arrive at a language C that satisfies the criterion even better but which cannot be reached from A. (Conceptual progress like any other kind of progress depends on psychological circumstances, and these may prohibit in one case what they encourage in another. Moreover, the psychological factors which come into play are never clear in advance.)

Nor can the demand for practicability and sensory content be regarded as a *conditio sine qua non*. We possess detecting mechanisms whose performance far outdistances our senses. Combining such detectors with a computer, we may test a theory directly, without intervention of a human observer. This would eliminate sensations and perceptions from the process of testing. Using hypnosis, one could eliminate them also from the transfer of the results into the human brain and thus arrive at a science, a physics, that is completely without experience. Considerations like these, which indicate possible paths of development, should cure us once and for all of the belief that judgments of progress, improvement, etc., are based on rules which can be revealed now and will remain in action for all the years to come. Therefore, my discussion of Galileo is not aimed at arriving at the "correct method." Instead, it has the aim to show that such a "correct method" does not and cannot exist. More specifically, it has the limited aim to show that counterinduction is often a reasonable move. Let us now proceed a step further with our analysis of Galileo's reasoning.

Galileo replaces one natural interpretation by a very different and as yet (1630!) at least partly unnatural interpretation. How does he proceed? How does he manage to overcome the strong connection between observation and the idea of the operativeness of all motion? How does he manage to introduce absurd and counterinductive assertions such as the assertion that the earth moves, and how does he succeed in getting them a just and attentive hearing? We anticipate that arguments will not suffice—an interesting and highly important limitation of rationalism— and Galileo's utterances are, indeed, arguments in appearance only. For Galileo uses propaganda; he uses psychological tricks, in addition to whatever intellectual reasons he has to offer. These tricks are very successful; they lead him to victory—but they obscure the new attitude toward experience that is in the making and postpone for centuries the possibility of a reasonable philosophy. They obscure the fact that the experience on which Galileo wants to base the Copernican view is nothing but the result of his own fertile imagination, that it has been invented. They obscure this fact by insinuating that the new results which emerge are known and conceded by all and need only be called to our attention to appear as the most obvious expression of the truth.

Galileo "reminds" us that there are situations in which the nonoperative character of shared motion is just as evident [153] and as firmly believed as the idea of the operative character of all motion is in other circumstances (this latter idea is, therefore, not the only natural interpretation of motion). The situations he uses are events in a ship and events in a smoothly moving carriage and in any other system that can contain an observer and permit him to carry out some simple operations. He discusses the events in a ship:

Sagredo: There just occurred to me a certain fantasy which passed through my imagination one day while I was sailing to Aleppo, where I was going as consul for our country. . . . If the point of a pen had been on the ship during my whole voyage from Venice to Alexandretta and had had the property of leaving visible marks of its whole trip, what trace—what mark—what line would it have left?

Simplicio: It would have left a line extending from Venice to there; not perfectly straight—or rather not lying in the perfect arc of a circle—but more or less fluctuating according as the vessel would now and again have rocked. But this bending in some places a yard or two to the right or left, up or down, in a length of many hundred miles, would have made little alteration in the whole extent of the line. These would scarcely be sensible, and without an error of any moment it could be called part of a perfect arc.

Sagredo: So that if the fluctuation of the waves were taken away and the motion of the vessel were calm and tranquil, the true and precise motion of that pen point would have been an arc of a perfect circle. Now if I had had that same pen continually in my hand, and had moved it only a little sometimes this way or that, what alteration should I have brought into the main extent of this line?

Simplicio: Less than that would be given to a straight line a thousand yards long which deviated from absolute straightness here and there by a flea's eye.

Sagredo: Then if an artist had begun drawing with that pen on a sheet of paper when he left the port and had continued doing so all the way to Alexandretta, he would have been able to derive from the pen's motion a whole narrative of many figures, completely traced and sketched in thousands of directions, with landscapes, buildings, animals, and other things. Yet the actual, real, essential movement marked by the pen point would have been only a line; long, indeed, but very simple. But as to the artist's own actions, these would have been conducted exactly the same as if the ship had been standing still. The reason that of the pen's long motion no trace would remain except the marks drawn upon the paper is that the gross motion from Venice to Alexandretta was common to the paper, the pen, and everything else in the ship. But the small motions back and forth, to right and left, communicated by the artist's fingers to the pen but not to the paper, and belonging to the former alone, could thereby leave a trace on the paper which remained stationary to those motions. (172–73)

Or:

Salviati: . . . imagine yourself in a boat with your eyes fixed on a point of the sail yard. Do you think that because the boat is moving along briskly, you will have to move your eyes in order to keep your vision always on that point of the sail yard and to follow its motion?

Simplicio: I am sure that I should not need to make any change at all; not just as to my vision, but if I had aimed a musket I should never have to move it a hairsbreadth to keep it aimed, no matter how the boat moved.

Salviati: And this comes about because the motion which the ship confers upon the sail yard, it confers also upon you and upon your eyes, so that you need not move them a bit in order to gaze at the top of the sail yard, which consequently appears motionless to you. [And the rays of vision go from the eye to the sail yard just as if a cord were tied between the two ends of the boat. Now a hundred cords are tied at different fixed points each of which keeps its place whether the ship moves or remains still [154]]. (249–50)

It is quite clear that these situations lead to a nonoperative concept of motion even within common sense.

On the other hand, common sense (and we now are referring to seventeenth-century common sense) also contains the idea of the operative character of all motion. This idea arises when a limited object that does not contain too many parts moves in vast and stable surroundings, for example, when a camel trots through the desert or when a stone descends from a tower.

Now, Galileo urges us to "remember" the conditions in which we assert the nonoperative character of shared motion in this case also and to subsume it under the first.

Thus the first of the two paradigms of nonoperative motion mentioned above is followed by the assertion that

it is likewise true that the earth being moved, the motion of the stone descending, is actually a long stretch of many hundred yards, or even many thousands; and had it been able to mark its course in motionless air or upon some other surface, it would have left a very long slanting line. But that part of all this motion which is common to the rock, the tower, and ourselves remains insensible and as if it did not exist. There remains observable only that part in which neither the tower nor we are participants; in a word, that with which the stone is falling measures the tower.

And the second paradigm precedes the exhortation to

transfer this argument to the whirling of the earth and to the rock placed on top of the tower, whose motion you cannot discern because in common with the rock you possess from the earth that motion which is required to follow the tower; you do not need to move your eyes. Next, if you add to the rock a downward motion which is peculiar to it and not shared by you, and which is mixed with the circular motion, the circular portion of the motion which is common to the stone and the eye continues to be imperceptible. The straight motion alone is sensible, for to follow that you must move your eyes downwards.

This is strong persuasion indeed. Yielding to this persuasion, we quite automatically start confounding the conditions of the two cases and become relativists (this is the essence of Galileo's trickery). As a result the clash between Copernicus and "the conditions affecting ourselves and those in the air above us" (see note 25) dissolves into thin air, and we realize "that all terrestrial events from which it is ordinarily held that the earth stands still and the sun and the fixed stars are moving would necessarily appear just the same to us if the earth moved and the others stood still" (416).

Let us now look at the situation from a more abstract point of view.

14. Inventing Experiences

We start with two conceptual subsystems of ordinary thought. One of them regards motion as an absolute process which always leads to effects, effects on our senses included. The description of this conceptual system given in the present paper may be somewhat idealized, but the arguments of the opponents of Copernicus which are quoted by Galileo himself and which according to him were "very plausible" (see note 12) show not only that there was a widespread tendency to think in its concepts but that this tendency was a serious obstacle for the discussion of alternative ideas. Occasionally one finds even more primitive ways of thinking in which concepts such as "up" and "down" are used absolutely, for example, in the assertion "that the earth is too heavy to climb up over the sun and then fall headlong back down again" (327) or in the assertion that "after a short time the mountains, sinking downward with the rotation of the terrestrial globe, would get into such a position that whereas a little earlier one would have had to climb steeply to their peaks, a few hours later one would have to stoop and descend in order to get there" (330). In his marginal notes Galileo calls these "utterly childish reasons [which] suffice to keep imbeciles believing in the fixity of the earth" (327), and he thinks it unnecessary "to bother about such men as these, *whose name is legion,* or to take notice of their fooleries" (327, my italics).[155] Yet we see that the absolute idea of motion was well entrenched and that the attempt to replace it was bound to encounter strong resistance.[156]

The second conceptual system entails the relativity of motion and is also well entrenched in its own domain of application. Galileo aims at replacing the first system by the second in all cases, terrestrial as well as celestial. Naïve realism with respect to motion is to be eliminated completely.

Now, we have seen that this naïve realism is on occasion an essential part of our observational vocabulary. In these occasions (paradigm i in figure 3) the observation language contains the idea of the efficacy of all motion. Or, to express it in the material mode of speech, our experience in these situations is the experience of objects which move absolutely. Taking this into consideration, we see that Galileo's proposal amounts to a partial revision of our observation language or of our experience. An experience which partly contradicts the idea of the motion of the earth is turned into an experience that confirms it, at least as far as "terrestrial things" are concerned (132; 416). This is what actually happens. But

Galileo wants to persuade us that no change has taken place, that the second conceptual system was already universally known, even though it was not universally used. Both Salviati (his representative in the dialogue) and his opponents (Simplicio and also Sagredo, the intelligent layman) connect Galileo's method of argumentation with Plato's theory of *anamnesis* [157]—a clever tactical move, typical of Galileo, one is inclined to say, which, however, must not deceive us about the revolutionary development that is actually taking place.

Paradigm i: Motion of compact objects in stable surroundings of great spatial extension. Deer observed by the hunter.		*Paradigm ii:* Motion of objects in boats, coaches, and other moving systems.	
Natural Interpretation: All motion is operative.		*Natural Interpretation:* Only relative motion is operative.	
Falling stone *proves* ↓ earth at rest.	Motion of earth *predicts* ↓ oblique motion of stone.	Falling stone *proves* ↓ no relative motion between starting point and earth.	Motion of earth *predicts* ↓ no relative motion between starting point and stone.

Figure 3

The resistance against the assumption that shared motion is nonoperative (171) is equated with the resistance exhibited by forgotten ideas toward the attempt to make them known. Let us accept this interpretation of the resistance! But let us not forget its existence. We must then admit that it restricts the use of the relativistic ideas, confining them to part of our everyday experience. Outside this part, and that means in interstellar space, they are forgotten and, therefore, not active. But outside this part there is not complete chaos either. Other concepts are used, among them those very same absolutistic concepts which derive from the first paradigm. We not only use them, but we must admit that they are empirically entirely adequate. No difficulties arise as long as one remains within the limits of the first paradigm. "Experience," that is, the totality of all facts from all domains described with the concepts which are appropriate in these domains, *this* experience cannot force us to carry out the change which Galileo wants to introduce. The motive for a change must come from a different source. [158]

It comes, first, from the desire to see "the whole correspond to its parts with wonderful simplicity" (341), as already expressed by Copernicus himself.[159] It comes from the "typically metaphysical urge" for unity of understanding and of conceptual presentation.[160] And, secondly, the motive for a change is connected with the intention to make room for the motion of the earth which Galileo has accepted and is not prepared to give up.[161] The idea of the motion of the earth is closer to the first paradigm than to the second—or at least it was at the time of Galileo. This gave great strength to the Aristotelian arguments and made them very plausible. To eliminate this plausibility, it becomes necessary to subsume the first paradigm under the second and to extend the relative notions to all phenomena. The idea of anamnesis functions here as a psychological crutch or as a psychological lever which smoothes the process of subsumption by concealing its existence. We are now ready to apply the relative notions not only to boats, coaches, and birds, but also to the "solid and well established earth" as a whole, and we have the impression that this readiness was in us all the time, although it took some effort to make it conscious. This impression is most certainly erroneous—it is the result of Galileo's propagandistic machinations. We would do better to describe the situation in a different way, as a change of our conceptual system, or, since we are dealing with concepts which belong to natural interpretations and are, therefore, connected with sensations in a very direct way, as a change of experience [162] that allows us to accommodate the Copernican doctrine. The change corresponds perfectly to the pattern outlined in section 7; an inadequate view, the Copernican theory, is supported by another inadequate view, the idea of the nonoperative character of shared motion, and each receives and gives strength in the process (the inadequacy of the second view is, of course, covered up by Galileo's tricks, but it can easily be revealed [see notes 12, 155, 156]). It is this change which constitutes the transition from the Aristotelian point of view to the epistemology of modern science.

For experience now ceases to be that unchangeable fundament which it is both in common-sense and in the Aristotelian philosophy. The attempt to support Copernicus makes experience fluid in the very same manner in which it makes the heavens fluid, "so that each star roves around by itself" (120). An empiricist who starts from experience and builds on it without ever looking back now loses the very ground on which he stands. Neither the earth, "the solid, well established earth," nor the facts on which he usually relies can be trusted any longer. It is clear that a philosophy that uses such a fluid and changing experience needs new methodological principles which do not insist on an asymmetric judgment of theories by experience. Classical physics intuitively adopts such principles; at least the great and independent thinkers, such as Newton, Faraday, and Boltzmann, proceed in this way. But its official doctrine still clings to the idea of a stable and unchanging experience. The clash between this doctrine and the actual procedure is concealed by a tendentious presentation of the results of research that hides their revolutionary origin and suggests that they have flown from the stable and unchanging experience praised by the doctrine. These methods of concealment start with Galileo's attempt to introduce new ideas under the cover of anamnesis, and they culminate in Newton.[163] They must be exposed if we want to arrive at a better account of the progressive elements in science.

15. Dynamics

Our discussion of the anti-Copernican argument is not yet complete. So far we have tried to discover what assumption will make a stone that moves alongside a moving tower appear to fall straight down instead of being seen to move in an arc. The assumption that our senses notice only relative motion—an assumption which we shall call the *relativity principle*—has been seen to do the trick. What remains to be explained is why the stone stays with the tower and why it is not left behind. If we want to save the Copernican view and if we want to discuss it in a general way that is not restricted by the ideology contained in the customary natural interpretations, then we must explain not only why a motion that preserves the relation among visible objects remains unnoticed but also why a common motion of various objects does not affect their relation or why such a motion is not a causal agent. Turning the question around in the manner illustrated in section 13, we now see that the anti-Copernican argument rests on at least two natural interpretations, namely, (i) the epistemological assumption that absolute motion is always noticed, as well as (ii) the dynamical principle that objects (such as the falling stone) which are not interfered with remain in a state of absolute rest. Furthermore, there is (iii) the cosmological idea that the universe contains absolute places in addition to absolute directions and absolute motions. Our present problem is to supplement the relativity principle with a new law of inertia in such a fashion that the motion of the earth can still be upheld.

The law of inertia which Galileo adopts asserts the constancy of frictionless circular motion around a center and, more specifically, around the center of the earth:

It has already been said . . . that circular motion is natural for the whole and for the parts when they are in the optimum arrangement; straight motion is to restore disorderly parts to order. Though it would be better to say that they never move in a straight motion, whether ordered or disordered, but in a mixed motion, which might even be a plain circle. But only a part of this mixed motion is visible and observable to us, which is the straight part; the circular remainder stays imperceptible because we also share in it. This applies to rockets, which do move up and around, but we cannot distinguish the circular motion because we are also moving with it. . . . But motion in a horizontal line which is tilted neither up nor down is circular motion about the center; therefore circular motion is never acquired naturally without straight motion to precede it; but, *being once acquired, it will continue perpetually with uniform velocity.* (242–43, 28, my italics)

Circular motion is not only conserved; it is also one of the two most fundamental motions in the universe:

Nothing, I assert, moves in a straight line by nature. The motion of all celestial objects is in a circle; ships, coaches, horses, birds, all move in a circle around the earth; the motions of the parts of animals are all circular; in sum, we are forced to assume that only *gravia deorsum* and *levia sursum* move apparently in a straight line; but even that is not certain as long as it has not been proven that the earth is at rest.[164]

Once the earth has been shown to be a star, it participates in the circular motion that is essential to all celestial objects (this is one of the basic principles of the Aristotelian philosophy), and the same is then true of all celestial objects. So much about circular motion. The fundamental character which it assumes in Galileo's system explains why shared motion, which will be circular for all participants, will not lead to dynamical effects.

Now, Galileo often acts as if no motion whatever could lead to dynamical effects:

Motion, in so far as it is and acts as motion, to that extent exists relatively to things that lack it; and among things which all share equally in any motion, it does not act, and is as if it did not exist. . . . *Whatever* motion comes to be attributed to the earth must necessarily remain imperceptible to us and as if nonexistent, so long as we look only at terrestrial objects. . . . It is obvious, then, that motion which is common to many moving things is idle and inconsequential to the relation of these movables among themselves, nothing being changed among them, and that it is operative only in the relation that they have with other bodies lacking that motion, among which their location is changed. Now, having divided the universe into two parts, one of which is necessarily movable and the other motionless, it is the same thing to make the earth alone move, and to move all the rest of the universe, so far as concerns any result which may depend upon such movement. For the action of such movement *is only in the relation* between the celestial bodies and the earth, which relation alone is changed. (116, 114, 116, my italics)

This last assertion goes far beyond the statement of a circular law of inertia, and it may be compared with the views of Berkeley (*de motu*) and Mach: joint motion has no effect not because "space" does not act upon objects which move together, but because there is no space over and above the relations between such individual material objects. It would be of interest to inquire whether this is an isolated aside or whether the relational theory of space plays an essential role in Galileo's philosophy. In the present paper I shall restrict myself to an examination of the consequences of the circular law.

The relativity principle was defended in two ways: first, by showing how it helps Copernicus; secondly, by exhibiting its use within common sense and by surreptitiously generalizing that use. Each defense has a well-determined function. The first, though perfectly ad hoc, indicates the direction in which support for Copernicus may be found and in this way prepares the invention of auxiliary sciences (cf. section 7). The second defense defuses the power of common sense by extending it to implausible ideas (this is the essence of Galileo's method of anamnesis [165]). Independent arguments are provided by neither.

Galileo's defense of the principle of circular inertia is of the exact same kind. It is again introduced not by reference to experiment, or to independent observation, but by reference to what everyone is already supposed to know:

Simplicio: So you have not made a hundred tests or even one? And yet you so freely declare it to be certain? . . .
Salviati: Without experiment, I am sure that the effect will happen as I tell you, because it must happen that way; and I might add that you yourself also know that it cannot happen otherwise, no matter how you may pretend not to know it. . . . But I am so handy at picking peoples' brains that I shall make you confess this in spite of yourself. (145)

Step by step Simplicio is forced to admit that a body that moves without friction on a sphere concentric with the center of the earth will carry out a "boundless" and a "perpetual" motion (147). This assertion, of course, is neither based on experiment nor on corroborated theory; it is a daring new suggestion involving a tremendous leap of the imagination.[166] A little more analysis then shows, in perfect accordance with what has been found in the case of the telescope, that the suggestion is connected with experiments by ad hoc hypotheses; for the amount of friction to be eliminated follows not from independent investigations—such investigations commence only much later, in the eighteenth century—but from the very result to be achieved, namely, the circular law of inertia.[167] The considerations of sections 7 and 8 finally convince us that this is not at all a drawback, that the ad hoc hypotheses are but temporary [168] placeholders for meteorological theories to be invented at some time in the future, and that their function is to preserve a cosmological view that is endangered by its being combined with inadequate auxiliary sciences. Viewing natural phenomena in this way leads to a complete reevaluation of all experience, or, as we can say now, it leads to the invention of a new kind of experience that is more sophisticated but also far more speculative than is the experience of Aristotle or of common sense.[169] Speaking paradoxically (but not incorrectly), we may say that Galileo's "experience" has metaphysical ingredients.

16. Conclusion

When the "Pythagorean idea" of the motion of the earth was revived by Copernicus, it met with difficulties which exceeded the difficulties encountered by contemporary Ptolemaic astronomy. Strictly speaking, one had to regard it as refuted. Galileo, who was convinced of the truth of the Copernican view and who did not share the common, though by no means universal belief in a stable experience, looked for new kinds of fact which might support Copernicus and which might still be acceptable to all. Such facts he obtained in two different ways: first, by the invention of his telescope which changed the sensory core of everyday experience and replaced it by very puzzling and unexplained phenomena; then by his principle of relativity and his dynamics which changed its conceptual components. Neither the telescopic phenomena nor the new ideas of motion were acceptable to common sense (or to the Aristotelians). Besides, the associated theories could easily be shown to be false. But these false theories, these unacceptable phenomena, are distorted by Galileo and are converted into strong support of Copernicus. The whole rich reservoir of the everyday experience and of the intuition of his readers is utilized in the argument; but the facts which they are invited to recall are arranged in a new way, approximations are made, known effects are omitted, and different conceptual lines are drawn so that a new kind of experience arises, manufactured almost out of thin air. This new experience is then solidified by insinuating that the reader has been familiar with it all the time. It is solidified and soon accepted as the gospel truth despite the fact that its conceptual component is incomparably more speculative than is the conceptual component of common-sense experience (we may, therefore, say that Galilean science rests on an *illustrated metaphysics*). The distortion permits Galileo to advance, but it prevents almost everyone else from making his effort the basis of a critical philosophy (even today emphasis is put on his mathematics, on his alleged experiments,[170] or on his frequent appeal to the Truth—and his propagandistic moves are altogether neglected [171]).

I suggest that what Galileo did was to let refuted theories support one another, that he built in this way a new world view which was only loosely (if at all!) connected with the preceding cosmology (everyday experience included), that he established fake connections with the perceptual elements of this cosmology which are only now being replaced by genuine theories (physiological optics, theory of continua), and that whenever possible he replaced old facts by a new type of experience which he simply invented for the purpose of supporting Copernicus. Let it be noticed, incidentally, that Galileo's procedure drastically reduces the content of dynamics: Aristotelian dynamics was a general theory of change comprising locomotion, qualitative change, generation, and corruption. Galileo's dynamics and its successors deal with locomotion only, other kinds of motion being pushed aside with the promissory note (due to Democritus) that locomotion will eventually be capable of comprehending *all* motion. Thus a comprehensive empirical theory of motion is replaced by a much narrower theory plus a metaphysics of motion, just as an empirical experience is replaced by an experience that contains speculative elements. This, I suggest, was the actual procedure followed by Galileo. Proceeding in this way, he exhibited a sense of style, a sense of humor, an elasticity and elegance, and an awareness of the valuable weaknesses of human thinking which has never been equalled in the history of science. Here is an almost inexhaustible source of material for methodological speculation and, much more importantly, for the recovery of those features of knowledge which do not only inform but which also delight us.

NOTES

1. The two quotations at the beginning of the paper are from Descartes, letter to Mersenne, October 11, 1638, in *Oeuvres*, II, p. 380, and from Galileo, letter to Leopold of Toscana, 1640 (usually quoted under the title *Sul Candor Lunare*, VIII, ed. naz., p. 491). For a detailed discussion of Galileo's style and its

connection with his natural philosophy, cf. L. Olschki, "Geschichte der Neu-sprachlichen Wissenschaftlichen Literatur," *Galilei und seine Zeit*, III (Halle, 1927; reprinted, Vaduz, 1965). The letter to Leopold is quoted and discussed on pp. 455 ff.

2. "Problems of Empiricism," in *Beyond the Edge of Certainty*, ed. Robert Colodny (Englewood Cliffs, N.J., 1965).

3. I am thinking here of writers such as Koestler, Ardrey, and Sperry. Ardrey's *African Genesis* is an especially valuable source of material for methodological discussion. Small wonder it keeps irritating the experts.

4. "The gist of the matter lies in this, that the different aspects of the historical process—economics, politics, the state, the growth of the working class—do not develop simultaneously along parallel lines" (L. Trotsky, "The School of Revolutionary Strategy," from a speech given at a general party membership meeting of the Moscow Organization, July 1921, in *The First Five Years of the Communist International*, II, [New York, 1953], p. 5). The same is true of the relation between observation, theory, auxiliary subjects, and so on. For the existence of multiple causes which may be out of phase, cf. also Lenin, *"Left Wing" Communism, an Infantile Disorder* (Peking, 1965), p. 59. Cf. also the following quotation from Lenin: "We can (and must) begin to build Socialism [or, in the present context, the science of the future] not with imaginary human material [as does the Aldwych doctrine of critical rationalism] nor with human material specially prepared by us [as do all Stalinists, in politics as well as in the philosophy of science] *but with the quite specific human material* bequeathed to us by capitalism [by the 'normal' science of yesteryear]. True, that is very 'difficult'; but no other approach to this task is serious enough to warrant discussion" (*"Left Wing" Communism*, pp. 40–41).

5. Cf. sec. 4 of Feyerabend, "Against Method," in *Minnesota Studies in the Philosophy of Science*, IV, eds. Radner and Winokur (Minneapolis, Minn., 1970).

6. "It sometimes happens that at a new turning point of a movement, theoretical absurdities cover up some practical truth" (Lenin, diary note at the Stuttgart Conference of the Second International, in Bertram D. Wolfe, *Three Who Made a Revolution* [New York, 1948], p. 599).

7. *Philosophical Review*, 73 (1964), pp. 264–66. Cf. also my correction in "Reply to Criticism," in *Boston Studies in the Philosophy of Science*, II, eds. R. Cohen and M. Wartowsky (New York, 1965), n. 34.

8. *Theory of Spectra and Atomic Constitution* (Cambridge, 1922), p. 114. Cf. also the detailed analysis of this principle in chap. I of K. M. Meyer-Abich, *Korrespondenz, Individualitaet, und Komplementaritaet* (Wiesbaden, 1965).

9. In sec. 10 of my paper "Bemerkungen zur Geschichte und Systematik des Empirismus," in *Die Wissenschaft und Ihre Wurzeln in der Metaphysik*, ed. Weingartner (Salzburg-Munich, 1967), I made the same historical material support a very different and much more orderly (or "rationalistic") conclusion. This does not mean that I have changed my mind. It means that every case—and, therefore, also the present case—has different aspects and that an intelligent person can always win by concentrating on one side of the issue. In other words it means that there is no room, within a critical theory of knowledge, for an "approach to the truth." The aims of a critical methodology of the kind described in the text have been stated with unsurpassed clarity by K. R. Popper, *Open Society and Its Enemies*, 5th ed. (Princeton, 1969). Cf. also sec. 12 of Feyerabend, "Against Method."

10. For tentative accounts, cf. my essay "The Theatre as an Instrument of the Criticism of Ideologies," *Inquiry*, 10 (1967), pp. 298–312, as well as "On the

Improvement of the Sciences and the Arts and the Possible Identity of the Two," in *Boston Studies in the Philosophy of Science,* III, eds. R. Cohen and M. Wartowsky (New York, 1968). Cf. also secs. 14 and 15 of Feyerabend, "Against Method."

11. Galileo Galilei, *Dialogue Concerning the Two Chief World Systems,* trans. Stillman Drake (Berkeley and Los Angeles, 1953).
12. Ibid., pp. 131, 256.
13. Ibid., p. 328.
14. Ibid., p. 335.
15. Ibid., p. 339. This quotation is discussed by K. R. Popper, "Science, Problems, Aims, and Responsibilities," *Federal Proceedings of the American Societies for Experimental Biology,* 22 (1963), p. 962.
16. This objection was raised by Tycho Brahe among others. Cf. J. L. E. Dreyer, *Tycho Brahe* (New York, 1963), p. 177.
17. This proof is closely connected with the doctrine of natural places. According to this doctrine, which has strong observational support, the elements of the universe are distinguished by the places to which they tend to move: the earth moves toward the center, fire moves toward the circumference, water and air move toward intermediate places. Occasionally these purely dynamical properties are regarded as the *sole defining properties* of the elements. Earth is distinguished from fire not by its appearance, nor by the fact that the latter burns whereas the former cools, nor by its constituents, *but solely by the fact that it moves down whereas fire moves toward the circumference* (for further details cf. F. Solmsen, *Aristotle's System of the Physical World* [New York, 1960], chaps. 11 ff). The doctrine gives physical content to the notion of position and, thereby, of (a finite) absolute space. The proof against the motion of the earth which derives from it runs as follows:

> So far as the composite objects in the universe and their motion on their own account and in their own nature are concerned, those objects which are light, being composed of fine particles, fly towards the outside, that is, towards the circumference, though their impulse seems to be towards what is for individuals "up," because with all of us what is over our heads, and is also called "up," points towards the bounding surface; but all things which are heavy, being composed of denser particles, are carried towards the middle, that is to the center, though they seem to fall "down," because, again, with all of us the place at our feet, called "down," itself points towards the center of the earth, and they naturally settle in a position about the center, under the action of mutual resistance and pressure which is equal and similar from all directions. Thus it is easy to conceive that the whole mass of earth is of huge size in comparison with the things that are carried down to it, and that the earth remains unaffected by the impact of the quite small weights (falling on it), seeing that these fall from all sides alike. . . . But, of course, if as a whole it had a common motion, one and the same with that of the weights, it would, as it was carried down, have got ahead of every other falling body, in virtue of its enormous excess of size, and the animals and all separate weights would have been left behind floating in the air, while the earth, for its part, at its great speed, would have fallen completely out of the universe itself. But indeed this sort of suggestion has only to be thought of in order to be seen to be utterly ridiculous. (Ptolemy, *Syntaxis;* quoted from *Source Book in Greek Science,* eds. M. R. Cohen and I. E. Drabkin [New York, 1948], p. 126)

It would be unhistorical at this place to refer to the relativity of location, velocity, and, perhaps, of all motion. Place, or position, in Aristotle has physical properties. "The typical locomotions of the elementary bodies . . . show not only that place is something, but also that it exerts a certain influence. Each is carried to its own place, if it is not hindered, the one up, the other down" (Aristotle, *Physics*, trans. Ross [Oxford, 1930], sec. 208b). These different properties of different locations enable us to distinguish them absolutely and not only in relation to objects occupying them. The idea that the observed motions are prompted by objects in space (such as the earth) rather than by places (such as the center of the closed universe) is an alternative theory whose physical advantages were realized only after the Copernican point of view had been generally accepted. To a certain extent the theory of general relativity implies a return to the Aristotelian views. Cf. n. 149.

18. This argument is mentioned in Galileo's "Trattato della Sfera," in *Opere*, ed. naz., II, p. 224. It had convinced Tycho that the earth must be at rest and was the starting point of his own system (see his letter to Rothmann, 1587, as reported in Dreyer, *Tycho Brahe*, pp. 176, 208).

19. This argument was frequently discussed in the Middle Ages. As an example, cf. the following quotation from Buridan:

> But the last appearance [that must be adduced against a rotation of the earth] is more demonstrative in the question at hand. This is that an arrow projected from a bow directly upward falls again in the same spot of the earth from which it was projected. This would not be so if the earth were moved with such velocity. Rather before the arrow falls the part of the earth from which the arrow was projected would be a league's distance away. But still the supporters would respond that it happens so because the air, moved with the earth, carries the arrow, although the arrow appears to us to be moved simply in a straight line motion because it is being carried along with us. Therefore, we do not perceive that motion by which it is carried with the air. But this evasion is not sufficient because the violent impetus of the arrow in ascending would resist the lateral motion of the air so that it would not be moved as much as the air. This is similar to the occasion when the air is moved by a high wind. For then an arrow projected upward is not moved as much laterally as the wind is moved, although it would be moved somewhat. (*Questions on the Four Books on the Heavens and the World of Aristotle*, bk. II, sec. 9, ques. 22; quoted from *The Science of Mechanics in the Middle Ages* by M. Clagett [Madison, Wis., 1957], doc. 101)

Can there be any doubt of the empirical character of this argument?

20. Galileo, *Dialogue*, pp. 124 ff.

21. Ibid., p. 334.

22. Ibid. Copernicus himself does not seem to mention this difficulty. He refers to it neither in the *Commentariolus* nor in his major work (at least I have not found any reference of this kind). Rheticus (*Narratio Prima*; reprinted in E. Rosen, ed. and trans., *Three Copernican Treatises*, 2nd ed. [New York, 1959], p. 137) mentions the variability of Mars as an argument against Ptolemy but remains silent about the similar problem in Copernicus. It is only Osiander who discusses the difficulty in his introduction to the *De Revolutionibus*, turning it into an argument for the "hypothetical," i.e., instrumentalistic (in Popper's sense), character of Copernicus and of all other astronomical hypotheses. He writes:

It is not necessary that these hypotheses be true; they need not even be like the truth; it suffices when they lead to calculations which agree with the results of observation; *except* some one should be so ignorant in matters of geometry and of optics that he is prepared to regard the epicycle of Venus as being like the truth and to assume that it is the cause of its being now forty (or more) degrees ahead of the sun, now the same amount behind it. For who does not see that this assumption necessarily implies that the diameter of the planet when close to the earth must be four times as large than when it is in the point most remote from the earth, and its body more than 60 times as large—a fact which is contradicted by the experience of all ages.

In summary the Copernican system, regarded as a true description of the world, is inconsistent with highly corroborated basic statements and must, therefore, be regarded as refuted. From this Osiander concludes that *all* astronomical hypotheses have the same character and are instruments for the prediction of observable events only. (This point is not mentioned in Popper's account of the matter where Osiander is quoted but only up to the "except" that introduces the reasons for his instrumentalism. As a result, Popper's Osiander appears much more naïve than the actual Osiander seems to have been; cf. also my essay "Realism and Instrumentalism," in *The Critical Approach to Science and Philosophy*, ed. M. A. Bunge [New York, 1964].)

The actual variations of Mars and Venus (*Nautical Almanach,* 1957–58) are four magnitudes and one magnitude respectively.

23. *The Assayer;* quoted from Drake and O'Malley, eds., *The Controversy on the Comets of 1618* (Philadelphia, 1960), pp. 184–85.

24. This refers to the period before the end of the sixteenth century. At that time Copernicanism was, empirically at least, as doubtful as was the Ptolemaic point of view. This is shown very clearly in Derek J. de S. Price, "Contra-Copernicus: A Critical Re-Estimation of the Mathematical Planetary Theory of Ptolemy, Copernicus, and Kepler," in *Critical Problems in the History of Science,* ed. M. Clagett (Madison, Wis., 1959), pp. 197–218. Price deals only with the *kinematic* and the *optical* difficulties of the new views. (Consideration of the *dynamical* difficulties would further strengthen his case.) He points out that "under the best conditions a geostatic or heliostatic system using eccentric circles (or their equivalents) with central epicycles can account for all angular motions of the planets to an accuracy better than 6′ . . . excepting only the special theory needed to account for . . . Mercury and excepting also the planet Mars which shows deviations up to 30′ from such a theory. [This is] certainly better than the accuracy of 10′ which Copernicus himself stated as a satisfactory goal for his own theory." This was also difficult to test especially in view of the fact that refraction (almost 1° on the horizon) was not taken into account at the time of Copernicus and the observational basis of the predictions was unsatisfactory.

Carl Schumacher (*Untersuchungen ueber die ptolemaeische Theorie der unteren Planeten* [Muenster, 1917]) has found that the predictions concerning Mercury and Venus made by Ptolemy differ at most by an amount of 30′ from those of Copernicus. The deviations found between modern predictions and those of Ptolemy (and Copernicus), which in the case of Mercury may be as large as 7°, are due mainly to faulty observations, including an incorrect value of the constant of precession. For the value of Ptolemaic astronomy, cf. also N. R. Hanson, "The Mathematical Power of Epicyclical Astronomy," *Isis,* 51 (1960), pp. 150–58.

It seems that the situation described in the text is a common one in the sciences. It would, therefore, seem to be important to develop methodologies which advise us *how to choose between refuted theories* rather than *how to choose between theories on the basis of refutations*.

25. This is Ptolemy's general characterization of the dynamical arguments "against the whirling of the earth." He says:

> Certain thinkers though they have nothing to oppose to the above arguments [for the stationary earth], have concocted a scheme which they consider more acceptable, and they think that no evidence can be brought against them if they suggest for the sake of argument that the heaven is motionless, but that the earth rotates about one and the same axis from west to east, completing one revolution approximately every day, or alternatively that both the heaven and the earth have a rotation of a certain amount, whatever it is, about the same axis, as we said, but such as to maintain their *relative* situations.
>
> These persons forget however that, while, so far as appearances in the stellar world are concerned, there might, perhaps, be no objection to this theory in the simpler form, yet, to judge by the conditions affecting ourselves and those in the air above us, such a hypothesis must seem to be quite ridiculous. (*Syntaxis*, I. 7, pp. 126–27)

26. Galileo, *Dialogue*, p. 328.

27. For this view cf. Ludovico Geymonat, *Galileo Galilei*, trans. Stillman Drake (New York, 1965), p. 184.

28. *The Sidereal Messenger of Galileo Galilei*, trans. E. St. Carlos (London, 1960), p. 10 (originally published in London in 1880).

29. Galileo, *Opere*, X, p. 441.

30. *Ad Vitellionem Paralipomena quibus Astronomiae Pars Optica Traditur* (Frankfurt, 1604); to be quoted from *Johannes Kepler, Gesammelte Werke*, II, ed. Franz Hammer (Munich, 1939). This particular work will be referred to as the "Optics of 1604." It was the only useful optics that existed at the time. Cf. n. 89.

31. Galileo's curiosity was most likely stimulated by the many references to this work in Kepler's reply to the *Sidereus Nuncius*. For the history of this reply as well as a translation, cf. *Kepler's Conversation with Galileo's Sidereal Messenger*, trans. E. Rosen (New York, 1965). The many references in the *Conversation* to earlier works were interpreted by some of Galileo's enemies as a sign that "his mask had been torn from his face" (G. Fugger to Kepler, May 28, 1610, in Galileo, *Opere*, X, p. 361) and that Kepler "had well plucked him" (Maestlin to Kepler, August 7, 1610, in ibid., p. 428). Galileo must have received Kepler's *Conversation* before May 7 (ibid., p. 349), and he acknowledges receipt of the printed *Conversation* in a letter to Kepler, August 19, 1610 (ibid., p. 421).

32. *Dioptrice* (Augsburg, 1611); reprinted in *Werke*, IV (Munich, 1941). This work was written after Galileo's discoveries. Kepler's reference to them in the preface has been translated by E. St. Carlos, *Sidereal Messenger*, pp. 79 ff. The problem referred to by Tarde is treated in Kepler's *Dioptrics* of 1611.

33. Geymonat, *Galileo Galilei*, p. 37.

34. Galileo, *Opere*, VIII, p. 208.

35. *Die Geschichte der Optik* (Leipzig, 1926), p. 32. Various writers, whose lack of imagination and temperament are properly matched by their high moral standards, have been put off by such signs of worldliness on the part of Galileo, and they have tried their best to explain his actions as the result of high (and dry) motives. A much less important episode, viz., Galileo's almost complete silence

about Copernicus in his "Trattato della Sfera" (pp. 211 ff) at a time when according to some he had already accepted the Copernican creed, has led to much soul-searching and to some convenient ad hoc hypotheses even on the part of so worldly an author as L. Geymonat (*Galileo Galilei*, p. 23). However, there is no reason why a man, and an extremely intelligent man at that, should conform to the standards of the academic squares of today, and why he should not try in his own way to further his interests. Moreover, what strange moral principle is this that requires a great thinker to be a blabbermouth who "expresses" only what he believes to be "the truth" and who never mentions what he does not believe? (Is that what the contemporary search for "authenticity" demands?) A puritanic view like this is surely too naïve a background for understanding a man of the late Renaissance and early Baroque periods. Moreover, Galileo the mountebank is a much more interesting character than the moralistic searcher for the truth we are usually invited to revere. *Finally, it was only through sleights of hand such as these that progress could be made at this particular time, as we shall see.*

(Kepler, the most knowledgeable and most lovable of Galileo's contemporaries, gives a clear account of the reasons why, despite his superior knowledge of optical matters, he "refrained from attempting to construct the device," i.e., the telescope [*Conversation*, p. 18]. "You however," he addresses Galileo, "deserve my praise. Putting aside all misgivings you turned directly to visual experimentation." It remains to be added that Galileo, due to his lack of knowledge in optics, had no "misgivings" to overcome: "Galileo . . . was totally ignorant of the science of optics, and it is not too bold to assume that this was a most happy accident both for him and for humanity at large" [Ronchi, "Complexities, advances, and misconceptions in the development of the science of vision: what is being discovered?" in *Scientific Change*, ed. Alistair Crombie (New York, 1963), p. 550].)

When considering the aberrations that are implied in the above account of Galileo, one might object that Kepler could do very well without them, that he managed to advance without the tricks to which Galileo occasionally liked to resort, and that these tricks are, therefore, not really essential to science. The reply is that Kepler could well afford to be honest. One does not need to break what some regard as the moral rules of science if one breaks a sufficient number of methodological rules, has a sufficiently strong faith, and is able to invent, abandon, and reinvent wild ideas abundantly.

Hoppe's judgment concerning the invention of the telescope is shared by Wolf, Zinner, and others. Huyghens points out that superhuman intelligence would have been needed to invent the telescope on the basis of the available physics and geometry. After all, he says, we still do not understand the workings of the telescope. ("Dioptrica," *Hugenii Opuscula postuma* [Lugd. Bat., 1903], p. 163, paraphrased from A. G. Kästner, *Geschichte der Mathematik*, IV [Göttingen, 1800], p. 60).

36. Geymonat, *Galileo Galilei*, p. 39.
37. Letter to Carioso, May 24, 1616, in *Opere*, X, p. 357. Cf. also his letter to P. Dini, May 12, 1611, in ibid., p. 106: "Nor can it be doubted that I, over a period of two years now, have tested my instrument (or rather dozens of my instruments) by hundreds and thousands of objects, near and far, large and small, bright and dark; hence I do not see how it can enter the mind of anyone that I have simplemindedly remained deceived in my observations." The "hundreds of thousands" of experiments remind one of Hooke and are most likely equally as spurious. Cf. n. 49.

38. Lagalla, *De phaenomenis in orbe lunae novi telescopii usa a D. Galileo Galilei nunc iterum suscitatis physica disputatio* (Venice, 1612), p. 8; quoted from E. Rosen, *The Naming of the Telescope* (New York, 1947), p. 54. The regular reports (*Avvisi*) of the Duchy of Urbino on events and gossip in Rome contain the following notice of the event:

> Galileo Galilei, the mathematician, arrived here from Florence before Easter. Formerly a Professor at Padua, he is at present retained by the Grand Duke [of Tuscany] at a salary of 1,000 scudi. He has observed the motion of the stars with the *occhiali*, which he invented or rather improved. Against the opinion of all the ancient philosophers, he declares that there are four more stars or planets, which are satellites of Jupiter and which he calls the Medicean bodies, as well as two companions of Saturn. He has here discussed this opinion of his with Father Clavius, the Jesuit. Thursday evening, at Monsignor Malvasia's estate outside the St. Pancratius gate, a high and open place, a banquet was given for him by [Frederick Cesi], the marquis of Monticelli and nephew of Cardinal Cesi, who was accompanied by his kinsman, Paul Monaldesco. In the gathering there were Galileo; a Fleming named Terrentius; Persio, of Cardinal Cesi's retinue; [La]galla, Professor at the University here; the Greek, who is Cardinal Gonzaga's mathematician; Piffari, professor at Siena; and as many as eight others. Some of them went out expressly to perform this observation, and even though they stayed until one o'clock in the morning, they still did not reach an agreement in their views. (Quoted from Rosen, *Naming of the Telescope*, p. 31)

39. Jerome Sirturi writes as follows about the event reported by Lagalla and the *Avvisi* in n. 38:

> Galileo was there with his telescope of imperishable fame. It chanced that on a certain day Prince Frederick Cesi, marquis of Monticelli, a learned man and patron of letters, had invited Galileo and some other writers to dinner on the Malvasia estate, as it is called. Having arrived there before sunset, they began to look through the telescope at the inscription of Pope Sixtus V over the entrance to a building in the Lateran, which is about a mile away. Taking my turn, I too saw it, and I read the inscription to my heart's content. Then at nightfall, after dinner, we observed Jupiter and the motions of its companion bodies. Being quite stirred by the vision of such immense light, and by the desire to understand the matter, they retired to inspect the telescope. To satisfy their curiosity, Galileo himself took out the lens and the concave eyepiece, and showed them plainly. (*Telescopium sive ars perficiendi novum illud Galilaei visorium instrumentum ad sidera* [Frankfurt, 1618], p. 27; quoted from Rosen, *Naming of the Telescope*, p. 53)

40. Carlos, *Sidereal Messenger*, p. 11. According to Borellus, *De Vero Telescopii Inventore* (Hague, 1655), p. 4, Prince Moritz immediately realized the military use of the telescope and ordered that its invention—which Borellus attributes to Zacharias Jansen—be kept a secret. Thus the telescope commenced as a secret weapon and was turned to astronomical use only later.

41. This is hardly ever realized by those who argue with Kästner, *Geschichte der Mathematik*, IV, p. 133, that "one does not see how a telescope can be good and useful on the earth and yet deceive on the sky." Kästner's comment is directed against Horky. See text to nn. 49 and 50.

42. The stability of the heavens, the success of astronomical predictions over

meteorological predictions, and the erratic behavior of material on the surface of the earth were the main evidence.

In the Aristotelian theory of the elements, the nature of an element is determined not (only) by its internal constitution, but (mainly) by its dynamical behavior. Earth is occasionally distinguished from fire not by its appearance, nor by the fact that it has a cooling effect whereas fire burns, but solely by the fact that it moves down whereas fire moves toward the circumference (cf. also. n. 17). This theory (which has much in common with a similar principle in contemporary particle physics where particles are distinguished not by their composition but by their behavior) leads to postulating a fifth element, the ether, which was assumed to be entirely changeless and whose properties were confirmed by the evidence of all ages up to the discussion of the first Nova of 1572.

43. For this theory cf. G. E. L. Owen, "ΤΙΘΕΝΑΙ ΤΑ ΦΑΙΝΟΜΕΝΑ," in *Aristote et les problèmes de la méthode* (Louvain, 1961), pp. 83–103. For the development of the Aristotelian thought in the Middle Ages, cf. A. C. Crombie, *Robert Grosseteste and the Origins of Experimental Science* (Oxford, 1953). Relevant works of Aristotle are *Analytica Priora, Analytica Posteriora, De Anima,* and *De Sensu.* Concerning the motion of the earth, cf. Aristotle, *De Coelo,* 293a28 f: "But there are many others who would agree that it is wrong to give the earth the central position, *looking for confirmation rather to theory, than to the facts of observation*" [my italics]. As we shall see in sections 12 ff, this was precisely the manner in which Galileo introduced Copernicanism, *changing* experience in order to fit this favorite theory of his. That the senses are acquainted with our everyday surroundings but are liable to give misleading reports about objects outside this domain is proved at once by the *appearance of the moon.* Large but distant objects on the earth are seen as being large but far away. The appearance of the moon, however, does not give us any idea of its size.

44. It is not too difficult to separate the letters of a familiar alphabet from a background of unfamiliar lines, even if they should happen to have been written with an almost illegible hand. No such separation is possible with letters which belong to an unfamiliar alphabet. The parts of such letters do not drift together and form distinct patterns which stick out from the background of general (optical) noise (in the manner described by K. Koffka, "Perception: An Introduction to the *Gestalt-Theorie*," *Psychological Bulletin,* 19 (1922), pp. 551 ff, partly reprinted in M. D. Vernon, ed., *Experiments in Visual Perception* [Baltimore, Md., 1966]; cf. also the article by Gottschaldt in the same volume).

45. For the importance of cues, such as diaphragms, crossed wires, and background, in the localization and shape of the telescopic image and the strange situations arising when no cues are present, cf. chap. IV of Ronchi, *Optics, the Science of Vision* (New York, 1957), esp. pp. 151, 174, 189, 191. Cf. also R. L. Gregory, *Eye and Brain* (World University Library, 1966), p. 99 and passim (on the autokinetic phenomenon). F. P. Kilpatrick, ed., *Explorations in Transactional Psychology* (New York, 1961) contains ample material on what happens in the absence of familiar cues. Any historian dealing with the history of the telescope should consult this material—especially Professor Ronchi's account—before embarking on a criticism of the opponents of Galileo.

46. It is for this reason that the "deep study of the theory of refraction" which Galileo pretended to have carried out would have been quite insufficient for establishing the usefulness of the telescope. Cf. text to n. 28 and also the last paragraph of n. 57.

47. This is how the ring of Saturn was seen at the time. Cf. also R. L. Gregory, *The Intelligent Eye* (London, 1970), p. 120.

48. The "hundreds" and "thousands" of observations and trials which we find here are part of contemporary rhetorics (corresponding to our "I have told you a thousand times") which cannot be used to infer a life of incessant observation. Cf. also n. 37.

49. Here again is a case where external clues are missing. Cf. Ronchi, *Optics*, concerning the representation of flames, small lights, etc.

50. Galileo, *Opere*, X, p. 342 (my italics, which refer to the difference, commented upon above, between celestial and terrestrial observation).

51. Galileo, *Opere*, III, letter, May 16.

52. Ibid., p. 196.

53. Kepler suffered from polyopia. ("Instead of a single small object at a great distance, two or three are seen by those who suffer from this defect. Hence, instead of a single moon, ten or more present themselves to me". [*Gesammelte Werke*, II, p. 180; also in *Kepler's Conversation*, n. 94]. Cf. also the remainder of the note for further quotations.) He was familiar with Platter's anatomical investigations (cf. S. L. Polyak, *The Retina* [Chicago, 1942], p. 134–35, for details and literature). As a consequence he was well aware of the need for a *physiological criticism of astronomical observations*.

54. Letter, August 9, 1610; quoted from Kaspar and Dyck, eds., *Johannes Kepler in seinen Briefen*, I, (Munich, 1930), p. 349.

55. Kaspar and Dyck, *Kepler Briefen*, p. 352.

56. Thus Emil Wohlwill writes: "No doubt the unpleasant results were due to the lack of training in telescopic observation, the restricted field of vision of the Galilean telescope as well to the absence of any possibility for changing the distance of the glasses in order to make them fit the peculiarities of the eyes of the learned men" (*Galileo und sein Kampf fuer die Kopernikanische Lehre*, I [Hamburg, 1909], p. 288). A similar but more dramatically expressed judgment is found in Arthur Koestler, *Sleepwalkers* (New York, 1968), p. 369.

57. Cf. Ronchi, *Optics; Histoire de la Lumière* (Paris, 1956); *Storia del Cannochiale* (Vatican City, 1964); *Critica dei Fondamenti dell' Acustica e del' Ottica* (Rome, 1964). Cf. also E. Cantore's summary in *Archives Internationales d'histoire des sciences* (December 1966), pp. 333 ff. I acknowledge that Professor Ronchi's investigations have greatly influenced my thinking on scientific method. For a brief historical account of Galileo's work, cf. Ronchi's article in Crombie, *Scientific Change*, pp. 542–61.

How little this field is explored becomes clear from S. Tolansky's *Optical Illusions* (London, 1964). Tolansky is a physicist who in his microscopic research (on crystals and metals) was distracted by one optical illusion after another. He writes: "This turned our interest to the analysis of other situations, with the ultimate unexpected discovery that optical illusions can and do play a very real part in affecting many daily scientific observations. This warned me to be on the lookout and as a result I met more illusions than I had bargained for."

The illusions of direct vision, which are slowly being rediscovered today, were well known to medieval writers on optics and were discussed in special chapters of their textbooks. Moreover, these writers treated lens images as *psychological* phenomena, the results of a misapprehension, since an image "is merely the appearance of an object outside its place" as we read in John Pecham (cf. David Lindberg, "The 'Perspectiva Communis' of John Pecham," *Archives Internationales d'histoire des Sciences* [1965], p. 51). All this was further argument for the illusory character of telescopic images.

58. Ronchi, *Optics*, p. 189.

59. This may explain the frequently uttered desire to look inside the telescope. Cf.,

e.g., the end of Sirturi's report, n. 39 above. No such problems arise in the case of terrestrial objects whose images are regularly placed "in the plane of the object" (Ronchi, *Optics*, p. 182).

In this connection the following phenomenon is of interest. In his second letter from England (October 10, 1775), Lichtenberg reports: "Once I showed the increasing moon through a telescope of great magnification to a group of people who knew nothing of astronomy. Some asked me whether there were droplets on the glass. Now in the first quarter the spots in the moon do indeed appear like raindrops on a windowpane . . ." (G. C. Lichtenberg, *Aphorismen, Briefe, Schriften*, ed. Paul Requardt [Stuttgart, 1939], p. 416). Phenomena such as these have much in common with the ambiguous pictures of psychological research (figure-ground and others) and like them they depend on the attitude of the observer. Observers expecting visual *illusions* may have been especially prone to perceiving the nonobjective aspects of telescopic images (and perhaps even of observation with the naked eye, cf. text to n. 82). Galileo's vision, on the other hand, must have been firmly objective.

60. For the magnification of Galileo's telescope cf. Carlos, *The Sidereal Messenger*, p. 11. The old rule "that the size, position and arrangement according to which a thing is seen depends on the size of the angle through which it is seen" (R. Grosseteste, *De Iride;* quoted from Crombie, *Robert Grosseteste*, p. 120) is almost always wrong. I still remember my disappointment when, having built a reflector with an alleged linear magnification of about 150, I found that the moon was only about five times enlarged and situated quite close to the ocular (1935).

61. The image remains sharp and unchanged over a considerable interval—the lack of focusing may show itself in a doubling, however.

62. The first usable telescope which Kepler received from elector Ernst of Köln (who, in turn, had received it from Galileo) and on which he based his *Narratio de observatis a se quartuor Jovis satellibus* (Frankfurt, 1611) showed the stars as *squares* and intensely *colored* (*Gesammelte Werke*, IV, p. 461). Francesco Fontana who, from 1643 on, observed the phases of Venus notes an unevenness of the boundary (and infers mountains). Cf. R. Wolf, *Geschichte der Astronomie* (Munich, 1877), p. 398. For the idiosyncrasies of contemporary telescopes and descriptive literature, cf. Ernst Zinner, *Deutsche und Niederlaendische Astronomische Instrumente des 11. bis 18. Jahrhunderts* (Muenchen, 1956), pp. 216–21. Refer also to the author catalogue in the second part of the book.

63. Father Clavius, the astronomer of the powerful Jesuit Collegium Romanum, praises Galileo as the first "who has observed them [the moons of Jupiter]" (letter, December 17, 1610; *Opere*, X, p. 485) and completely recognizes their reality. Magini, Grienberger, and others soon followed suit. It is clear that in doing so they did not proceed according to the methods prescribed by their own philosophy, or else they were very lax in the investigation of the matter. One should remember, incidentally, that contemporary optics (such as the optics of Witelo and Peckham) devoted long discussions to the illusions of the natural senses and to the additional illusions caused by instruments. Cf. also the beginning of n. 85.

64. Carlos, *Sidereal Messenger*, p. 8.

65. Ibid., p. 24. Cf. fig. 1 in this essay, which is taken from Galileo's publication. Kepler in his "Optics of 1604" writes on the basis of observations with the naked eye that "it seemed . . . as though something was missing in the circularity of the outmost periphery" (*Gesammelte Werke*, II, p. 219). He returns to this assertion in his *Conversation*, pp. 28–29, criticizing Galileo's telescopic results by what he himself had seen with the unaided eye: "You ask why the moon's outermost circle

334 : Paul K. Feyerabend

does not also appear irregular. I do not know how carefully you have thought about this subject or whether your query, as is more likely, is based on popular impression. For in my book ["Optics of 1604"] I state that there was surely some imperfection in that outermost circle during full moon. Study the matter, and once again tell us, how it looks to you." Here the results of observation with the naked eye are quoted against Galileo's telescopic reports—and with perfectly good reason, as we shall see below. The reader who remembers Kepler's polyopia (see n. 53) may wonder how he could trust his senses to such an extent. His reply is contained in the following quotation: "When eclipses of the moon begin, I, who suffer from this defect, become aware of the eclipse before all the other observers. Long before the eclipse starts, I even detect the direction from which the shadow is approaching, while the others, who have very acute vision, are still in doubt. . . . The aforementioned waviness of the moon [see the above quotation] stops for me when the moon approaches the shadow, and the strongest part of the sun's rays is cut off" (Kepler, *Gesammelte Werke*, II, pp. 194–95).

Galileo has two explanations for the contradictory appearance of the moon. The one involves a lunar atmosphere (Carlos, *Sidereal Messenger*, pp. 26–27). The other explanation (ibid., pp. 25–26), which involves the tangential appearance of series of mountains lying behind one another, is not really very plausible, for the distribution of mountains near the visible side of the lunar globe does not show the arrangement that would be needed. (This distribution is now even better established by the publication of the Russian moon photograph of October 7, 1959; cf. Zdenek Kopal, *An Introduction to the Study of the Moon* [Reidel, Holland, 1966], p. 242.)

66. Carlos, *Sidereal Messenger*, p. 38. Cf. also the more detailed account in Galileo, *Dialogue*, pp. 336 ff. "The telescope, as it were, removes the heavens from us," writes A. Chwalina in his edition of *Kleomedes. Die Kreisbewegung der Gestirne* (Leipzig, 1927), p. 90, commenting on the decrease of the apparent diameter of *all* stars with the sole exception of the sun and the moon.

Later on the different magnification of planets (or comets) and fixed stars was used as a means of distinguishing them. "From experience I knew," writes Herschel in his paper that reports his first observation of Uranus, "that the diameters of the fixed stars are not proportionally magnified with higher powers, as the planets are; therefore I now put on the powers of 460 and 932, and found the diameter of the comet increased in proportion to the power, as it ought to be" (*Philosophical Transactions*, 71 [1781], p. 492 f—the planet is here identified as a *comet*). However, it is noteworthy that the rule did not invariably apply to the telescopes which were in use at Galileo's time. Thus, commenting on a comet of November 1618, Horatio Grassi points out "that when the comet was observed through a telescope it suffered scarcely any enlargement" and he infers, in perfect accordance with Herschel's "experience," that "it will have to be said that it is more remote from us than the moon" (*On the Three Comets of 1618;* quoted from Drake and O'Malley, *The Controversy on the Comets of 1618*, p. 17). In his *Astronomical Balance*, Grassi repeats that according to the common experience of "illustrious astronomers, . . . from many parts of Europe, . . . the comet observed with a very extended telescope received scarcely any increment" (quoted from Drake and O'Malley, *The Controversy on the Comets*, p. 80). (Galileo, in his *Dialogue*, pp. 177 ff, accepts this statement as fact, although he criticizes the conclusions which Grassi wants to draw from it.) All these phenomena refute Galileo's assertion that the telescope "works always in the same way" (*The Assayer;* quoted from Drake and O'Malley, *The Controversy on the Comets*, p. 204), and they also undermine his theory of irradiation. Cf. n. 97.

67. Kopal, *Study of the Moon*, p. 207.

68. Ibid. R. Wolf in *Geschichte der Astronomie*, p. 396, remarks on the poor quality of Galileo's drawings of the moon ("seine Abbildung des Mondes kann man . . . kaum . . . eine Karte nennen"); and Zinner, in *Geschichte der Sternkunde* (Berlin, 1931), p. 473, calls Galileo's observations of the moon and of Venus "typical for the observations of a beginner." His picture of the moon, in particular, "has no similarity with the moon" (ibid., p. 472). Zinner also mentions the much better quality of the almost simultaneous observations of the Jesuits (ibid., p. 473), and he finally asks whether Galileo's observations of the moon and of Venus were not the result of a fertile brain rather than of a careful eye ("sollte dabei . . . der Wunsch der Vater der Beobachtung gewesen sein?" [ibid.])—a just question especially in view of the phenomena we briefly described in n. 59.

69. His discovery and identification of the moons of Jupiter were no mean achievements, especially as a useful stable support for the telescope had not yet been developed.

70. The reason, among other things, is the great variation of telescopic vision from one observer to the next. Cf. again Ronchi, *Optics*, chap. IV.

71. For a survey and some introductory literature, cf. chap. 11 of Gregory, *Eye and Brain*. For a more detailed discussion and more literature, cf. K. W. Smith and W. M. Smith, *Perception and Motion* (Philadelphia, 1962, reprinted in part in Vernon, *Experiments in Visual Perception*). The reader should also consult Ames's chapter "Aniseikonic Glasses," in Kilpatrick, *Explorations in Transactional Psychology*. Ames deals with the change of normal vision caused by abnormal optical conditions, which are sometimes quite slight. A comprehensive recent account is I. Rock, *The Nature of Perceptual Adaptation* (New York, 1966).

72. Many of the old instruments and excellent descriptions of them are still available. Cf. Zinner, *Deutsche und Niederlaendische Astronomische Instrumente*.

73. For interesting information the reader should consult the relevant passages of Kepler, *Conversation* as well as of *Somnium*. (The latter is now available in a new translation by E. Rosen, who has added a considerable amount of background material. See *Kepler's Somnium* [Madison, Wis., 1967].) The standard work for the beliefs of the time is still Plutarch's *Face on the Moon*. It will be quoted from *Moralia, XII*, trans. H. Cherniss (London, 1967).

74. Note the following: "One describes the moon after objects one thinks one can perceive on its surface" (Kästner, *Geschichte der Mathematik*, IV, p. 167, describing Fontana's observational reports of 1646). "Maestlin even saw rain on the moon" (Kepler, *Conversation*, pp. 29–30, presenting Maestlin's own observational report). For the instability of the images of unknown objects and their dependence on belief (or "knowledge"), cf. again Ronchi, *Optics*, chap. IV.

75. Chapter 15 of Kopal, *Study of the Moon* contains an interesting collection of exactly this kind.

76. Of course, one must also investigate the dependence of what is seen on the current methods of *pictorial representation*. Outside the field of astronomy this has been done by E. H. Gombrich, *Art and Illusion* (Princeton, N.J., 1961), who deals with general art history, and L. Choulant, *A History and Bibliography of Anatomical Illustration*, trans. Singer et al. (New York, 1945), who deals with anatomy.

   Astronomy has the advantage that one side of the puzzle, viz., the stars, is fairly simple (much simpler than the uterus, for example) and relatively well known.

77. For these theories and further literature, cf. J. L. E. Dreyer, *A History of Astronomy from Thales to Kepler* (New York, 1953).

78. For Berossos's theory cf. Stephen Toulmin, "The Astrophysics of Berossos the Chaldean," *Isis*, 58 (1967), pp. 65 ff. Lucretius writes: "Again, she may revolve upon herself / like to a ball's sphere—if perchance that be—/ one half of her dyed o'er with glowing light / and by the revolution of that sphere / she may beget for us her varying shapes / until she turns that fiery part of her / full to the sight and open eyes of men" (*On the Nature of Things*, trans. Leonard [New York, 1957], pp. 216–17).

79. Cf. text to notes 50 ff of my "Reply to Criticism," pp. 246 ff.

80. Simplicius in Aristotle's *De Coelo*, II, p. 12. Here Polemarchus considers the difficulties of Eudoxos's theory of homocentric spheres, viz., that Venus and Mars "appear in the midst of the retrograde movement many times brighter, so that [Venus] on moonless nights causes bodies to throw shadows" (objection of Autolycus, reported by Simplicius, ibid., p. 504), and he may well be appealing to the possibility of a deception of the senses (which was frequently discussed by ancient schools). Aristotle, who must have been familiar with all these facts, does not mention them in *De Coelo* or the *Metaphysics*, though he gives an account of Eudoxos's system and of the improvements of Polemarchus and Kalippus.

81. *De Coelo*, pp. 290a25 ff.

82. Ibid., p. 37. Cf. also S. Sambursky, *The Physical World of the Greeks* (New York, 1962), pp. 244 ff.

83. Ibid. Cf., however, n. 59 as well as Pliny's remark that the moon is "now spotted and then suddenly shining clear" (*Natural History*, II, pp. 43, 46).

84. Sambursky, *Physical World of the Greeks*, p. 50.

85. Cf. also n. 96 and text. All this requires further research, especially in view of the general distrust in vision as expressed in the principle *non potest fieri scientia per visum solum.* ("No scientific value should be attached to anything observed by sight alone. Visual observation could never be considered valid unless confirmation was available by touch. [As a consequence] no one used [the] enlarged images [created by concave mirrors] as the basis of a microscope. The reason for this essential fact is clear: nobody believed what he saw in a mirror, once he realized that he could not confirm it by touch" [Ronchi, "Complexities, advances, and misconceptions," p. 544].) Distrust in vision may be raised also by the surprising changes of normal terrestrial perception which can perhaps be inferred from the results of Snell and Dodds. Perhaps it is also a little unreasonable to assume that phenomena will be unaffected by one's views about their relation to the world. (Afterimages may be bright and disturbing for someone who has just obtained his sight. However, later on they become almost unnoticeable and must be studied by special methods.)

    The hypothesis in the text is developed in one particular direction not so much because I am convinced that it is true, but in order to indicate possible avenues of research and also to give a clear impression of the complexity of the situation at Galileo's time. In this connection let me briefly mention Alhazen's "On the Light of the Moon," German trans. K. Kohl, *Sitzungsberichte Phys. Med. Soc. Erlangen*, 55/56 (1924/1925), pp. 305–98. (I am using the discussion and translation of parts of this work from M. Schramm, *Ibn Al-Haythams Weg zur Physik* [Wiesbaden, 1963], pp. 70 ff.) All the early hypotheses about phases and eclipses (such as turning of a partly luminous, partly darkened sphere; rotation of an opaque semisphere; change of distance; intervening bodies of different sizes) are reviewed in this work but without any reference to the "face." (Schramm, *Ibn Al-Haythams*, p. 88, n. 3 makes Alhazen's hypothesis that "the spots are caused by a body external to the moon and with varying power of absorption" responsible for this. In Alhazen's later work on the spots, the absorptive power of different media is discussed. This suggests that we are again dealing not with different *phenomena* but with different *explanations* of phenomena which are otherwise well known.)

One of the main tasks of Alhazen's treatise is to establish that the disk of the moon itself, which is surely seen by everybody, cats, dogs, and vampires included, is not a subjective glare caused by a very different objective structure but corresponds *point for point* to objective light traveling from the surface of the moon to the eye of the observer. This is shown in a very simple and ingenious fashion by a projective device that subdivides the moon and projects it point for point on a screen. Each point that is seen with the naked eye also leaves its mark on the screen and is thus proved to be an objective phenomenon. (The very same method is used to show that Aristotle's distinction between light and color—light belongs to the medium and affects the eye whereas color belongs to the object and has no effect at all upon the eye—cannot be upheld: colored objects shining in secondary light can also be projected, point for point, on a white screen.)

86. One of the main arguments in favor of this contention is Kepler's description of the moon in his "Optics of 1604." He comments on the broken character of the boundary between light and shadow (*Gesammelte Werke*, II, p. 218); he describes the dark part of the moon during an eclipse as looking like torn flesh or broken wood (ibid., p. 219). He returns to those passages in the *Conversation*, p. 27, where he tells Galileo that "these very acute observations of yours do not lack the support of even my own testimony. For [in my] 'Optics' you have the half moon divided by a wavy line. From this fact I deduced peaks and depressions in the body of the moon. [Later on] I describe the moon during eclipse as looking like torn flesh or broken wood, with bright streaks penetrating into the region of the shadow." Remember also that Kepler criticizes Galileo's telescopic reports on the basis of his own observations with the naked eye (see n. 65).

87. There is one other point which I must on no account forget, which I have noticed and rather wondered at it. It is this: The middle of the Moon, as it seems, is occupied by a certain cavity larger than all the rest, and in shape perfectly round. I have looked at this depression near both the first and the third quarters, and I have represented it as well as I can in the second illustration already given. It produces the same appearance as to effects of light and shade as a tract like Bohemia would produce on the Earth, if it were shut in on all sides by very lofty mountains arranged on the circumference of a perfect circle; for the tract in the moon is walled in with peaks of such enormous height that the furthest side adjacent to the dark portion of the moon is seen bathed in sunlight before the boundary between light and shade reaches half way across the circular space. (Carlos, *Sidereal Messenger*, pp. 21–22)

This description, I think, definitely refutes Kopal's conjecture of observational laxity. It also refutes Professor Agassi's conjecture (in a private communication) that Galileo simply could not draw. Kopal tentatively identifies Galileo's monster with Ptolemy—which seems somewhat unlikely.

It is interesting to note the difference between the woodcuts in the *Sidereus Nuncius* and Galileo's original drawing, which are shown in Wolf, *Geschichte der Astronomie*. The woodcut corresponds closely to the description whereas the original drawing with its impressionistic features is vague enough to escape the accusation of gross observational error ("kaum . . . eine Karte," says Wolf—see

n. 68). This fact presents a second difficulty for Professor Agassi's conjecture that Galileo simply could not draw.

For the difficulties of lunar observation, cf. also the quotation from Lichtenberg in n. 59. Galileo and Kepler, both object-oriented, would, of course, never see the spots as droplets on the glass of the lens (though a double effect of this kind, returning the droplets to the moon, might account for Galileo's monster).

88. "I cannot help wondering about the meaning of that large circular cavity in what I usually call the left corner of the mouth," writes Kepler (*Conversation,* p. 28), who then proceeds to make conjectures as to its origin (conscious efforts by intelligent beings included).

89. So far I have disregarded here the work of della Porta (*De Refractione*) and Maurolycus, both of whom anticipated Kepler in certain respects (and are duly mentioned by him). Maurolycus makes the important step to consider only the cusp of the caustic (see *Photismi de Lumine,* trans. Henry Crew [New York, 1940], p. 45 [mirrors] and p. 74 [lenses]). However, a connection with what is seen on direct vision is still not established. For the difficulties which were removed by Kepler's simple and ingenious hypothesis, cf. Ronchi, *Histoire de la Lumière,* ch. iii.

90. *Gesammelte Werke,* II, p. 72. Part of the "Optics of 1604" has been translated into German in *Klassiker der Exakten Wissenschaften,* ed. Ostwald (Leipzig, 1922), *J. Keplers Grundlagen der geometrischen Optik,* trans. F. Plehn, chap. 3, sec. 2, pp. 38–48.

91. Kepler, *Gesammelte Werke,* II, p. 67.

92. Ibid., p. 64.

93. Ibid., p. 66. "In visione tenet sensus communis oculorum suorum distantiam ex assuefactione, angulos verò ad illam distantiam notat ex sensu contortionis oculorum."

94. Ibid., p. 67.

95. Ronchi, *Optics,* p. 44. One should also consult the second chapter of this book for a history of pre-Keplerian optics.

96. Ronchi, *Optics,* pp. 182, 202. This phenomenon was known to everyone who had used a magnifying glass only once, Kepler himself included. This fact only shows that disregard of familiar phenomena does not entail that the phenomena are seen differently (cf. text to n. 85).

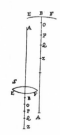

In *Lectiones XVIII Cantabrigiae in Scholis publicis habitae in quibus Opticorum Phenomenon genuinae Rationes investigantur ac exponentur* ([London, 1669], pp. 125–26), Isaac Barrow asks where "the point A" which is seen through a lens or in a mirror that unites its rays at Z "ought to appear" to an eye situated near O. (See diagram.) He responds:

From our tenets it should seem to follow that it would appear before the eye at a vast distance off, so great as should in some sort surpass all sensible distance. . . . But on the contrary, we are assured by experience, that the point A appears variously distant, according to the different situations of the eye between the points B and Z. And that it does almost never seem farther off than it would if it were beheld by the naked eye; but, on the contrary, it does sometimes appear much nearer. . . . All which does seem repugnant to our principles; at least, not rightly to agree with them [the "principles" are those of "Kepler, Scheinerus, Des Cartes," etc.]. (Quoted from the English translation in Berkeley, "An Essay Towards a New Theory of Vision" in *Works,* ed. Frazer, I [London, 1901], pp. 137 f)

"But as for me," Barrow continues, exhibiting the general attitude of all opticians up to the present time,

> neither this nor any other difficulty shall have so great an influence on me, as to make me renounce that which I know to be manifestly agreeable to reason. Especially when, as it here falls out, the difficulty is founded in the particular nature of certain odd cases. (Ibid.)

(This last remark ought to be compared with the quotation from Galileo that introduces section 2 and with my considerations in the text to n. 85, concerning the possible changes of [celestial] phenomena as seen with the naked eye.) According to Berkeley, "This phenomenon . . . entirely subverts the opinion of those who will have us judge of distances by lines and angles" ("New Theory of Vision," p. 141). Berkeley replaces this opinion by his own theory according to which the mind judges distance from the clarity or confusion of the primary impressions. Kepler's idea of the telemetric triangle was adopted at once by almost all thinkers in the field. It was given a fundamental position by Descartes according to whom "distantian . . . discimus, per mutuam quandam conspirationem oculorum" (*Dioptrice;* quoted from *Renati Descartes Specima Philosophiae* [Amsterdam, 1657], p. 87).

"But," says Barrow, "neither this nor any other difficulty shall . . . make me renounce that which I know to be manifestly agreeable to reason." This attitude was responsible for the slow advance of a scientific theory of eyeglasses and of optics in general. "It can hardly be denied," writes Moritz von Rohr, "that up to the most recent times no optical instrument has been treated more rarely in a scientific spirit. Even today one frequently lacks understanding of what can be expected from eyeglasses. The reason for this peculiar phenomenon is to be sought in the close connexion between the eyeglass and the eye and it is impossible to give an acceptable theory of eyeglasses without understanding what happens in the process of vision itself" (*Das Brillenglas als Optisches Instrument* [Berlin, 1934], p. 1). The telemetric triangle omits precisely this process or, rather, it gives a simplistic and entirely false account of it. (It is interesting to note here that modern optics to some extent repeats the optics of the thirteenth to the sixteenth centuries which did not deal with lenses and eyeglasses despite the fact that they were widely used. Witelo nowhere refers to lenses, and even the author of the *Margerita Philosophica,* that *epitome omnis philosophiae, tractans de omni genere scibili,* [no less!] entirely refrains from mentioning them.)

The state of optics at the beginning of the twentieth century is described well by A. Gullstrand. He explains how a return to the psycho-physiological process of vision has enabled physicists to arrive at a more reasonable account of even the *physics* of optical imagery:

> The reason why the laws of actual optical imagery have been, so to speak, summoned to life by the requirements of physiological optics is due partly to the fact that by means of trigonometrical calculations, tedious to be sure, but easy to perform, it has been possible for the optical engineer to get closer to the realities of his problem. Thus, thanks to the labours of such men as Abbe and his school, technical optics has attained its present splendid

development; whereas with the scientific means available a comprehensive grasp of the intricate relations in the case of the imagery in the eye has been actually impossible. ("Appendices to Part I" in Helmholtz, *Treatise on Physiological Optics*, trans. Southall (New York, 1962), pp. 261 ff)

97. "O Nicholas Copernicus, what a pleasure it would have been for you to see this part of your system confirmed by so clear an experiment!" writes Galileo, implying that the new telescopic phenomena are additional support for Copernicus (*Dialogue*, p. 339). The difference in the appearance of planets and fixed stars (cf. n. 66) he explains by the hypothesis that "the very instrument of seeing introduces a hindrance of its own" (ibid., p. 335) and that the telescope removes this hindrance, viz., irradiation, permitting the eye to see the stars and the planets as they really are. (In the "Discourse on the Comets" of Galileo's follower Mario Giuducci, irradiation is ascribed to refraction by the moisture on the surface of the eye. See Drake and O'Malley, *Controversy on the Comets*, p. 47.)

This explanation, plausible as it may seem (especially in view of Galileo's attempt to show how irradiation can be removed by means other than the telescope) is not as straightforward as one might wish. Gullstrand says that "owing to the properties of the wave surface of the bundle of rays refracted in the eye . . . it is a mathematical impossibility for any cross section to cut the caustic surface in a smooth curve in the form of a circle concentric with the pupil" ("Appendices to Part I," p. 426). Other authors point to "inhomogeneities in the various humours, and above all in the crystalline lens" (Ronchi, *Optics*, p. 104). Kepler gives this account: "Point sources of light transmit their cones to the crystalline lens. There refraction takes place, and behind the lens the cones again contract to a point. But this point does not reach as far as the retina. Therefore, the light is dispersed once more, and spreads over a small area of the retina, whereas it should impinge on a point. Hence the telescope, by introducing another refraction, makes this point coincide with the retina" (*Conversation*, pp. 33–34). Polyak, in his classical work *The Retina*, attributes irradiation partly to "defects of the dioptrical media and to the imperfect accomodation [but] chiefly [to the] peculiar structural constitution of the retina itself" (p. 176), adding that it may be a function of the brain also (p. 429).

None of these hypotheses covers *all* the facts known about irradiation. Gullstrand, Ronchi, and Polyak (if we omit his reference to the brain, which can be made to explain anything we want) cannot explain the disappearance of irradiation in the telescope. Kepler, Gullstrand, and Ronchi also fail to give an account of the fact, emphasized by Ronchi, that large objects show no irradiation at their edges. ("Anyone undertaking to account for the phenomenon of irradiation must admit that when he looks at an electrical bulb from afar so that it seems like a point, he sees it surrounded by an immense crown of rays whereas from nearby he sees nothing at all around it" [*Optics*, p. 105].) We know now that large objects are made definite by the lateral inhibitory interaction of retinal elements (which is further increased by brain function). Cf. Ratliff, *Mach Bands*, pp. 146 ff. But the variation of the phenomenon with the diameter of the object and under the conditions of telescopic vision is entirely unexplored. Galileo's hypothesis received support mainly from its agreement with the Copernican point of view and was therefore largely ad hoc.

98. The actual variations of Mars and Venus are four magnitudes and one magnitude respectively.

99. Galileo, *Dialogue*, p. 328.

100. Geymonat, *Galileo Galilei*, n. 36, pp. 38–39.

101. Kepler, *Gesammelte Werke*, IV, p. 447. Kepler (*Conversation*, p. 14) speaks of "mutually self supporting evidence." Remember, however, that what is "mutually self supporting" are two refuted hypotheses (or two hypotheses which may even be incommensurable with the available basic statements) and not two hypotheses which have independent support in the domain of basic statements. In a letter to Herwarth, March 26, 1598, Kepler speaks of the "many reasons" that he wants to adduce for the motion of the earth adding that "each of these reasons, taken for itself, would find only scant belief" (Caspar and Dyck, *Johannes Kepler in seinen Briefen*, I, p. 68).

102. Cf. the article by Price referred to in n. 24.

103. For some of these techniques cf. n. 157 and secs. 13 and 14. For Galileo's style, cf. Olschki (in n. 1). Galileo's style is closely connected with the situation described in the text. Its vividness not only expresses Galileo's strong and colorful personality, but also functions to attract and to persuade in a situation in which conclusive argument is not yet available and in which progress depends on the ability to remove temporarily one's scruples (one's "professional integrity," as Professor Lakatos is fond of saying) and to take the Copernican view for granted. As will be shown below, one had to proceed in this way, for one needed the Copernican view as a starting point for the development of appropriate auxiliary sciences (cf. sec. 8, the quotation from Mill in n. 146, and the text to n. 146). Now, a situation such as the one in which Galileo found himself can arise at any time. It can be created at once by magnifying the problems which beset our current theories, thus turning them into refuting instances (Bohr and Einstein were masters in this art of "testing an idea in its furthest consequences by exaggeration," as Bohr's method has been described). But then the current theories need additional argumentative support just as much as Copernicus's did at the time of Galileo. Style is always important—Boltzmann, Einstein, and their less gifted imitators notwithstanding.

104. In all fields today it is customary to strive for sophistication, to use quantitative methods and, if possible, at least *one* formula. Even philosophy—which was once a welcome refuge for a healthy, entertaining, and intelligent dilettantism, where one could see specialties in perspective and put them in their place—has been corrupted by this craving for minutiae and formalization. In the philosophy of science this trend has led to various *theories of confirmation*, and so great is the pressure and the desire not to be left out that the most reasonable members of this strange profession have become unbalanced and have started developing their own formalisms of *corroboration*. Now if the ideas in the present paper are correct, that is, if one can and should proceed by counterinduction also, then all these formalisms are as relevant as are the properties of the number of the Great Beast or of magical squares. Observations and theories, then, are on the same footing so that it is vain to try to construct numerical functions for evaluating the latter on the basis of the former. For details, cf. my essay "Against Method," especially secs. 12 and 13.

105. Apparent exceptions are Popper, Agassi, Lakatos, and Hanson.
    Popper's *background knowledge* seems to incorporate precisely that part of the transient historical situation that affects the development of our theories. For Agassi one should study his article in the *Australian Journal for Philosophy*, 39, pp. 87–88. For Lakatos the reader should consult "Changes in the Problem of Inductive Logic," in *The Problem of Inductive Logic*, ed. Lakatos (Amsterdam, 1968), pp. 387 ff. In this paper Lakatos gives an excellent survey of earlier attempts at historization and concludes with an account of his own dialectical point of view. In a lecture in Berkeley (May 1967) he pointed out, in complete

agreement with the principles of dialectical materialism (cf. n. 23), that evidence for auxiliary theories (which he calls "touchstone theories") may arrive long after the refutation of the cosmology to be saved. He has also emphasized the possibility of eliminating theories known to be true and of accepting theories known to be false. Regarding the last point, my position is different, for I do not intend to make use of the theological term *true* (cf. n. 9). Lakatos also seems to be more confident than I concerning the value of a reasonable theory of corroboration. (For Hanson cf. his numerous articles, but especially *Patterns of Discovery* [New York, 1958].)

Despite this temporary show of reason on the part of some Popperians, there was always the tendency to argue *sub specie aeternitatis*. Since Popper's papers on the "Third World," this tendency has become outright orthodoxy, and so we are back again with good old Plato, only it is a warmed-over version of him which we receive, entirely void of the excitement created by the original discovery, but considerably more pompous.

106. Galileo, *Dialogue*, p. 171.
107. Galileo Galilei, *Sidereus Nuncius, Nachricht von neuen Sternen*, ed. Sammlung Insel, I (Frankfurt, 1965), p. 13. Aristotle was more open-minded: "The evidence [concerning celestial phenomena] is furnished but scantily by sensations; whereas respecting perishable plants and animals we have abundant information, living as we do in their midst, and ample data may be collected concerning all their various kinds, if only we are willing to take sufficient pains. Both departments, however, have their special charms" (*De Partibus Animalium*, 644b26 ff).
108. The existence of such phase differences has been repeatedly emphasized by dialectical materialists, though they usually restrict themselves to a political context. Cf. nn. 4 and 6. For a more recent application to problems of science, cf. Lakatos, "Criticism and the Method of Scientific Research Programs," in *Criticism and the Growth of Knowledge*, eds. Lakatos and Musgrave (Cambridge, 1970). Cf. also my paper in the same volume for a criticism of Kuhn and Lakatos.
109. In the case of the dynamical arguments, there is the additional feature that concepts are changed, although the attempt is made to keep this change from sight and to create the impression that one just remembers what one knows already and develops it further. Cf. sec. 13.
110. For a brief account of this philosophy, cf. my paper "On a Recent Critique of Complementarity," *Philosophy of Science*, 35/36 (1968/1969). It must be admitted, however, that Bohr seemed to change his view after 1927, though the extent of the change is still not too well known.
111. H. Margenau, "The Philosophical Legacy of the Quantum Theory," *Mind and Cosmos*, ed. R. Colodny (Pittsburgh, 1966), p. 349. This seems to be the standard opinion of many physicists.
112. E. P. Wigner, "The Problem of Measurement," *American Journal of Physics*, 31 (1963), p. 14.
113. Schrödinger (*Nature*, 173 [1954], p. 442) has shown the difficulties of defining the angle between two mutually inclined surfaces of a crystal.
114. Cf. his essay "Geloeste und ungeloeste Probleme des Messprozesses in der Quantenmechanik," in *Werner Heisenberg und die Physik unserer Zeit* (Braunschweig, 1961), and the literature given there.
115. Cf., e.g., Daneri, Loinger, and Prosperi, *Nuclear Physics*, 33 (1962), pp. 287 ff.
116. Ludwig, *Werner Heisenberg*, p. 160.

117. Ibid., p. 159.
118. This is overlooked, for example, by Margenau ("Philosophial Legacy") who cites the behavior of certain wave packets as proof of the existence of a smooth classical limit. What he does not mention is that there exist other descriptions of macroobjects which have no such convenient properties. For the effects of linearity, cf. N. Rosen, *American Journal of Physics*, 32 (1964), pp. 597 ff as well as the instructive analysis by E. L. Hill in Feyerabend and Maxwell, eds., *Mind, Matter, and Method, Essays in Honor of Herbert Feigl* (Minneapolis, Minn., 1966), pp. 430 ff.
119. *Naturwissenschaften*, 23 (1935), p. 812.
120. *Scientific Papers Presented to Max Born* (Edinburgh, 1953).
121. Wigner, "The Problem of Measurement," p. 11.
122. Dirac, *The Principles of Quantum Mechanics* (Oxford, 1947), par. 3. Of course, here we do not regard the electron as a perennial particle (as does Popper, cf. M. Bunge, ed., *Quantum Theory and Reality* [New York, 1967], p. 27. For arguments against such a point of view, cf. my essay "On a Recent Critique of Complementarity").
123. For these points cf. my essay "On the Quantum Theory of Measurement," in *Observation and Interpretation*, ed. Koerner (London, 1957), p. 125. Professor Wigner is against extending the argument to the subjective impressions of the observer which, according to him, contain no element of uncertainty whatever. But, of course, we must also know that the impressions have something to do with the case (the electron)—and here certainty can no longer be achieved. Besides, Professor Wigner's point shows what we want to establish, viz., that the application of the quantum theory to the macrolevel (sensations included) is impossible without additional assumptions.
124. This was asserted in ibid., p. 122. For mathematical details cf. Jauch, *Helvetica Physica Acta*, 37 (1964), pp. 193 ff. The same idea seems to inspire the work of the Italian school. For an analysis of the latter which, despite its brevity, is full of extremely useful observations, cf. Jauch, Wigner, and Yanase, *Nuovo Cimento*, 48 (1967), pp. 144 ff.
125. Cf. the brief report in my "Problems of Microphysics," *Frontiers of Science and Philosophy*, ed. R. Colodny, n. 203 and text. A report of the conversations can be obtained from the files of the Minnesota Center.
126. "When [the wave packet of the electron] arrives at the film it begins to interact with the wave-functions of the atoms of the film. In the interaction it is 'broken up' into many very small packets which cease to interfere coherently. What is still unexplained, however, is that only one of these packets contains the electron" (D. Bohm and J. Bub, "A Proposed Solution of the Measurement Problem in Quantum Mechanics," *Reviews of Modern Physics*, 38 [1966], p. 459).
127. The situation is further aggravated by the fact that a mixture does not uniquely determine its elements unless the weights are all different. Cf. von Neumann, *Mathematische Grundlagen der Quantenmechanik* (Berlin, 1932), pp. 175–76.
128. For part of the mathematics cf. von Neumann's proof that the "cut" between observer and observed object can be moved back and forth in an arbitrary manner (*Mathematische Grundlagen*, pp. 225 ff).
129. The rest of this section has been included to deal with an objection by Dr. Heinz Post of the Chelsea College of Technology.
130. *Atomic Theory and the Description of Nature* (Cambridge, 1932), p. 75.

131. Ibid., p. 79.
132. An excellent summary of these results is contained in chap. 23 of Bohm's *Quantum Theory* (Princeton, 1951):

> Quantum theory presupposes the classical level and the general correctness of classical concepts in describing this level; it does not deduce classical concepts as limiting cases of quantum concepts (as, for example, one deduces Newtonian mechanics as a limiting case of special relativity). . . . The necessity for presupposing a classical level and the appropriate classical concepts implies that the large scale behaviour of a system is not completely expressible in terms of concepts that are appropriate at the small scale level. . . . As we go from small scale to large scale level, new (classical) properties . . . appear which cannot be deduced from the quantum description in terms of the wave function alone. . . . These new properties manifest themselves . . . in the appearance of definite objects and events which cannot exist at the quantum level.

What Bohm does not mention in his summary is that the transition to the classical level involves not only additions (definite objects) but also eliminations (interference terms). The latter are dealt with in chap. 22 of his book, which describes the quantum theory of measurement.

133. "Bohr has expressed himself in discussions somewhat as follows: classical physics and the quantum theory, taken as descriptions of nature, are both caricatures; they allow us, so to speak, to asymptotically represent actual events in two extreme regions of phenomena" (H. A. Kramers, *Naturwissenschaften*, 11 [1923], p. 559).

134. It is interesting to see to what extent the presence of classical concepts determines the structure of the theory; direct measurement is possible only of those situations which (like position, momentum, and angular momentum) have a classical analogon. There is no direct way (involving, for example, trajectories) of measuring the intrinsic angular momentum of the electron. The Stern-Gerlach experiment always measures the total angular momentum of atoms, and a Stern-Gerlach experiment with isolated electrons will fail as the expected effects are hidden by the wave properties of the electron. For a more detailed account cf. W. Pauli, "L'électron magnétique," *Proceedings of the Sixth Solvay Conference* (Paris, 1930), pp. 217 ff. The discussion of the Stern-Gerlach experiment occurs on p. 220, the general discussion of the feature just mentioned (with comments by Bohr) on pp. 276 ff. Unfortunately, the formalism does not mirror this important difference between variables. (The same is true of the quantum field theory.)

135. Interest in classical physics has been considerably revived by R. H. Dicke's investigations. These investigations make it clear that one was premature in believing that the explanation of, say, the advance of the perihelion of Mercury was beyond its reach. Cf. "The Observational Basis of General Relativity," in *Gravitation and Relativity*, eds. Chiu and Hoffmann (New York, 1964), pp. 1–16. The experimental basis of Dicke's investigations was reported at the January 1967 meeting of the American Physical Society and was made public in the press.

The foregoing is not supposed to imply that Professor Dicke's *positive theory* is classical. All I want to point out is that he is using classical physics when criticizing Einstein's theory of relativity. We have here, therefore, a very interesting relation between the theory to be criticized, the theory that does the criticizing, and the theory that is supposed to take over after the criticism has been effective. The theory that does the criticizing is not thought to be adequate as a description of reality. Still, having played a decisive role in the acceptance of general relativity (the 43″ which support general relativity were, after all, obtained by observation in conjunction with classical perturbation theory), it can now play an equally decisive role in its rejection. The reader should also be aware of the fact that classical mechanics has recently undergone a "renascence [which] broadened and deepened it but rather elevated than annulled the older parts of the subject" (C. Truesdell, *The Elements of Continuum Mechanics* [New York, 1966], p. 1; the reference is to continuum mechanics as well as to the science of general mechanics which was founded by Birkhoff).

136. "The ephemerides are calculated in accordance with the Newtonian law of gravitation, modified by the theory of general relativity" (*Explanatory Supplement to the Astronomical Ephemeris and the American Ephemeris and Nautical Almanac* [London, 1961], p. 11). Cf. also D. Brouwer and G. M. Clemence, *Methods of Celestial Mechanics* (New York, 1961), p. 3. "This mixture of the theories and Newton and Einstein is intellectually repellent, since the two theories are based on such different fundamental concepts. The situation will be made clear only when the many body problem has been handled relativistically in a rational and mathematically satisfactory way" (J. L. Synge, *Relativity, the General Theory* [Amsterdam, 1964], pp. 296–97). For a first step toward a more comprehensive treatment, cf. Levi-Civita, *The N-Body Problem in General Relativity* (Amsterdam, 1964), chap. III.

137. Born, *Annalen der Physik*, 30 (1909), pp. 1 ff has given a definition of rigid motion which demands that the distance between adjacent world lines of the motion remain constant. However, a rigid object in this sense cannot be made to rotate (Ehrenfest, *Physikalische Zeitschrift*, 10 [1909], p. 918) and has only three degrees of freedom (Herglotz, *Annalen der Physik*, 31 [1910], p. 393, and Noether, *Annalen der Physik*, 31 [1909], pp. 919 ff). Rotating solid objects must be dealt with by relativistic *fluid mechanics* which has not yet led to a simple and concise characterization of isolated bodies.

138. Dicke's investigations (cf. n. 135) have done much to separate the classical and the relativistic elements of the allegedly purely relativistic predictions. One might be inclined to deny the conclusion in the text by referring to the numerous "derivations" of classical mechanics from the general theory of relativity. However, such derivations are but formal exercises unless it is shown that not only momentary effects but also long-term effects are excluded and this for the whole period for which useful astronomical observations are available (2,800 years); one would have to show that the minute deviations neglected in the usual approximations have no cumulative effects which might endanger the stability of the planetary system. No such proof is available, as far as I know. Considering the difficulties of the relativistic many-body problem (see n. 136), it is not likely that it will soon be found. But without it the "derivations" referred to are useless for the purpose of unified prediction. That they are useless for an even wider purpose, viz., for showing the conceptual continuity of the theories of Einstein and Newton, is demonstrated by more detailed examinations, such as those by Peter Havas, "Four-Dimensional Formulations of Newtonian Mechanics and Their Relation to the Special and the General Theory of Relativity," *Reviews of Modern Physics*, 36 (1964), pp. 938 ff. Concerning this paper, the reader is invited not to remain content with the introduction which promises to show continuity between Newton and Einstein but also to consult the remaining sections of the essay, which, he will find, tell a very different story.

139. This is, of course, realized in physics where the so-called evidence for relativity is now subjected to a critical reexamination that separates the classical and the relativistic elements in the derivation. Cf. nn. 135 and 138 and the literature given there.
140. Letter to Karl Seelig, January 1953; quoted from C. Seelig, *Albert Einstein* (Zurich, 1960), p. 271. Cf. also Gerald Holton's illuminating article "Influence on Einstein's Early Work," *Organon*, 3 (1966), p. 242.
141. Unfortunately, we do not possess any account of this philosophy in the English language. This is regrettable, for possibilism seems to be a much better framework for understanding science than does any other existent methodology (perhaps excluding dialectical materialism in the form given to it by Lakatos). I hope that the readers of the present essay will join me in urging Professor Naess to throw caution to the winds and give us what we have been wanting to see for now over a decade, a brief outline, in English, of the main principles of his philosophy.
142. *Advancement of Learning*, 1605 ed. (New York, 1944), p. 21. Cf. also *Novum Organum*, Aphorisms 79, 86, as well as J. W. N. Watkins, *Hobbes's System of Ideas* (London, 1965), p. 169.
143. Cf. A. Koyré, *Etudes Galileénnes*, III (Paris, 1939), pp. 53 ff.
144. In a letter to Max Brod, Einstein (who had just read "about one third" of Brod's *Galilei in Gefangenschaft*), considering Galileo's propagandistic efforts, comments:

> It is difficult for me to imagine that he [Galileo] should have found it worth while, as a mature man, and against so much resistance, to try to incorporate the found truth into the consciousness of a multitude caught up in superficial and petty interests. Was this really so important for him to waste his last years? . . . His arguments were accessible for everyone searching for knowledge. . . . At any rate, I cannot possibly imagine that I could undertake similar things in order to defend relativity. I should think: truth is so much stronger than I and it would appear to me ridiculous and donquichotesque to try to protect it with sword and Rosinante. (Letter of July 4, 1949; quoted from Seelig, *Albert Einstein*, p. 210)

Concerning this letter, let me point out, first, that the idea of the strength of the truth allows for an antihumanitarian inversion which the antiheretical missionary Peter von Pilichsdorf (fourteenth century) used against new religious views: "I say boldly: if thy doctrine were true, it would be easy for you to preach it in every place. Now, forger!, you must needs hide the false coin with greater caution" (quoted from Coulton, *Inquisition and Liberty* [Boston], p. 198). Secondly, it is overlooked that arguments presuppose an attitude, more especially that Galileo's arguments presupposed a completely new and metaphysical attitude toward the facts of observation, and that an attitude can be brought about by propaganda only.
145. "The Methodological Character of Theoretical Concepts," in *Minnesota Studies in the Philosophy of Science*, I, eds. H. Feigl and M. Scriven (Minneapolis, Minn., 1956), p. 47.
146. *Autobiography*; quoted from *Essential Works of John Stuart Mill*, ed. Lerner (New York, 1965), p. 21.
147. *Novum Organum*, introduction.
148. Hereafter numbers in parenthesis refer to pages of the *Dialogue*.
149. The idea of absolute position, motion, and direction made excellent sense in the Aristotelian cosmology where space was a well-structured entity, very much like the *Bezugsmolluske* of Einstein.

150. Quoted from Chiaramonti, *De Tribus Novis Stellis,* an anti-Copernican treatise of 1572, with new editions in 1600 and 1609.
151. Feyerabend, "Problems of Empiricism," pp. 204 ff.
152. John L. Austin, *Sense and Sensibilia* (New York, 1962), p. 74.
153. "This is good sound doctrine," says Simplicio when the matter is brought up, "and entirely Peripatetic" (116).
154. The passage in brackets was added by Galileo after publication of the first edition. It makes it very clear that the discussion is purely kinematic and that dynamical effects are simply assumed not to exist. Cf. also sec. 15.
155. The idea that there is an absolute direction in the universe has a very interesting history. It evidently rests on the structure of the gravitational field on the surface of the earth or of that part of the earth which the observer knows, and it generalizes the experiences made there. The generalization is only rarely regarded as a separate hypothesis; it rather enters the grammar (in the modern sense of linguistic philosophy) of common sense and gives the terms "up" and "down" an absolute meaning. Lactantius, a church father of the fourth century appeals to this meaning when asking: "Is one really going to be so confused as to assume the existence of humans whose feet are above their heads? Or of regions where the objects which fall with us rise instead? Where trees and fruit grow not upwards, but downwards?" (*Divinae Institutiones,* III, de falsa sapientia). The same use of language is presupposed by that "mass of untutored men" who raise the question why the antipodes are not falling off the earth (Pliny, *Natural History,* II, pp. 161–66). The attempts of the pre-Socratics (Thales, Anaximenes, Xenophanes) to find a support for the earth which prevents it from falling "down" (Aristotle, *De Coelo,* 294a12 ff) shows that almost all early philosophers (with the only exception of Anaximander) have shared this way of thinking. (For the atomists who assume that the atoms originally all fell in one direction, viz., "down," cf. Jammer, *Concepts of Space* [Cambridge, Mass., 1953], p. 11.) Even Galileo, who thoroughly ridicules the idea of the falling antipodes (331), occasionally speaks of "the upper half of the moon" (65) when he means that part of the moon "which is invisible to us." And let us not forget that some linguistic philosophers of today "who are too stupid to recognise their own limitations" (327) want to revive the absolute meaning of the pair "up-down" at least locally. Thus the power of a primitive conceptual frame assuming an anisotropic world which also held the minds of Galileo's contemporaries and which was fought by Galileo himself must not be underestimated. For an examination of some aspects of common sense at the time of Galileo, cf. E. M. W. Tillyard, *The Elizabethan World Picture* (Baltimore, Md., 1963).
156. A more detailed investigation would have to distinguish between the Latin of the time and the language of the common people, Italian in the case of Galileo. The former had degenerated; it was rigid, perfectly adapted to the barren conceptual distinctions of the schoolmen, and absolutely immovable. At the time it was also an object of considerable merriment among intelligent laymen (cf. Olschki, "Geschichte der Neusprachlichen Wissenschaftlichen Literatur," pp. 64 ff). Italian was not adapted to any particular conceptual system; it was elastic and perfectly suited for arguments from remembrance such as those outlined in n. 157. Copernicanism could be introduced into it with relative ease. This is one of the reasons why Galileo preferred writing in Italian. Another reason is the vividness of the language, which allowed him to use phrases almost as if they were pictures and in this way to appeal to everyone's imagination. He also wished to make science known to all, so that it might profit from the still unbroken intelligence of the common man.

157.   *Simplicio:* I have frequently studied your manner of arguing, which gives me the impression that you lean towards Plato's opinion that *nostrum scire sit quoddam reminisci.* So please remove all questions for me by telling me your idea of this.

   Salviati: How I feel about Plato's opinion I can indicate to you by means of words and also by deeds. In my previous arguments I have more than once explained myself with deeds. I shall pursue the same method in the matter at hand. (190)

A closer look at the "method" indicates that Galileo uses the idea of anamnesis as a practical tool in the course of argument which connects various parts of everyday thinking and everyday knowledge, partly strengthening already existing connections, partly establishing new connections altogether, and which thereby turns common sense from a loose collection of memories, prejudices, hypotheses, and myths into a new and unified cosmology. However, the element of change implied in the use of this method is concealed and kept from view.

   Here are some examples:

   i) The argument showing that the earth shines as much as the moon: "You can see [now, Simplicio,] how you yourself really knew that the earth shone no less than the moon, and that not my instruction but merely the recollection of certain things already known to you have made you sure of it" (89–90).

   ii) The argument showing that a rotating earth will not "throw everything into the sky" (190). "The unraveling [of this problem]," says Salviati, "depends on some data well known and believed by you just as much as me, but because they do not strike you, you do not see the solution. Without teaching them to you then, since you already know them, I shall cause you to resolve the objection by merely recalling them" (190). There follows the quotation at the beginning of this note, and then the argument continues. Step by step Simplicio is made to approach the conclusion. First, the reason for the centrifugal motion is explained. "I believe I can say I am certain of that," says Simplicio at the end of this explanation, "but this new knowledge only increases my incredulity that the earth should revolve with such great speed and not throw to the skies all stones, animals, etc." (193). "In the same way that you knew what went before," replies Salviati, "you will know—or rather, do know—the rest too. And by thinking it over for yourself you would likewise recall it by yourself. But to save time, I shall help you to remember it" (193). At a later place (on the occasion of the discussion of the Copernican arrangement of the planets), Simplicio is asked to draw the arrangement himself. "You will see," says Salviati, "that however firmly you may believe yourself not to understand it, you do so perfectly, and just by answering my questions you will describe it exactly" (322). Sagredo expresses his admiration for this method and says: "Then, for once, I may be able to instruct both of you. And since proceeding by interrogations seems to me to shed much light upon things, in addition to the pleasure one may get out of pumping one's companion and making things drop from his lips which he never knew that he knew, I shall make use of that artifice" (251).

   iii) "*Sagredo:* I say to you that if one does not know the truth by himself, it is impossible for anyone to make him know it. I can indeed point out things to you, things being neither true nor false; but as for the true—that is the necessary; that which cannot be otherwise—every man of ordinary intelligence either knows this by himself or it is impossible for him ever to know it" (157 f).

iv) Occasionally, remembering is improved by the removal of contrary hypotheses: "*Salviati:* Meanwhile I wish to set forth some conjectures, not to teach you anything new, but to take away from you a certain contrary belief and to show you how matters may stand" (226).

These quotations would seem to make it clear that Galileo is not at all reluctant to explain his epistemological views. True, he does not give a direct reply to Simplicio's question as quoted at the beginning of this note. But this ground is much too slender for asserting, as Professor Watkins has done, that "at this point Galileo's epistemological caginess rather asserts itself" (*Hobbes*, p. 63). Galileo does not give a verbal reply, at this place at any rate. But he does at once proceed to show, by some piece of concrete analysis, what the correct answer is. And whatever hesitation occurs has an artistic rather than an epistemological function (fortunately, the *Dialogue* is not a *textbook* in the theory of knowledge).

158. It is simply untrue that the critical remedy against the [Aristotelian] philosophy of nature could have been an increase in *experience*. The kind of experience which led [Galileo] to break with Aristotle was already adapted to certain phenomena, it was collected and arranged in accordance with these phenomena, it was obtained under well determined conditions, i.e. it was *experimental* kind of experience. [Our considerations in the text seem to show that the relation to experiment was very tenuous. But the view that there was a "collection" and an "arrangement" in accordance with certain "phenomena"—i.e., the Copernican hypothesis—is perfectly sound.] This kind of experience is not immediately given, and it is not exhausted by what is evident to the senses: it confirms, or refutes, assumptions concerning a certain . . . aspect of the total phenomenon. But such regulated experience cannot stand at the beginning of a theoretical revolution. We rather start *by stepping back* from our normal everyday experience. (Blumenberg, *Sidereus Nuncius*, pp. 36–37)

For this point, one should also compare Herbert Butterfield, *The Origins of Modern Science* (London, 1957), p. 80:

It was commonly argued, even by the enemies of the Aristotelian system, that that system itself could never have been founded, except on the footing of observation and experiments—a reminder necessary perhaps in the case of those university teachers of the sixteenth and seventeenth century who still clung to the old routine and went on commenting too much (in what we might call a "literary" manner) upon the works of the ancient writers. We may be surprised to note, however, that in one of the dialogues of Galileo, it is Simplicio, the spokesman of the Aristotelians—the butt of the whole piece—who defends the experimental method of Aristotle against what is described as the mathematical method of Galileo. . . . What is more remarkable still is the fact that the science in which experiment reigned supreme—the science which was centered in laboratories even before the beginning of modern times—was remarkably slow, if not the slowest of all, in reaching its modern form. It was long before alchemy became chemistry.

159. *De Revolutionibus*, Introduction.
160. Cf. with this the still existing and even increasing resistance against universal understanding as it can be found in Professor Austin's "philosophy" (*Sense and Sensibilia*, remarks on Thales). What has advanced and even created science is

today ridiculed and pushed aside by men (and women) who have barely advanced beyond the multiplication table.

161. "He has never had the experience of a truth which forced itself upon him against his intentions and he was perhaps not even disposed for such an experience" (Blumenberg, *Sidereus Nuncius*, 41). This may be another reason why Galileo had to cheat, whereas Kepler could stay honest. Cf. n. 35 (however, one must point out, in view of the changes from the *de motu* to the *Discorsi*, that Blumenberg's judgment is somewhat harsh).

162. To this interpretation of Galileo's results one might object that the sense impressions do not change (they do not change in the present case, though they change in the case of the telescope, as we have seen) and that they alone are the foundation of our knowledge. This objection overlooks that sense impressions without an interpretation speak neither for a theory nor against it and that our present case is concerned with the combination sense datum plus interpretation. This opposition between fixed data which do not deceive us and changeable, partly illusory experience is not reflected in Galileo's terminology. However, the concepts are clear, even if they are not always easily separated from their background.

The possibility of an *illusion of the senses* is constantly emphasized by Galileo and occasionally confirmed by the Aristotelian Simplicio (257). It is also said that "reason can conquer sense so that in defiance of the latter the former" can determine what assumptions are finally accepted (328). This remark cannot refer to sense impressions which are mute and do not need to be "defied," but it is concerned with natural interpretations which are so closely linked to the senses that the data themselves seem to be speaking. Galileo makes it clear that this is exactly what he means; reflecting on the assertion of the nonoperative character of the motion of the earth in which we participate but which we do not feel, Simplicio offers the following quotation from Chiaramonti: "And from this opinion we must necessarily suspect our own senses as wholly fallible or stupid in judging sensible things which are very close at hand. Then what truth can we hope for, deriving its origin from so deceptive a faculty?" (De Tribus Novis Stellis, pp. 255–56). To this Salviati, the representative of the Copernicans, gives the following reply:

> Oh, I wish to derive still more useful and more certain precepts from it, learning to be more circumspect and less confident about that which the senses represent to us at first impression, for they may easily deceive us. And I wish that this author [Chiaramonti] would not put himself to such trouble trying to have us understand from our senses that this motion of falling bodies is simple straight motion and no other kind, nor get angry and complain because such a clear, obvious, and manifest thing should be called into question. For in this way he hints at believing that those who say that such motion is not straight at all, but rather circular, it seems they see the stone move visibly in an arc, since he calls upon their senses rather than their reason to clarify the effect. This is not the case, Simplicio; for just as I . . . have never seen, nor ever expect to see the rock fall any way but perpendicularly, just so do I believe that it appears to the eyes of everyone else. *It is therefore better to put aside the appearance, on which we all agree, and to use the power of reason either to confirm its reality, or to reveal its fallacy.* (256, my italics; this is excellent advice to those who try to solve the mind-body problem by quoting the results of

introspection. Nobody wants to deny these results. What is at stake is their interpretation and the inferences we can draw from them.)

To this Sagredo at once adds further proof of how easily we can be deceived by the senses:

> If I ever had a chance to meet this philosopher [Chiaramonti, from whom Simplicio has chosen his objection] who seems to me a cut above most of the followers of these doctrines, I should as a token of my esteem acquaint him with an event which he has surely seen many times, from which one may learn [in complete agreement with what we are saying] how easily one may be deceived by simple appearances, or, let us say, by the impressions of one's senses. This event is the appearance to those who travel along a street by night of being followed by the moon . . . [continued as already quoted in the text after n. 148].

Let us begin with the last example. The wanderer has the impression that the moon is following him, and the impression deceives unless he consults reason. Reason does not change the impression (just as it does not change the motion of the stone in the case of the Copernican argument—the stone still continues falling *straight down*). The deception, therefore, does not consist in the presence of the false impression. (There are deceptions of this kind too. Irradiation is an example, but it does not concern us in the present case.) It consists in the fact that the impression quite automatically leads to a judgment, to a natural interpretation, and it is this judgment which must be changed. The close psychological connection between impression and judgment makes us believe that the senses are speaking themselves (this exact phrase is used by Galileo to characterize the Aristotelians: "My senses tell me so" [56]), and the falsity of the judgment, therefore, seems to fall back on *them*. But—and with this we return to Salviati's comment—it is not the impression that is now being considered. Nobody denies that falling stones move in a straight line toward the center of the universe. What is in question, and what is to be decided by the "power of reason," is the judgment that is inferred from such phenomena. "It is therefore better to put aside the appearances."

We see that the reference to "the senses" involves two entirely different ideas. On the one side one means "the appearance" or even "the sensations" of later philosophy (though Galileo never pushes abstraction that far). This appearance is not denied, and the question of deception does not arise either. Then again the reference to the senses means the combination appearance plus judgment, i.e., it means what we now (after Kant) call *experience*. Deception is now possible, and it is in order to doubt the message of senses of this kind.

Now, when referring to the senses, an Aristotelian understands them in the second way, for he inextricably combines them with the principles and phrases of common sense (cf. again Owen's essay in n. 43). Galileo, on the other hand, does not want to remove the sensory judgments, for he needs "the senses accompanied by reasoning" (225) for supporting his own position. It is his intention to establish a connection between Copernicus and the senses that is as close and as forceful as is the already existing connection between the senses and the idea of the unmoved earth. We are, therefore, indeed dealing with a change of experience.

163. This schizophrenia is described in some detail in my essay, "Classical Empiricism," in *The Methodological Heritage of Newton*, eds. R. E. Butts and J. W.

Davis (Toronto, 1969). For Galileo's notion of experience, cf. also the articles by Dubarle and Gurwitsch in E. McMullin, ed., *Galileo, Man of Science* (New York, 1967).

164. Blumenberg, *Sidereus Nuncius*, p. 227, quoting one of Galileo's additions to his original text.

165. For details cf. n. 157. The reference to anamnesis is often used to support Galileo's Platonism (see, for example, A. Koyré, *Metaphysics and Measurement* [Cambridge, 1968], p. 42, quoting from an article the author published in 1943). The propagandistic function of anamnesis is hardly ever commented upon, not even by Koyré, who calls the *Dialogue* "a polemic, . . . a pedagogical, . . . a propagandistic book" (*Études Galileénnes*, III, p. 53) and who is aware of the fact that the defenders of Copernicus "had, to begin with, to reshape and to reform our intellect itself" (*Metaphysics and Measurement*, p. 3). This is another result of the belief (briefly commented upon in n. 35) that a great scientist cannot possibly try to cheat and that each argument he uses can have only one function: to approach the Truth with the help of the Truth.

166. The dialogue which contains this assertion deals with the question whether a stone falling along the mast of a moving ship will arrive at the foot of the mast or whether it will be deviated toward the poop. Salviati tries to convince Simplicio that the stone will arrive at the foot of the mast whether the ship moves or not, and he tries to convince him without carrying out the experiment. This is just as well, for the existing experiments (and there were such experiments) gave first one result and then another, and both Copernicans and Ptolemaics could quote experience in their favor. Nor is this divergence very surprising. Galileo raised questions and solved problems in a theoretical way long before there existed experiments of a precision sufficient for a decisive test. Just remember that on one of the few occasions when Galileo presents us with a numerical value (his value of the acceleration of free fall), he is wrong by a factor of two. For details cf. Koyré, *Metaphysics and Measurement*, chaps. IV and V.

167. While the ancients had been ready to make precise mathematical statements, right or wrong, about the doings of the planets, their approach to motions on the earth had been largely qualitative, using the language of cause and effect, and tendency as in biology and medicine. This separation of heavenly geometry from earthly mechanics had been maintained by Galileo who, while he had done much to publicize the kinematical properties of uniformly accelerated motion, derived correctly by the schoolmen 300 years before, had widened the gap between theoretical mechanics and practical mechanical phenomena, first by refusal to connect celestial and terrestrial motions in any way, and secondly by setting up as governing presumably near and familiar motions laws valid only in an ideal or vacuous medium, the unfortunate practicer who had to work on the real earth being put off with effects recognized by name as being due to resistance and friction but explained only as qualities, or tendencies in the style of the Aristotelian physics so contemned by Galileo. (C. Truesdell, "A Program Towards Rediscovering the Rational Mechanics of the Age of Reason," *Archive for the History of Exact Sciences*, I [1960], p. 6)

Truesdell's article contains an excellent account of the difficulties encountered in the attempt to construct a meteorology of the kind already referred to in the text, and he points out that these difficulties were gradually solved only by the invention of still another Parmenidean frame: "Euler's success in this most difficult matter lay in his *analysis of concepts*. After years of trial, sometimes adopting a semiempirical compromise with experimental data, Euler saw that the *experiments had to be set aside for a time;* some remain not fully understood today. By creating a simple field model for fluids, defined by a set of partial differential equations, Euler opened to us a new range of vision in physical science" (ibid., p. 28. Cf. also Truesdell's introductions to vols. 12 and 13 of *L. Euleri Opera Omnia* [1954, 1956 respectively]). For a detailed examination of the conceptual changes entailed in Galileo's procedure, as well as of his peculiar technique of persuasion, cf. my paper "Bemerkungen zur Geschichte und Systematik des Empirismus," in *Metaphysik und Wissenschaft*, ed. Paul Weingartner (Salzburg, 1968).

168. *Temporary* may mean *centuries*.
169. Another reason for the great distance between the new Galilean experience and the experience of common sense (that has not been discussed in this paper) is the use of mathematics: it is impossible to give a mathematical account of quality and, therefore, of common-sense experience. "And well we know that Galileo, like Descartes somewhat later, and just for the same reason, was forced to drop the notion of quality, to declare it subjective, to ban it from the realm of nature" (Koyré, *Metaphysics and Measurement*, p. 8). A similar comment is to be made abount the increased demand for precision. "Modern science . . . has precision for principle; it asserts that the real is, in its essence, geometrical and, consequently, subject of rigorous determination and measurement; . . . it discovers and formulates (mathematically) laws that allow it to deduce and to calculate the position and speed of a body at each point of its trajectory and at each moment of its motion, *and it is not able to use them because it has no ways to determine a moment, or to measure a speed*. Yet, without these measures the laws of the new dynamics *remain abstract and void*" (ibid., p. 95). This is another reason why Galileo frequently had to resort to arguments of plausibility. Cf. also n. 166.
170. For an evaluation of Galileo's experiments and the gradual improvement of experimental technique, cf. ibid., sec. IV.
171. This situation should be remembered when considering Popper's accusation that Bohr and his followers are guilty of a new betrayal of Galilean science ("Three Views of Human Knowledge," *Conjectures and Refutations* [New York, 1962], p. 97). For another objection against this very misleading article, cf. n. 22 of the present paper.

Index of Names

Index of Topics

Aerodynamics, 239, 256, 260, 263, 264, 265, 266, 267, 268, 269. *See also* Fluid mechanics; Magnus effect
Ancient science, xi, xiii, 64, 65, 79, 80, 82, 84, 98, 145, 160, 161, 261, 278, 279, 283, 284, 285, 287, 293, 294, 310, 317, 318, 320, 322, 323, 325, 326, 327, 328, 331, 336, 342, 347, 348, 349, 352. *See also* Aristotle; History of science; Plato; Thales
Astronomy. *See* Celestial mechanics

Bayes's theorem, 84, 85, 86, 91, 92, 96, 97, 102, 103, 107, 119, 133, 162, 165, 228. *See also* Probability theory
Behaviorism, 29, 65
Bertrand paradox, 152
Boyle's law, 237

Causality, 71, 197, 198, 199, 200, 209, 210, 212, 214, 219, 220. *See also* Determinism; Time
Celestial mechanics, 237, 254, 274, 279, 290, 292, 293, 320, 321, 325, 326, 327, 334, 347. *See also* Newtonian physics
Circularity and scientific inference, 139, 159, 167
Confirmation theory, 5, 6, 7, 9, 12, 14, 22, 27, 31, 34, 63, 83, 121, 130, 182, 221, 341. *See also* Frequency theory of probability; Inductive logic; Language and truth; Personalist theory of probability; Probability theory
Correspondence rules, 66, 68. *See also* Language and scientific discovery; Language and scientific theory
Cosmology. *See* Celestial mechanics

d'Alembert paradox, 261. *See also* Fluid mechanics

358

Deductive logic, 66, 67, 72, 85, 173, 175, 176, 179, 180, 193, 194, 204, 227, 302. *See also* Hempel, C.; Scientific laws; Theoretical entities
Determinism, 147, 208. *See also* Causality
Dialectics, 160, 161, 164, 276, 341, 342. *See also* Materialism; Socialism

Empiricism, 3, 14, 15, 18, 19, 31, 37, 39, 40, 50, 53, 55, 56, 62, 65, 66, 67, 68, 69, 73, 74, 76, 79, 84, 233, 275–353 *passim. See also* Positivism
Entropy, 58, 211, 212
Epistemologism: fallacy of, 16, 26. *See also* Epistemology
Epistemology, xiii, 28, 31, 53, 80, 82, 83, 84, 117, 135, 159, 161, 166, 188, 194, 223, 235, 306, 307, 319. *See also* Language and learning; Metaphysics; Rationalism; Scientific laws; Theoretical entities
Evolutionary theory, 81, 145, 147, 157, 229. *See also* Innate ideas; Learning theory

Fluid mechanics, 258, 259, 261, 262, 263, 264, 344. *See also* Aerodynamics; Euler, L.; Magnus effect
Frequency theory of probability, 123, 183–226 *passim. See also* Confirmation theory; Mathematics and probability theory; Probability theory; Statistics and induction; Statistics and probability theory

Gaussian distribution, 250
Geometry, 52, 55, 56, 62, 76, 87, 94, 141, 152, 153, 240, 250, 256, 297, 301; and aerodynamics, 270, 271, 274. *See also*

Graphs and representation; Magnus effect; Mathematics
Goethe's theory of color, 292
Graphs and representation, 247, 248, 249, 251, 271, 272. *See also* Geometry; Mapping and representation

History of science, xiii, xv, 43, 53, 55, 75, 94, 95, 119, 121, 138, 143, 144, 145, 150, 254, 256, 257, 276, 277, 280, 287, 291, 293, 295, 296, 301, 304, 312, 319, 321, 323, 327, 328, 329, 330, 331, 332, 334, 335, 336, 337, 338, 339, 341, 349, 352, 353. *See also* Ancient science; Optics; Scientific revolution of the seventeenth century
Hume's problem, 3, 123, 161, 166, 181. *See also* Inductive logic

Illusion, 350, 351. *See also* Optics
Innate ideas, 141, 144, 145. *See also* Evolutionary theory
Inductive and deductive differences, 174, 181, 191, 195. *See also* Inductive logic
Inductive logic, xiii, xiv, 3, 7, 8, 13, 27, 32, 52, 80, 89, 90, 110, 121, 122, 123, 125, 127, 143, 148, 149, 150, 152, 158, 164, 174, 185, 221, 222, 226. *See also* Confirmation theory; Hume's problem; Inductive and deductive differences; Scientific laws; Simplicity; Statistics and induction; Statistics and probability theory; Theoretical entities
Inquisition, 346. *See also* Magic; Witchcraft
Instrumentalism and realism, 10, 11, 81, 326, 327. *See also* Operationalism

Language
—and common sense, 157. *See also* Perception and common sense; Perception and language; Science and common sense
—and explanation, 38, 221. *See also* Correspondence rules; Theoretical entities
—and learning, 70, 76, 306. *See also* Epistemology
—and observation, 45, 165, 243, 253, 308
—and scientific discovery, 310, 311. *See also* Correspondence rules
—and scientific theory, 42, 47, 55, 65, 69, 82, 140, 252, 253, 255, 302, 303. *See also* Correspondence rules; Theoretical entities
—and symbols, 251
—and truth, 88. *See also* Confirmation theory; Learning theory
Learning theory, 141, 142, 144, 145, 147, 157, 305, 306. *See also* Evolutionary theory; Language and truth

Mach Bands, 340. *See also* Optics
Magic, 151, 305, 341, 346; and logic, 177, 178, 191, 203
Magnus effect, 261, 262, 265, 267, 269. *See also* Aerodynamics; Fluid mechanics; Geometry
Mapping and representation, 244, 245, 246. *See also* Graphs and representation
Materialism, 30, 79, 276, 303, 342. *See also* Dialectics; Socialism
Mathematics, 87, 112, 113, 155, 156, 252, 256, 259, 263, 271, 272, 330, 335, 343, 353. *See also* Geometry; Statistics and induction; Statistics and probability theory
—and probability theory, 104, 105, 106, 109, 114, 115, 116, 120, 163, 196
—and quantum theory, 299
—and statistics, 153
Measurement theory, 134, 297, 353. *See also* Quantum theory
Metaphysics, 127, 323. *See also* Epistemology; Realism
Models, 75, 158
Monism, 30. *See also* Materialism
Myth, 60. *See also* Magic

Newtonian physics, 54, 66, 67, 257, 261, 274. *See also* Celestial mechanics; *Principia Mathematica;* Scientific revolution of the seventeenth century

Observation language, 35–76 *passim*
Ockham's rule, 100
Ontology, 13. *See also* Realism
Operationalism, 47. *See also* Instrumentalism and realism
Optics, 62, 281, 282, 285, 286, 288, 289, 292, 329, 331, 333, 336, 337, 338, 339, 340. *See also* Illusion; Perception and illusion